Preface

This is the third book with this title. The first was published in 1975. Before the first renewal in 1982 the book had become out of date in a number of respects. Inevitably the book has again become out of date, even though demand has continued. This time there has been an interval of eleven years, during which the changes in data communications have been far greater than in the previous seven.

With a certain amount of information out of date, there was a temptation to completely rewrite the book. However, it was felt that the general structure of the previous edition was still sound.

Even with a large amount of the information still relevant, there was a considerable task in bringing sections up to date. In other cases information has been discarded, except where seemingly out of date material has historical interest or helps to put some later development into perspective.

The audience aimed at is computing and information technology professionals wanting to find out more about the transmission and telecommunications area and telecommunications specialists wishing to expand their knowledge of computer communications. In both cases it is likely that this desire has been fuelled further by experience of being a user of PCs. Inevitably the PCs will have been networked locally, over a wide area or both and the user will have first hand experience of what networks can provide as well as the problems.

Over the past ten years data communications have increased in complexity to provide networks that have switching, concentration and high connectivity. Packet Switching is the most widespread example of this in wide area networking. The two standard Packet Switching interfaces, X25 and Asynchronous, are on Host and Terminal equipment that can be provided by most manufacturers. Although Packet Switching started in the province of PTTs, this is no longer the case and large numbers of private networks have been installed. Local Area Networks (LANs) have been part of private networking from their introduction. They have achieved their success not only by providing relatively cheap connectivity, but also by developing interfaces to a wide variety of systems. Another important service, which is not yet widespread, is Integrated Services Digital Network (ISDN). As with Packet Switching and LANs, ISDN provides a network with inherent switching, concentration and high connectivity.

Networks, both public and private, have increased in complexity to meet the demands of data communications systems on which businesses have become dependent. It follows that the management of these networks is fundamental to business success. This has led to greater demands from users for systems for fault detection, performance monitoring, configuration management and inventory. This demand will continue to exert a great influence on the development of communications networks.

The most important development, however, has been the Reference Model for Open Systems Interconnection (OSI) and the standards that apply to it. The lower three layers of the Reference Model apply to networks. These three layers are conformed to by the interfaces for Packet Networks, ISDN and the lower layers of LANs. The higher layers apply to end user systems, that is Hosts and Terminals. OSI is of particular importance to users in helping them overcome their problems of systems interoperability and procurement.

The *Handbook of Data Communications* is structured so that the first part is concerned with transmission theory and how this has been developed to produce the analogue and digital services used for data and digitised voice transmission. This part also discusses the ways in which the services can be taken advantage of by the terminal equipment. The next part concentrates on the basic elements necessary to provide coherent data transmission; error control, flow control and protocols. The part following this covers the distributed connection services; Packet Switching, LANs and ISDN. Network Management issues are then covered. Finally, the last chapter takes us forward in time, predicting the developments and changes expected over the next few years.

The book is meant to be used as a reference to help in the understanding of a wide range of data communications topics and to encourage some readers to seek further knowledge in specific areas. It should prove extremely valuable to students planning to enter the communications industry.

Acknowledgements

The authors of the individual chapters within this Handbook are as follows:

1	Origins and Basic Concepts	Martin Gronow, Managing Consultant, NCC Communications Division
2	General Telecommunications Principles	Ian Benton, Senior Consultant,
3	Modulation and Multiplexing	Henry Brysh, Senior Consultant, NCC, OSI Communications
4	Analogue to Digital Conversion	Ian Benton, Senior Consultant
5	Telephone Networks	John Abrahams, Independent Consultant
6	Data Networks	Dave Crawford, Independent Consultant

Handbook of Data Communications

The Nat

 BLACKWELL

Copyright © NCC Blackwell 1982, 1993, 1995

First published 1982
Second edition published 1993
Paperback edition first published 1995

NCC Blackwell
108 Cowley Road, Oxford OX4 1JF, UK

Blackwell Publishers Inc.
238 Main Street, Cambridge, MA 02142, USA

British Library Cataloguin___ ___ ___ ___ata
A CIP catalogue record ___ ___ble from the British Library.

Library of Congress ___ ___ Data has been applied for.

ISBN 1–85554 ___ ___bk.)

Typeset i___ ___The National Computing Centre Ltd.
Printed ___ ___Bodmin, Cornwall

This b___ ___

Contents

1

Origins and basic concepts

Origins

One of the best historical yardsticks by which one can measure the progress of mankind has been his ability to communicate information, both over distance and through time. The efficient recording and dissemination of information had been a problem until the invention of the printing press.

Although this allowed effective preservation of records, telecommunication, communication over distance, still usually involved physical travel. Indeed, until the advent of true telecommunications, speed of information transfer over distance was governed formerly by man's ability to exploit his fellow animals and latterly to capitalise on new methods of transport, and physical transportation of hard copy is still very much in evidence. However, it is to the development of those early methods of rapid communication over distance, which did not involve physical travel, that we owe sophisticated systems of today.

Developments in Telecommunications

Our inability to make our voices heard beyond a very limited distance is something we all learn as children. Most of man's attempts at rapid communication beyond voice range have therefore necessitated using some form of telegraphy, which means the conveyance of messages using signalling symbols or codes. Fiery beacons, for example, used to be a very popular method of indicating danger. Homer, describing the fall of Troy in the 11th Century BC, spoke of a chain of flaring beacons which brought the news to Argos. In 1588, warning of the Spanish Armada was given by a chain of beacons throughout the length and breadth of Britain. Smoke signalling was used widely by the North American Indians.

How much information could be conveyed accurately by such means obviously depended on weather conditions and it is reasonable to conclude that the Indians had a few error control problems. The tom-tom is one of the most ancient methods employed and is still in use today by the natives of Africa, South America and Polynesia. Little is known about the codes used but it is interesting to reflect that such primitive peoples, by inventing suitable codes and the means to transmit and relay information, have communicated effectively and rapidly over considerable distances since before the time of Christ.

1

In more recent history, signalling systems have been developed using the movements of human or mechanical arms (semaphore) or reflected light (heliograph) and these systems are still effective today where no alternative form of communication is possible. Since electrical telegraphic communication began in 1837 there have been many ingenious machines using various codes and signals. Although the first reliable machine for sending letters and figures was invented by Wheatstone in 1840, we owe much to Baudot, Hughes, Morkrum and others for their pioneering work in electrical telegraphy.

Since the invention of the telephone by Bell in 1876 to completely different, but nevertheless complementary methods of communication have been developed, both reaching high levels of sophistication. Firstly, the telephone effectively removed the problem of the limited range of the voice; apart form the enormous social benefits and changes it has brought it has also enabled man to advance rapidly in removing barriers which had previously prevented the exchange of ideas over any significant distance. Secondly, telegraphic communication through telegram services, private telegraph circuits and the dialled telex system has enabled worldwide communication of textual information albeit on a more modest scale of development than the telephone. It may be regarded as remote typewriting using teleprinter machines; the transmission speed of telegraphy is closely allied to the keying speed of teleprinter operators.

An additional major thrust in the third and fourth quarters of this century, has been the rapid expansion of data communications services, largely as a result of developments in computer technologies, coupled with the thirst of modern business for information and the ability to transport and manipulate it. This form of communication, generally at much higher speeds than those employed in telegraphic applications has been closely associated with the development of the computer. Modern data communications practices now demand levels of interconnectivity between disparate systems unheard of only ten years ago. This push for open systems has become a major influence in the development of computer architectures and applications.

Finally the demand for open systems and networks capable of supporting multi-media systems with high information rates has spurred the development of a new generation of networks. These networks known as Integrated Services Digital Networks (ISDN) capitalise on the latest technologies to support systems and services and will take us into the next century and beyond.

The Electronic Computer

Charles Babbage's nineteenth century work, now regarded as the real precursor of modern computing, lay dormant for nearly a century until the technology emerged for the construction of the first automatic computer. In 1937 Claude Shannon first demonstrated the parallel between switching circuits and the algebra of logic. He also defined a universal unit of information, a *bit* being the amount of information needed to remove the uncertainty between yes and no (or between on or off).

Data and Information

The modern history of digital computers began in 1939 with the efforts of Howard Aiken and his associates at Harvard University. Work on their electromechanical Sequence Control Calculator began in 1939 and was completed in 1944.

The second key development was the construction of the ENIAC (Electronic Numerical Integrator And Calculator) as a joint project between the University of Pennsylvania and the United States Army. This machine, which was completed in 1946, used thermionic valves rather than electromagnetic relays and was the forerunner of the first generation of electronic digital computers. Until the discovery of the germanium transistor in 1947, electronic computers consumed enormous amounts of electrical power, generated much heat, and were unreliable. The early machines using valves were employed mainly for scientific and experimental work. There was then little desire to add to the problems of computing by passing information remotely over telegraph and telephone lines.

The next generation of electronic computers emerged in the mid-1950s. These were *stored program* machines, holding programming instructions in the main memory of the computer rather than externally on punched cards or tape. They used transistors instead of valves, had more efficient storage and consumed less power; they were faster and more reliable than their predecessors. Work at Birkbeck College, London, in the development of magnetic drum storage, contributed to this area of development. This reflected earlier contributions by Cambridge University and J Lyons & Co Ltd who introduced LEO (Lyons Electronic Office, 1951), the first computer for commercial use. Because they could store and rapidly process large amounts of information, computers were increasingly being developed for commercial data processing. When information for processing originated at sites remote from the central computer, the source documents were usually sent by post to the computer centre; here the information was converted into a machine-readable coded form and held on punched paper tape or punched cards for processing at some later time.

In the late 1950s, as the volume of information grew and as jobs became more time critical, people turned towards the existing telegraph circuits and the Telex system as faster alternatives to the post for collecting information *off-line*. By the 1960s and 1970s, as the demand for remote processing grew, slow-speed circuits could not cope with the volumes of data, and requirements emerged for lines capable of higher speeds. Networks, involving many users and computers, developed — locally, nationally, and internationally.

During the 1980s, the volume of information being processed has risen dramatically. The development of the microprocessor has brought Information Technology closer to the user, particularly with Personal Computers. With this processing requirement has come a huge demand for communications resulting in LANs and Packet Switching becoming commonplace. Value Added Networks are widespread as are lines running at Megabit speeds and ISDN is now on the threshold of a huge demand by many customers.

Communicating with Computers

Complementing the development of computer systems has been the utilisation of communications to extend the power of the computer beyond the computer room, thereby allowing the benefits of the system to be more widely available geographically. This mixture of computers and communications gives the benefits of computer facilities at one's fingertips regardless of location whilst preserving on one site the expertise needed to operate the system. A simple example is where a person dials a connection over the telephone network to interconnect a terminal with the facilities available at the computer. Other examples are central processing, distributed processing, being able to locate development staff remotely from a system.

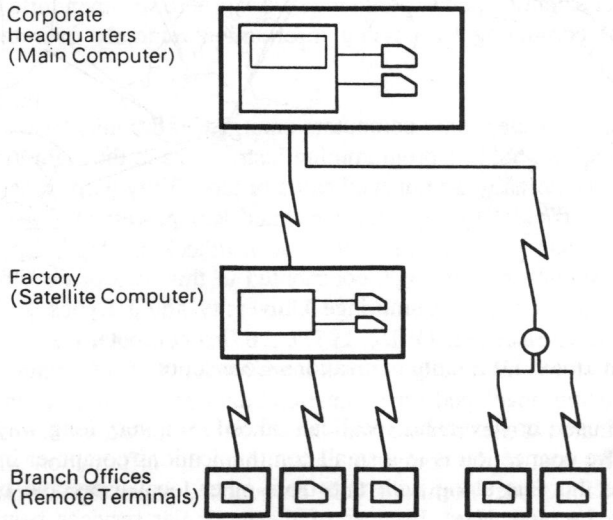

Corporate Headquarters (Main Computer)

Factory (Satellite Computer)

Branch Offices (Remote Terminals)

Figure 1.1 Simplified configuration of a business data communication network

A typical involvement with data communications is shown by the example of a large firm with numerous branch offices and several factories (*see* Figure 1.1). Each of the factories could have its own medium-sized computer which would be linked to terminals within the factory and also to adjacent branch offices. Further communications links could exist between these machines and a larger machine at the corporate headquarters. The individual terminals would be used for the collection and dissemination of the user data, with the *satellite* computer collating and editing this data and carrying out much of the local minor data processing. Major computation and corporate matters would be passed into the large machine at headquarters.

This blend of computers and communications is now taken for granted in a rapidly growing proportion of business organisations. Even the most unsophisticated of users may unconsciously be using very complex systems. For example, a small business in Manchester, England, may have a fairly simple terminal which is connected, via a local telephone call, into a computer service bureau to use one of the facilities offered by that bureau.

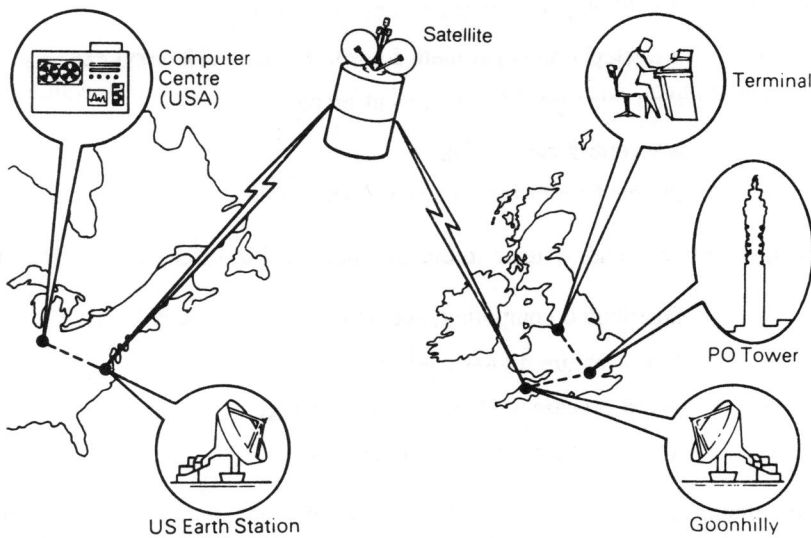

Figure 1.2 International Data Communications Connection

Unknown to the user, however, his local call takes him a very long way from home (*see* Figure 1.2). The connection is to a small communications computer in Manchester which concentrates the data along with that from other local users and passes it to a larger computer in London. Here, because of the particular services being used, it is passed via a communications satellite in orbit above the Atlantic, to the service company's main computer centre in the USA. The results come back over the same links giving the user the impression that the bureau is just next door.

Voice Data Integration

Developments in data communications to support information technology products have brought with it the opportunity to integrate systems from a wide range of sectors and applications. The primary impetus for the integration of not only voice and data, but also of computing and office automation has come from developments in personal office computing (for example, word processing, desktop computing, etc), coupled with the developments in digital communications, digital telephone networks, digital facsimile, etc.

Opportunities have been seized which take advantage of economies of scale and added cost efficiency by combining services wherever possible. Thus, the convergence of voice, computing, word processing and messaging systems is being capitalised upon to the advantage of the user. No longer does the business manager have to turn separately to:

- the DP department for computer services;
- the telephone department for voice and PABX services;
- the typing pool for document preparation;
- a secretary for messages;
- the mail room for document transmission.

Integrated networks and systems in the automated office offer such services as:

- Distributed computing, access to distributed data bases;
- Multi-feature digital telephones;
- Word processing/document interchange;
- Voice and electronic mail services.

Telecommunications link the systems both locally and remotely across company, national and indeed international boundaries, if necessary. Networks and services are already available to transmit information in its principle forms with equal faciles, be it speech, numerical information, written text or image.

Communicating Information

As you read these lines, you are forming part of an information system; the author being the message source (transmitter), the publication the message medium, and yourself the receiver. This is an example of a *simplex* system, communication being in one direction only. In a telephone conversation people do not generally speak at once but exchange roles as transmitter and receiver, maintaining a check on the understanding and accuracy of the messages they are each receiving. This type of information system where messages are transmitted in both directions but not at the same time is termed *half duplex*, and again, we find the three essential parts of an information system — a message source, a message medium (the telephone line) and a receiver. A great deal of information can, of course, be exchanged in both directions at the same time by two people gazing into one another's eyes. In their attempts to describe this simultaneous transmission of messages, poets have surprisingly failed to recognise that this is simply a *duplex* information system.

It is only recently that communication itself has been studied, man throughout the ages having concentrated on the *methods* of communicating. The study of communication theory has provided a better understanding of the factors which limit the rate at which information can be transferred. To the communications theorist, information is 'any organised signal'. This, of course, presupposes that these organised signals have some meaning which is understandable to a receiver. In speech conversation within a room, information in this sense would consist of the complex sound waves 'organised' by the vocal chords, pharynx and tongue of the speaker (transmitter) being transmitted via the medium of air to the listener (receiver). Information on a telephone circuit may consist of a series of tones, a group of tones, direct current pulses or any other signals which are organised. Collectively, the information constitutes the message. Noise also has a precise meaning in communication theory and can be defined as 'any signal which interferes with the message being sent'. Music, no matter how beautiful, could therefore be regarded as noise by two people wishing to talk to one another during a concert.

In addition to the problem of noise, the communication channel must have sufficient capacity for the intended purpose. Analogous to the transmission of a fluid down a pipe, a communications channel does have a finite capacity. This is not readily apparent in speech conversation, where the primary concern is with intelligibility, but it has very important implications for information transmission in its general sense. From this discussion it is evident that there are a number of concepts and principles which require a closer examination and a more precise definition.

The Nature of Data Communications

Data and Information

The term *data* can be defined as 'any representation, such as a figure or a letter, to which meaning can be ascribed'. *Information* has a number of meanings; to the communications theorist, for example, it means 'any organised signal'. Generally, however, the term is understood to describe something which is meaningful. Throughout this book we shall tend to use the terms interchangeably, bearing in mind that information can assume various significant forms meaningful to the user, as we have noted earlier.

Data Transmission

Data transmission can be defined as the movement of information over some physical medium, using some form of physical representation appropriate to the medium. Thus, we include:

— electrical signals carried along a wire;

— radio waves propagated through space; and

— thermal or infra-red signals transmitted through space from a laser source.

Data Communications

Data communications has a much wider meaning than data transmission and embraces not just the electrical transmission but many other factors involved in controlling, checking and handling the movement of information in a communications-based computer system. For example, it includes:

— the physical transmission circuits and networks;

— the hardware and software components required to support the data communications functions;

— procedures for detecting and recovering from errors;

— standards for interfacing user equipment to the transmission network; and

— a variety of rules or protocols for ensuring the disciplined exchange of information.

Transmission Concepts

Transmission Codes

The power of a computer lies in its ability to implement a predefined set of instructions or code in order to perform some task. Such systems are designed around architectures based upon digital logical circuits, each capable of making a yes or no decision. In mathematical parlance these logic states can be represented in *binary* notation. Each binary digit or *bit* can be represented mathematically by a 1 or 0 or electrically by two differing conditions, for example a positive (+Ve) voltage level for logical '1' and a negative (-Ve) voltage for logical '0'. It follows then that if a computer system operates on purely digital information, ie 1s and 0s, then for meaningful communication to take place some form of coding system must be present to convert alpha and numeric characters (letters and numbers) familiar to humans to a form acceptable by the computer.

A coding system with the capacity to represent, say, all of the letters of the alphabet and numerals between 0 and 9 (36 characters in all) would require a coding structure comprising some six binary digits. Such a coding scheme that will give a full numeric and alpha character set is the Binary Coded Decimal (BCD) system.

For data communications, however, an extended character set is usually required to cater for additional characters for punctuation as well as a considerable number of control codes which exist. These control codes govern the transmission of data, enable manipulation of message formats, separate information and switch on or off devices that are connected to the line.

The Extended Binary Coded Decimal Interchange Code (EBCDIC) is an extension of BCD code and uses eight bits instead of six. This code is useful where an application calls for a large number of different characters. Although there are only 109 assigned meanings, there are $256(2^8)$ possible combinations. The code is used mainly to transmit the eight-bit byte[1] of some computers and obviates the need for the code conversion which is often necessary between the transmission code and the code used by the computer.

International Alphabetic No 5

Due to the proliferation of data transmission codes throughout the world, serious attempts have been made to standardise the codes used. The International Telegraph and Telephone Consultative Committee (CCITT), the International Organization for Standardization (ISO) and national bodies have given much thought to a problem which had become increasingly serious as the need grew to communicate between devices of different manufacturers and between different countries.

The International Alphabet No 5 (IA 5) is a seven bit code developed to satisfy the need for a standard code capable of supporting both sophisticated telegraphic and data communications. The alphabet originated from the development by the American National Standards Association of a coding scheme known as the American Standard Code for Information Interchange (ASCII) in 1962. The Alphabet IA 5 was developed and subsequently ratified by the CCITT and the ISO bodies in June 1968.

Standardisation too early can adversely affect progress by impeding the introduction of new and better ideas. This argument is recognised by CCITT and ISO and the IA 5 code allows for a certain amount of flexibility enabling users to 'escape' from the normal conventions of the code by the use of special characters.

Telegraph Codes

A number of computer-based data communications systems still make use of the long established telegraph codes. In telegraphy, the need to transmit plain language text predominates and fewer characters are needed than in data communications. Separate characters for all decimal figures and letters of the alphabet (though not necessarily capital letters and small letters) must be available, which gives a required minimum of 36 characters. Although punctuation marks, 'space' etc, could be indicated in words or combinations of these characters, this would be cumbersome and tedious for both the operator and the reader. Separate characters are therefore provided in practice for these and other purposes. Figure 1.3 shows the keyboard of a traditional teleprinter which

---

[1] A byte is a 'group of consecutive binary digits operated on as a unit by a computer'. Although the eight-bit byte is common, the number of bits in a byte may vary between computers of different manufacturers up to 16- or 32-bit bytes.

provides for the entry of 58 different characters. The code, or alphabet, used is the five-bit International Alphabet No 2 (IA 2) has been widely employed for telegraphy throughout the world since it was ratified by the International Telecommunications Union (ITU) in 1932.

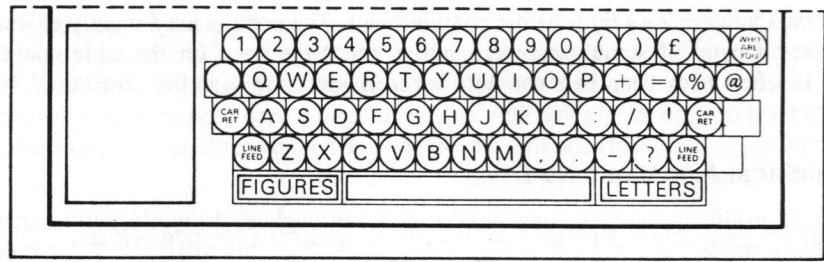

Figure 1.3   The keyboard of a traditional teleprinter

The number of different characters which can be derived from a code having two different states is normally given by a 2^n where n is the number of different units in the code; a five-unit code would therefore give 2^5 or 32 characters. However, IA 2 uses two 'shift' characters to extend the capacity of the code. The depression of a 'figures' key results in a unique character being reproduced which when transmitted informs the receiver that the characters which follow are to be interpreted as 'figures' or other 'secondary' characters. Similarly, 'letters' indicate that the following characters are to be letters or other 'primary' characters. In this way IA 2 offers 52 graphical (ie printable), two shift, three functional and one unallocated characters.

Efficiency of Codes

The efficiency of a two-condition code can be expressed by the formula:

$$E = \frac{\log_2 N}{M} \times 100$$

where   E = the efficiency of the code;
N = the number of characters or symbols required;
M = the number of bits in the code.

Let us assume that in a particular application 64 different characters are required and a seven-bit code is to be used. Applying the formula we have:

$$E = \frac{6}{7} \times 100 = 86 \text{ per cent}$$

If an eight-bit code were to be used then the efficiency would be 6/8 or 75 per cent.

Examined in this way, the telegraph codes such as IA 2 which employ 'letter shift' and 'figure shift' characters to extend the character set seem to be very efficient. IA 2 is only a five-unit code, yet 55 different characters are available to the user. Assuming that all these characters are required this gives a coding efficiency of approximately 116 per cent.

However, in practice a balance has to be struck between efficiency, ease of use within the system, acceptability to the human operator, and the range of characters required. The last two factors have become overwhelmingly important, and largely for these reasons are little used outside telegraphy. The requirement now is to send an even greater range of characters (eg graphical) and this can be done efficiently with data integrity protected by protocols that are 'data transparent'.

Coding by Statistical Probability and Information Compression

As the amount of information transmitted continues to grow, there is an increasing amount of attention paid to reducing transmission costs by using more efficient coding methods. The development of more efficient coding for data communications was accelerated by the pressure to reduce the escalating costs involved in storing and controlling the vast and ever increasing quantities of information held on computer files.

Considerable advances have been made in this area, and one application which is worth mentioning is the use of information compression techniques to reduce the bandwidth (see below) requirement for transmitting video signals. The techniques used either rely upon a coding scheme which reflects the statistical probability of occurrence or merely avoids transmitting information which is in some sense redundant. Examples of the latter are blank areas in printed text; and those parts of a video image which remain static within a given sampling interval.

Physical Transmission

In order to be transmitted, the coded representation must be converted into a physical representation or signal. This can be illustrated by reference to telegraphy or telex.

The method of entering information on a teleprinter is normally a keyboard, and the basic unit of information entered is a *character*. To arrange information entered by the keyboard into a form suitable for transmission, the depression of a key results in the conversion of a character into a code comprising a combination of five separate units. Each unit represents one or two possible conditions termed 'mark' or 'space'; these are equivalent to the binary symbols '1' and '0' respectively.

The code must be converted into electrical signals in order to be transmitted over a line. In practice, telegraph signals are transmitted 'serially' or one element at a time. Also, because a disconnection can provide a false signal if 'space' is represented by no current flow, the 'mark' and 'space' conditions are indicated by current flowing in different directions.

The teleprinter signals transmitted to line when the R or 4 key is depressed are shown in Figure 1.4. The signals transmitted are 'digital' ie they are transmitted at predetermined discrete levels; in this case 80 volt positive or negative, the polarity or direction of the current flow indicating whether a mark or a space was to be transmitted. Such high voltages for signal representation are now rare, modern telex systems now employ a tone signalling system using predefined frequencies to represent the 'mark' or 'space'. More complex signalling systems have been devised to take advantage of the ability to transmit data over the existing analogue networks. Digital networks were developed with the objective of sending large amounts of voice and data traffic on single links.

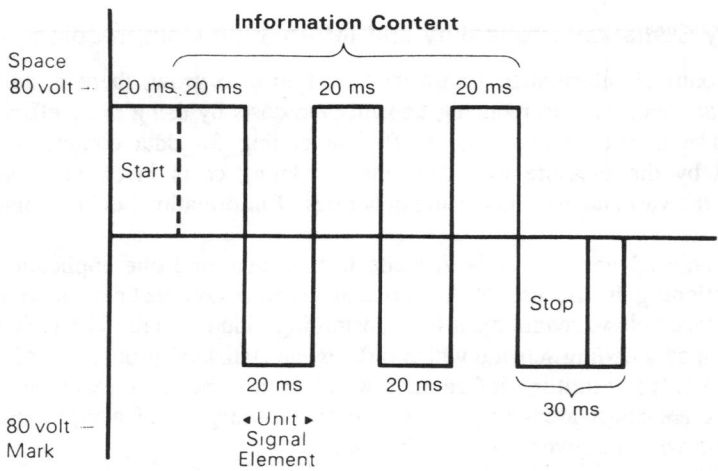

Figure 1.4 Signals transmitted to line from teleprinter

Transmission Signal Attributes

For telecommunication purposes the transmitted signal, and its behaviour during transmission, has three fundamental attributes. These are frequency, amplitude and phase. These are perhaps most readily appreciated in relation to sound or auditory signals.

Sound

Sound is the physical disturbance of air (or some other medium) which when received by the ear can be transmitted to the brain by the nervous system. Sound is produced by vibrations which cause compression and rarefaction of the air. The air vibrates with the sound source and the diaphragm of the ear will, if it is within range, vibrate in sympathy. The variations of air pressure which produce sound can be plotted graphically against time as shown in Figure 1.5. This example shows the sound wave generated by a single frequency sinusoidal source. Only one transition between compression and rarefaction is shown; this is termed a 'cycle'. In mathematical notation this would be represented by a trigonometrical expression. In the example shown, a pure sine wave, a complete cycle of the signal would involve a complete rotation through 360°.

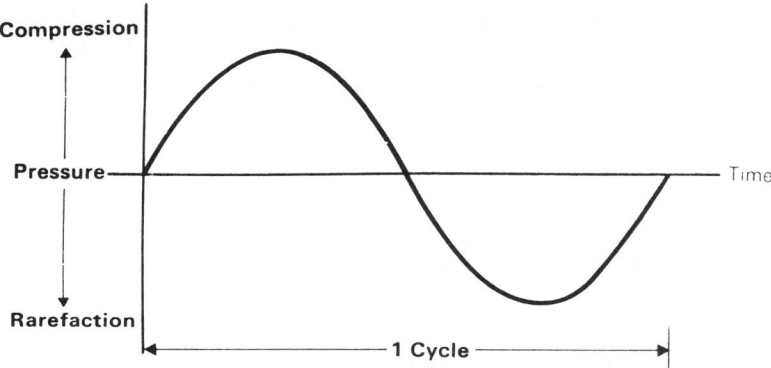

Figure 1.5 The pure sound wave

Frequency or Pitch

In music, notes are often referred to as being 'high' or 'low'. Perhaps this is because of the position of the notes on a musical scale or the tendency of some tenors to reach for a note by standing on their toes! The fact is that so-called 'high' notes have a greater number of cycles in a given time than do 'low' notes. Middle 'C' at concert pitch, for example, has a frequency of 270 cycles per second (270 Hertz), while the 'C' above middle 'C' has a frequency of 540 Hertz (Hz). The relationship between pitch and wavelength can be best illustrated by an example. If in Figure 1.5 the cycle shown took 0.001 of a second to complete and were to be continually repeated, a note of 1000 cycles per second (1000 Hz) would be produced. If this took place at normal room temperature, the velocity of sound would be 340 metres/s; the 'length' of each cycle would therefore be 340/1000 or 0.34 metres; the 'wavelength' of a 1000 Hz note therefore would be 0.34 metres through air at room temperature.

Amplitude or Volume

Volume or loudness is determined by the 'amplitude' of the waveform or the height of the peaks of compression and rarefaction. In Figure 1.6, the two sound waves have the same frequency or pitch but B has a greater volume than A.

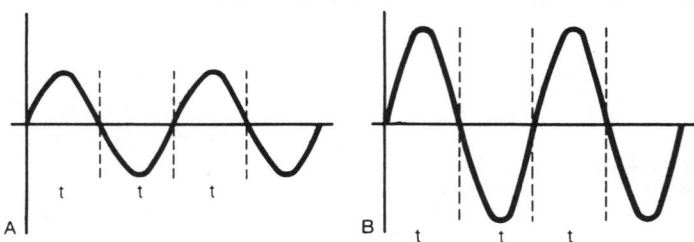

Figure 1.6 Frequency with two levels of amplitude

Tone

A person who is tone deaf might have difficult in hearing or reproducing correct pitch. This is a fairly common malady which in no way affects the afflicted person's ability to speak or sing. Indeed, tone deaf people are usually blissfully unaware of any problem and seem gifted with painfully robust voices. In contrast, the vast majority of people are able to distinguish between the characteristic sounds of, say, a dinner gong and a violin, or a male and female voice. This ability to tell the difference between 'tones' is of more fundamental importance to human beings for it is only in doing so that we are able to communicate.

Differences in the quality of sound are produced by variations in the fundamental waveform (the pitch). These variations are produced by the introduction of additional frequencies known as 'harmonics' or 'overtones'. For example, if a middle 'C' is produced by a piano, the fundamental frequency is caused by the middle 'C' string vibrating 270 times a second. However, the outer edges of the string, the frame and wooden components also vibrate at different frequencies to produce harmonics providing the characteristic tone, not only of the type of instrument, but of the particular piano being played.

The effect of harmonics upon what would otherwise be considered a pure sine wave is to produce a highly complex waveform in the form shown in Figure 1.7. In fact many signals can be proved both diagrammatically as well as mathematically to be of complex form, comprising a fundamental frequency and a series of harmonics.

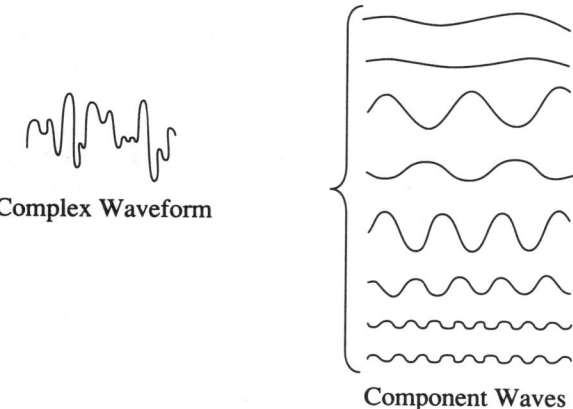

Complex Waveform

Component Waves

Figure 1.7   Components of a complex waveform

Phase

Figure 1.8 shows two sinusoidal waveforms in which B lags behind A. Each is said to be out of phase with the other. Whether one lags or the other leads depends upon the point chosen for the time origin. In the diagram, 0 is regarded as the origin and B therefore lags A by 90° since the amount of lag corresponds to a quarter of a revolution.

Phase is particularly important for modulating theory, and is the basis of several commonly used modulation techniques discussed in a later chapter.

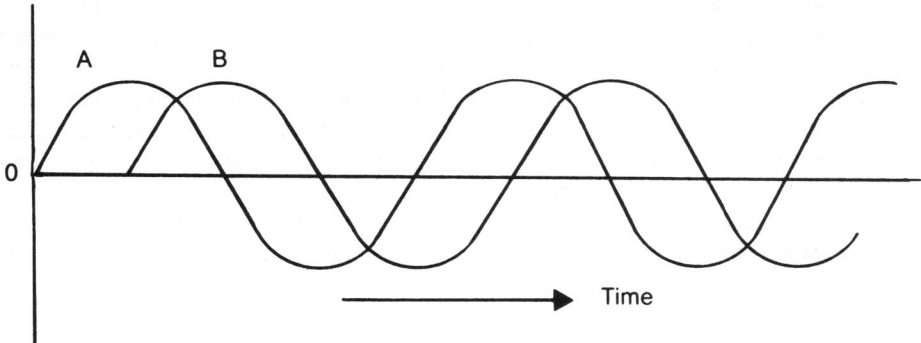

Figure 1.8   Phase relationships

Digital and Analogue Transmission

Although less fundamental from the foregoing in terms of providing a complete description of the signal, whether it is digital or analogue constitutes a very important qualitative difference.

Digital computers and associated equipment generate digital signals, and the distinction assumes significance when we have to consider the feasibility and efficiency with which digital signals can be transmitted through a network which is designed primarily to transmit analogue signals exemplified by the traditional telephone network.

The telegraphy example in Figure 1.4 illustrates the form of digital signals whilst the figures accompanying the discussion of frequency and amplitude illustrate the essential character of analogue signals. The distinguishing feature is that digital signals are discontinuous, and the analogue signals are continuous.

Noise and Attenuation

Noise

In our earlier discussion we noted that the quality of communication can be impaired if some concurrent extraneous communication interferes with it. This introduces the problem of noise.

Gardeners are well aware of the fact that man's view of orderliness does not coincide with that of nature. Reversion to nature is not a problem exclusive to gardeners, however, and all man's attempts to impose his own version of order are opposed by nature's. This strong natural force is evident in the oxidisation of metals and is most familiar in the form of rust. This force is at work on telecommunications channels in opposition to organised signals (information), producing weakening and distortion of the signals. It also manifests itself in the background noise which can sometimes be heard as a hiss on telephone or radio channels. This type of 'white noise ' (or gaussian noise) is inevitable on telecommunications channels, being produced by a natural movement of electrons which varies with temperature. The term 'white noise' is used because just as white light contains all the colours in the spectrum, white noise is purely random and can be of any frequency. Impulsive noise is also a serious problem in the transfer of information on telecommunications links. This can often be heard as clicks in a telephone conversation and is less of a natural phenomenon, being produced as a result of interference from other telephone circuits.

The full impact of noise depends upon whether speech or data is being transmitted. With speech communications, we are primarily concerned with intelligibility. Intelligibility can be defined as 'a percentage of simple ideas correctly received over the system used for transmitting or reproducing speech'. There are problems in measuring intelligibility; this is due to the fact that the human brain has an error correction capability and, although we may miss sections from words or sentences in a telephone conversation, our brains have the ability to fill in some of the gaps and interpret the meaning. Let us assume that the following sentence is spoken during a telephone conversation — 'Now is Tom foot all goot men toot compt taid party'. This sentence would probably be corrected by the listener without too much difficulty and perhaps unconsciously into the familiar sentence — 'Now is the time for all good men to come to the aid of the party'. Many interesting tests have been made using 'logatoms' which are specially constructed sentences using words or syllables which contain no information whatsoever but are extremely useful in measuring intelligibility.

The intelligibility of the human voice is contained within the harmonics produced; most of the intelligence in human speech occurs between 125 and 2000 Hz — a 'bandwidth' of about 2000 Hz. The CCITT Recommendations are a circuit responding to frequencies between 300 Hz and 3400 Hz as being adequate for the purposes of telephony, giving a high degree of speech intelligibility.

Considering information (or data) to be processed by a computer or where content is not immediately meaningful to a human being, the situation is rather different. As far as the physical representation of the signal is concerned, a major effect of noise is to distort the shape of the signal; and the information received following transmission could either be complete gibberish or contain errors which are not necessarily readily apparent.

Whether noise does interfere significantly with information is obviously connected with the power of the signal relative to the power of the noise and to the sensitivity of the receiving apparatus. This leads us into the topic of attenuation.

Attenuation

It is a fundamental physical fact that a signal loses power during transmission, and this loss of signal strength is referred to as attenuation. The extent of the loss depends upon the transmission medium, and for a specific medium is generally a function of distance.

For example, an audible sound very quickly becomes inaudible in still air, and the same applies to electrical signals, although to a much lesser extent. Other factors, such as frequency, also affect attenuation, and these are discussed in later chapters.

In practice, it is therefore necessary when attenuation reaches a certain level to restore the power of the signal by amplification. However, in the case of analogue channels this has an adverse side effect, in that besides amplifying the signal, the power of the noise is also increased at the same time. This problem does not arise with digital transmission. Although the digital waveform or pulse eventually loses power, and its shape may be disturbed due to noise, a different technique is available for restoring the quality of the signal. Since the pulse corresponding to the binary value can be standardised in advance, all that is necessary is to introduce at intervals a much simpler device which, instead of amplifying the signal, compares it with the reference signal and reshapes the transmitted pulse accordingly.

Data Transmission Rates

There are a number of different ways in which the rate of transmission can legitimately be expressed: 'modulation rate', 'data signalling rate' and 'data (or information transfer) rate'. A great deal of confusion can be caused by the misuse of these terms and they are explained separately below.

Modulation Rate

This is a term used by the communication engineer to describe the performance of a circuit in terms of the rate at which changes in the condition of the circuit can be made in a given time. More precisely it is the reciprocal of the duration of the unit signal element. The unit used in expressing modulation rate is the 'baud' which is equal to one unit signal element per second. For example, in Figure 1.9 each unit signal element is 20 ms in duration. The modulation rate is therefore:

$$\frac{1}{0.020} = 50 \text{ baud}$$

It should be noted that the expression of modulation rate in bauds does not necessarily indicate the rate at which data is transmitted.

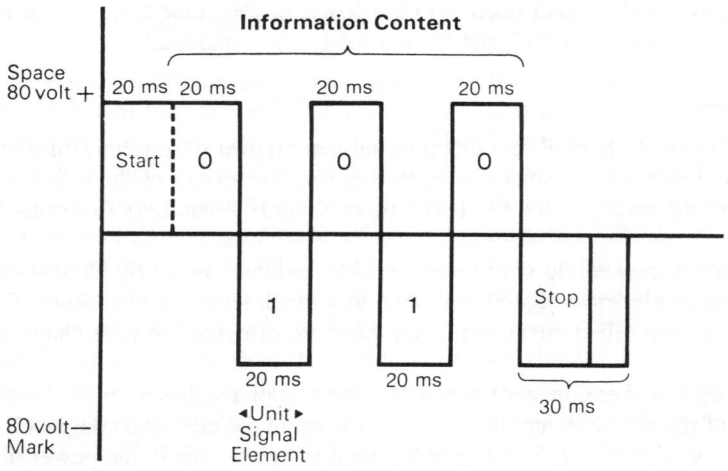

Figure 1.9 Data signalling using two voltage levels

Data Signalling Rate

The data signalling rate is used to express the rate at which information can be transmitted.

It is expressed in bits per second (bit/s) and for serial transmission it is defined as:

$$\left(\frac{1}{T}\right)\log_2 n$$

where T = the duration of the unit signal element in seconds and n = the number of signalling conditions.

Again, referring to Figure 1.9 the data signalling rate would be:

$$\left(\frac{1}{0.020}\right) \times 1 = 50 \text{ bit/s}$$

It would, however, be wrong to conclude from this that a baud is the same as 1 bit/s, for if more than two signalling states (multi-state signalling) were used we would have completely different answers.

Multi-state signalling can be explained by using a simple analogy.

Suppose the following data has to be passed between two persons in the same room without speaking or using written communication: 100001011001. There are, of course, many ways in which this might be achieved using two 'signal states', eg a white flag could be waved to represent binary '1' and a red flag for binary '0'.

However, if four different coloured flags were to be used, say white, red, green and yellow, further possibilities would be available. There are only four different ways of combining two binary digits, ie 00, 01, 10 and 11, therefore the white flag could be used to represent 00, the red 01, the green 10 and the yellow 11. Thus, in the example the data 100001011001 could be sent using flags in the following order: green, white, red, red, green, red. It will be seen that with this 'four-state signalling', the number of signals necessary to transmit the information is only half that necessary with two-state signalling. Eight different coloured flags could be used to convey three bits of information at a time (there are eight different ways of combining three binary digits), sixteen flags for four bits and so on. Theoretically an increase in the number of flags used should result in more information being transferred with the same number of signals. There is, however, an obvious problem with coding and decoding which progressively increases; there are also technical problems which increase with the number of different states employed when multi-state electrical signalling is employed for data transmission.

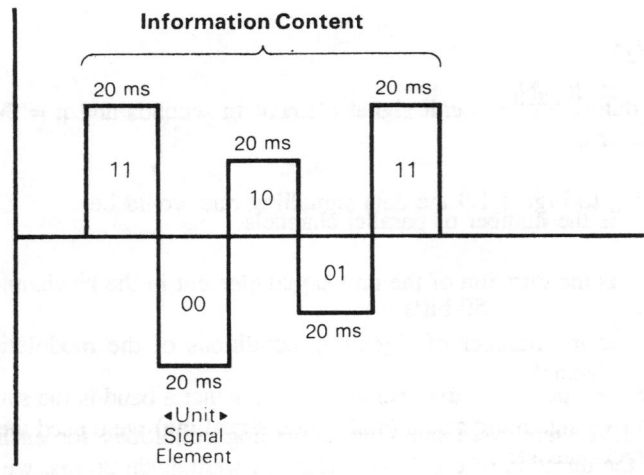

Figure 1.10 Data signalling using four voltage values

Figure 1.10 shows a simple four-state signalling system, each 20 ms unit signal element representing two binary digits. The data signalling rate would therefore be:

$$\left(\frac{1}{0.020}\right) \quad \log_2 4 = 50 \times 2 \text{ or } 100 \text{ bit/s}$$

Note that the modulation rate is still 1/0.020 or 50 bauds.

Although serial transmission is most common in telegraphy and data transmission, parallel transmission is sometimes used, complete characters rather than separate bits being transmitted at the same time. If the simple hypothetical example in Figure 1.11 is considered, it will be seen that the data signalling rate in bit/s transmitted will be a summation of the bits transmitted on each transmission path or 'channel'.

$$\left.\begin{array}{cc} 1 & 0 \\ 0 & 1 \\ 1 & 0 \\ 0 & 1 \\ 1 & 0 \end{array}\right\} \rightarrow$$

Figure 1.11 Serial transmission

The data signalling rate in a parallel system can therefore be expressed by:

$$\sum_{i=1}^{i=M} \frac{1}{T} \log_2 N_i$$

where   M   is the number of parallel channels

Ti   is the duration of the unit signal element in the ith channel in seconds

and   Ni   is the number of signalling conditions of the modulation in the ith channel.

If in Figure 1.11 there were only two signalling conditions for each of the five channels and if the duration of each signal element were again 20 ms, we would have:

$$\frac{1}{0.020} + \frac{1}{0.020} + \frac{1}{0.020} + \frac{1}{0.020} + \frac{1}{0.020} = 250 \text{ bit/s}$$

Data (Information) Transfer Rate

Unlike data signalling rate which is used to describe the rate at which data is transmitted, 'data transfer rate' describes the rate at which data actually arrives after transmission. It is defined by CCITT as 'the average number of bits, characters or blocks per unit time passing between corresponding equipment in a data transmission system'. It is expressed in terms of bits, characters or blocks per second, minute or hour.

Frequently the corresponding equipment referred to in this definition are a *data source* and a *data sink*. Consider the example of a terminal connected to one computer system spooling data via a dial up communications link to a printer on another (*see* Figure 1.12). For the purpose of this example, we will regard the data held on the intelligent terminal as the data source and the printer at location B as the data sink. Let us assume the communications link will transfer data to the printer at a data signalling rate of 1200 bit/s start/stop mode using 1 start bit and 1 stop bit. In one minute 72,000 bits will have passed across the link; however, in an asynchronous (start/stop) mode some 2 signal units per character are purely for delimiting each block of information. Thus, at the 'data sink' (the printer in this example) only 57,600 bits of user information will have been received, one fifth of the total data transmitted being redundant data.

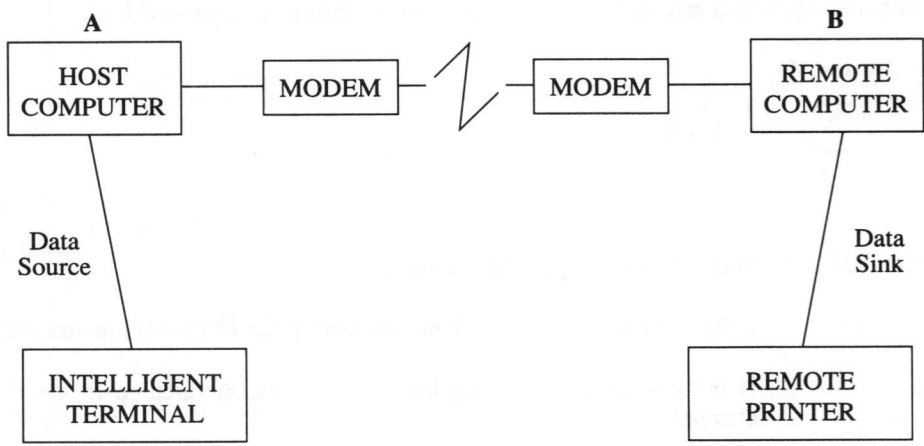

Figure 1.12 Remote data base access

The terminology used to describe this condition is referred to as the Data Transfer Rate (DTR). This can be expressed as follows:

$$\text{DTR} = \frac{N}{T} \quad \text{bits/second}$$

where N = the number of information bits accepted by the 'data sink'.

T = the time required (in seconds) to transmit N bits of data.

However, as will be seen in Chapter 14, the calculation of information transfer rate is not usually as simple as this and will depend upon the error rate on a particular transmission, the type of error control and link protocol employed and not only the number of bits, but the number of complete characters which are redundant.

Although DTR is a much more accurate way of describing the movement of usable information in a data communications system than any other, it cannot be stated without calculation based on knowledge of a particular system. A line or a piece of equipment cannot therefore be said to have a data transfer rate of a certain number of bits per second (bit/s). For this reason, bit/s in this book will be used to indicate data signalling rate unless stated otherwise.

Bandwidth

The bandwidth of a channel is the major determinant of the channel's information-carrying capacity. The bandwidth has a precise definition: it is the range of frequencies that the channel is capable of transmitting. There is therefore an intimate relationship between bandwidth and frequency, and as we have already seen, a signal in general can be analysed into a number of component frequencies. The analysis is performed using the technique of Fourier analysis. (Fourier analysis and its associated principles play a fundamental role in communications theory.)

The use of range in the above definition is important because not only are a number of frequencies present — so that the signal has a frequency spectrum — but the lowest frequency to be catered for is likely to be something different from zero, and the higher frequency can be either finite or theoretically infinite, as we shall see. Thus, the PSTN is designed to transmit frequencies in the bandwidth 300-3400 Hertz. Therefore, a number of the higher frequencies or harmonics present in the human voice are not transmitted, and this is the reason why people's telephone conversation often sounds 'flat' and relatively low-pitched.

With speech conversation the main problem is conveying intelligibility, and the bandwidth provided is generally adequate for the purpose. Also the rate at which analogue speech information is transmitted is not usually constrained by the telephone channel but by the natural rate of the conversion.

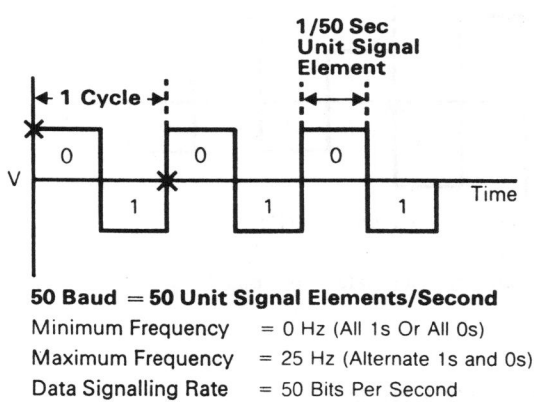

50 Baud = 50 Unit Signal Elements/Second

Minimum Frequency   = 0 Hz (All 1s Or All 0s)
Maximum Frequency   = 25 Hz (Alternate 1s and 0s)
Data Signalling Rate = 50 Bits Per Second

Figure 1.13 Relationship between bandwidth and information rate

However, bandwidth is a major constraint on the rate at which digital information can be transmitted. This can be seen more clearly if we examine the capacity of the human ear, with a maximum bandwidth of 20,000 Hz, to accept digital information. We often receive what might loosely be described as digital signals and an example of this can be heard whenever we sit in a train compartment. The typical two-state di-di-di-dum sound rises with the speed of the train. The pitch of this sound continues to rise with speed and if the train were to travel fast enough, the frequency of the signal would rise to a point where we could no longer hear it. The relationship between the rate at which information can be transmitted and bandwidth can perhaps be seen more clearly if we examine a telegraph type signal, ignoring for convenience the usual start/stop elements. We can see from Figure 1.13 that this 50 baud signal would carry 50 bits per second of information. However, it is also apparent that the fundamental frequency produced by sending this information would vary between 0 Hz (when all ones or zeros were transmitted) to a maximum of 25 Hz if alternate zeros and ones were transmitted. One bit of information is therefore represented in each half cycle of the 25 Hz square waveform. Nyquist[1] showed that, in a channel without noise, a signal with no frequencies greater than W could carry 2W voltage values. This means that in the idealised situation of a noiseless channel and using only two voltage values serially (to represent '0' and '1') the maximum theoretical capacity of the channel C would be 2W, where W is the bandwidth of the channel.

If more than 2 voltage levels were used, the maximum frequency would be unchanged. Figure 1.14 shows a system employing four states.

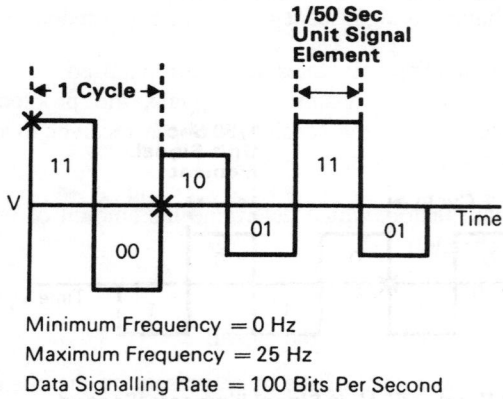

Minimum Frequency = 0 Hz
Maximum Frequency = 25 Hz
Data Signalling Rate = 100 Bits Per Second

Figure 1.14 Signalling using 4 levels

In this example, we again have fundamental frequencies which may vary between 0 Hz and 25 Hz (varying over a 25 Hz bandwidth), but two bits of information can be derived from each half cycle.

We can, therefore, extend the theoretical channel capacity in the absence of white noise to include provision for multi-state signalling:

---

[1] Nyquist, H 'Certain Factors Affecting Telegraph Speed' (1924) and 'Certain Topics in Telegraph Transmission Theory' (1928). Trans. AIEE

$$C = 2W \log_2 L$$

where $\quad$ C $\quad$ = the channel capacity in bit/s

$\qquad\quad$ W $\quad$ = the bandwidth

and $\qquad$ L $\quad$ = the number of states

eg: $\qquad$ bandwidth (W) = 2000 Hz

$\qquad\quad$ number of states (L) = 8, then

$\qquad\qquad$ C $\quad$ = (2 x 2000) x ($\log_2 8$)

$\qquad\qquad\qquad$ = 4000 x 3 = 12,000 bit/s

It would seem from this that channel capacity could be increased *ad infinitum* by increasing the number of signalling levels. Unfortunately, this is not true in practice; although these formulae are useful in showing the relationship between bandwidth, signalling levels and channel capacity, there are very real snags:

- There are no telecommunications channels which are completely free of white noise or other disturbances. However, fully-digital transmission of speech or data, combined with solid state electronics, is having a significant impact on noise reduction.

- The number of states that can be used is limited by the power available to transmit the signals, the problems of encoding and decoding and the sensitivity of the receiver to interpret the different signalling levels.

There are three main factors which determine the amount of information which can be transmitted on a channel:

- the bandwidth available;

- the power level of the signal;

- the power level of the noise present on the channel.

The work of Claude E Shannon[1] is of fundamental importance in providing mathematically that a communication channel has a finite capacity. The Shannon/Hartley law is now recognised as representing the theoretical maximum capacity of a channel in the presence of white noise and is given by:

$$C = W \log_2 \left(1 + \frac{S}{N} \right)$$

where $\quad$ C $\quad$ = channel capacity

$\qquad\quad$ W $\quad$ = bandwidth

$\qquad\quad$ $\dfrac{S}{N}$ = signal to noise ratio

---

[1] 'Mathematical Theory of Communication', Bell Systems Technical Journal (July and October 1948).

The ratio of signal to noise is a dimensionless quantity; however it is often expressed in decibels using the formulae:

$$\frac{S}{N} \quad \text{in dB} \ = \ 10\log_{10}\left(\frac{S}{N}\right)$$

The term decibel is a logarithmic unit of measure for comparing two power levels. It is also used to indicate absolute power levels and when used in this way a third letter is added to the notation, ie dBW meaning if the reference power is one watt, the power P is expressed in decibels above one watt.

NOTE: It is important to remember that the noise referred to here is 'white noise'. Other forms of disturbance, which are present in practice, include 'impulse noise', or noise peaks which are critical in determining the error performance of a channel.

Applying this formula to a fairly good quality speech channel with a bandwidth of 3000 Hz and a signal to noise ratio of -30 dB we have:

$$C \ = \ 3000\log_2\left(1 \ + \ \frac{1000}{1}\right)$$
$$C \ = \ 30{,}000 \ \text{bit/s}$$

This is very much more than could be achieved in practice with a circuit of this kind. The actual rates achieved on analogue channels are not solely dependent on the channel but on the modems used. Although much ingenuity has gone into the design of modems, and the introduction of microprocessor technology has enabled remarkable levels of sophistication and performance to be achieved, it is most unlikely that they could be designed to approach closely the theoretical maximum capacity of a channel.

Bandwidth Requirements for Digital Transmission

We have discussed bandwidth in fairly general terms, and we should now look more specifically at the bandwidth requirements for transmitting a typical digital signal along an analogue channel. In doing this we should present the Nyquist theorem more formally.

The digital data signal generated by a typical data terminal is a square wave, like the one depicted in Figure 1.15. This actually shows a *polar* signal, since the two states have equal positive and negative values. Since time increases towards the right, the bits in Figure 1.15 would be transmitted in the order 0, 1, 1, 0, 1, 0, 0. By convention in data transmission, the low-order bit is always transmitted first.

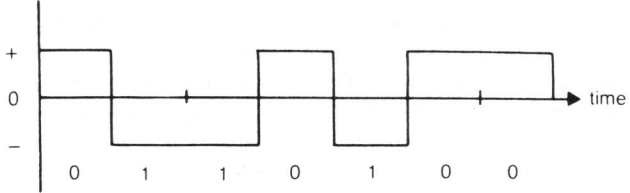

Figure 1.15 Polar binary signal

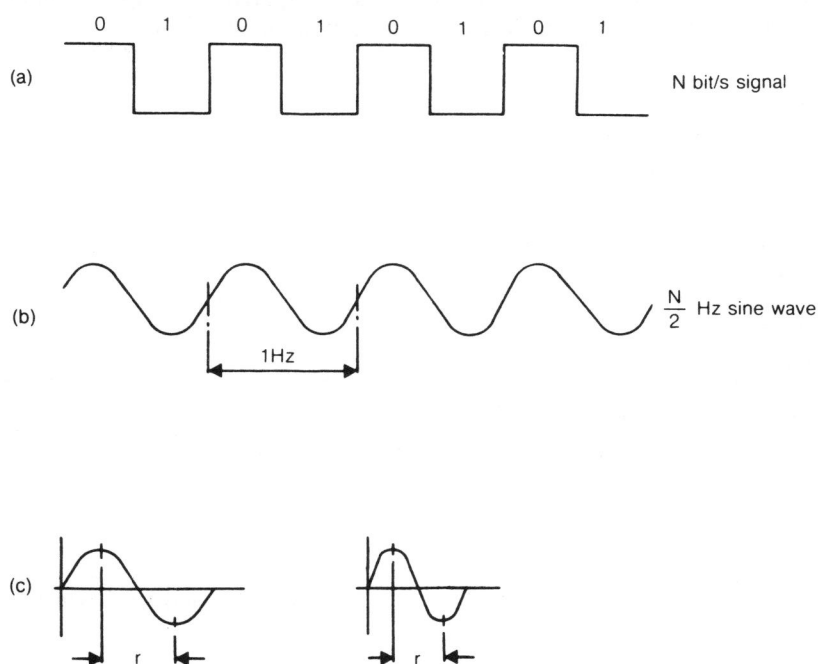

Figure 1.16 (a) Square wave, alternate 0s and 1s; (b) Sine wave equivalent; (c) Risetime of waves of different frequency

As we noted earlier, the greatest rate of change of information in a binary signal occurs when alternate 0s and 1s are transmitted. If the bit rate is N bit/s, it can be seen from Figure 1.16 that the binary signal 0 1 0 1 0 1... carries information at the same rate as a sine wave of frequency N/2 Hertz. However, if we were to transmit this data as a square wave through a channel with an upper frequency limit of N/2 Hz, we would find that the square wave was considerably rounded off. If we look at the wave forms in

Figure 1.16(c) we can begin to see why. The time taken for the signal to rise from its minimum to its maximum value is shown as 'r'. The higher the frequency of the wave, the shorter 'r' is. Now a square wave changes state almost instantaneously (ie 'r' is small), and this implies the presence of very high frequencies. In fact, a full analysis of a continuous square wave reveals that it is composed not of a single waveform at the given frequency but of a series of harmonically related sine or cosine waveforms of differing amplitudes, as shown in Figure 1.16. Such a series is known as a fourier series after its founder Baron de Fourier who in 1807 astounded his contemporaries by asserting that an arbitrary function could be expressed as a linear combination of sines and cosines.

Thus, to transmit a true square wave, that is one with a sharp transition from 0 to 1 (or cut-off), a channel of infinite bandwidth would be required in order to accommodate all the harmonic components which together constitute the waveform.

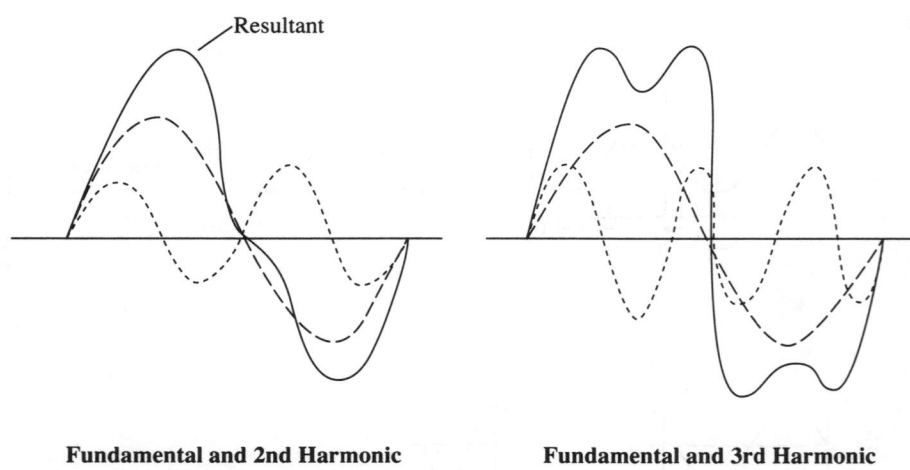

Fundamental and 2nd Harmonic   Fundamental and 3rd Harmonic

Figure 1.17 Complex waveforms showing the effects of harmonic components

Fourier also dealt with the analysis of a single square pulse. In this case analysis can show that the component frequencies are not a series of discrete waveforms but an envelope formed by a continuous spectrum of frequencies. This envelope is called a Fourier Transform and enables mathematical interpretation of period signals in a frequency domain. The Fourier Transform of a single square pulse is a curve in the form sin x/x, as shown in Figure 1.18 and gives the amplitude of all component frequencies of the pulse. Note that for a pulse of duration T, certain frequencies f, 2f, 3f, etc, have zero magnitude (F = 1/T).

Suppose, however, we had not started with a square pulse, but one of (sin x)/x shape (Figure 1.19(a)). What would be the component frequencies of that pulse? It turns out that the Fourier Transform of such a pulse has a spectrum with a sharp cut-off at one particular frequency as shown in Figure 1.18(b)).

Since a transmission channel always has a finite bandwidth, this would seem to be an ideal pulse to use for data transmission since the effects of distortion on bandwidth limitation are very much reduced.

Studies by Nyquist in the early part of this century proved that the maximum repetition rate of such pulses over a perfect channel is 2f pulses/second, where f is the bandwidth in Hertz. The time between pulses, given by $\frac{1}{2}$f pulses per second is known as the Nyquist interval.

The signalling rate on 2f pulses per second is known as the Nyquist rate.

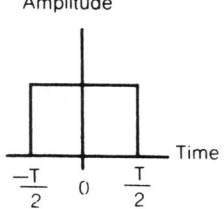

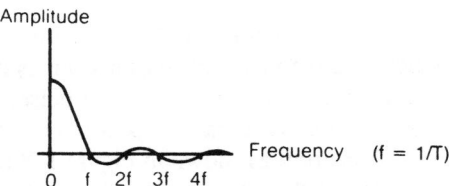

Figure 1.18   Single square pulse (a), and its Fourier Transform (b)

(a)   Amplitude

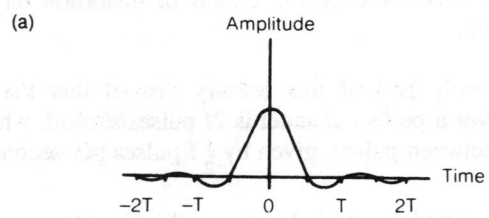

(b)   Amplitude

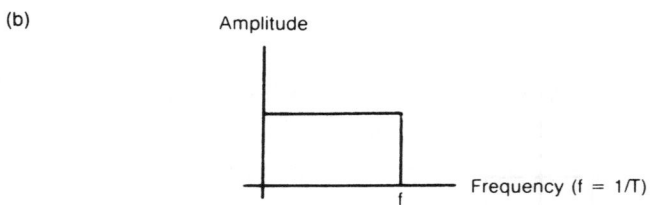

Figure 1.19   (Sin x)/x pulse (a), and its Fourier Transform (b)

Although pulses of (sin x)/x shape are ideal in theory, there are other shapes which are more tolerant of the deficiencies of practical transmission systems. One of these is based on the spectrum of what is known as a raised cosine pulse. A raised cosine pulse is sketched in Figure 1.20(a). A pulse based on the spectrum of the raised cosine pulse (Figure 1.20(c)) has a Fourier Transform as shown in Figure 1.20(d). It can be seen that the penalty paid in using a pulse of this shape is that twice the bandwidth is needed compared to the (sin x)/x pulse shape. However, one of the important properties of this pulse shape is that the level is only above the halfway point for half the pulse duration, and pulses of duration T can be sent at intervals of T/2 seconds.

Figure 1.21 shows how the bit pattern 0 1 1 0 1 0 0 would look in raised cosine spectrum form, together with the sampling instants.

These more sophisticated pulse shapes are not seen at the output of a data terminal, but may be generated by the modem or other transmission equipment.

(a)

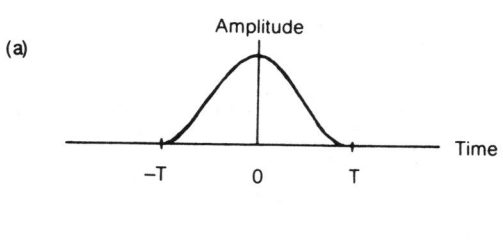

(b)

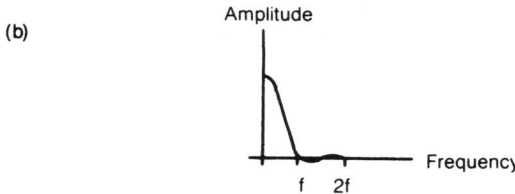

(c)

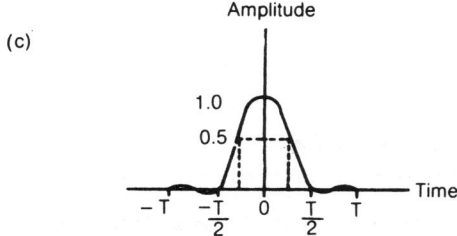

(d)

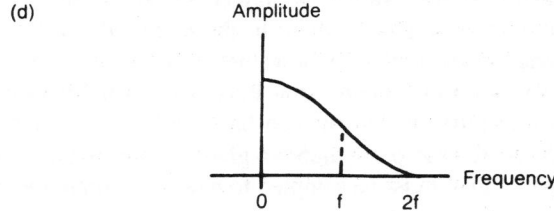

Figure 1.20 (a) Sketch of raised cosine pulse; (b) its Fourier Transform; (c) Pulse based on FT of raised cosine pulse(b); and (d) its Fourier Transform

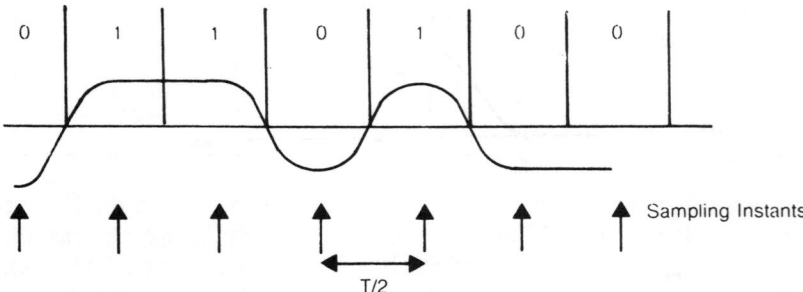

Figure 1.21   Binary waveform using a pulse shape based on the FT of a raised cosine pulse

Digital Transmission of Speech

The digital transmission of analogue signals is possible by virtue of the sampling theorem (sometimes referred to as the Nyquist theorem) which demonstrates that an analogue signal can be reproduced from an appropriate set of samples. Therefore, a system need only transmit sample values as they occur rather than the analogue signal itself using pulse modulation schemes where in the amplitude, width or position of a pulse waveform is varied in proportion to the values of the samples. This is an important application to the digitisation of speech using Pulse Code Modulation (PCM). Further explanation of PCM transmission systems can be found in Chapter 4 on Analogue to Digital Conversion. Nyquist's sampling theorem gives a lower bound to the rate at which a continuous (analogue) signal must be sampled to enable accurate reconstruction of the original waveform to be 2f pulses per second.

Thus, if we take a typical speech circuit with a bandwidth of 4 kHz, then for digitisation of the analogue waveform the voice signal must be sampled at least 8000 times per second to ensure that after transmission across a network, the receiving terminal can reconstruct an accurate representation of the original waveform.

2

General telecommunications principles

Introduction

When the first telephone network was designed and implemented in the 19th century, computers and other electronic devices did not exist. Since speech was the only communication over these original networks, the network was finely tuned to analogue transmission of the human voice. Any deviation from this fine tuning resulted in a loss of performance, making speech conversation and the ability to transmit data an awkward proposition.

However, as telephone subscribers began to purchase computers which needed to communicate over long distances, it was natural that they should examine the potential of the telephone network. Economic reasons made it hard to justify a reconstruction of the network, and so it was left to industry to devise products that could transmit data over the exiting network. This proved successful, and although not the ideal media for the purpose, telephone networks all over the world have been adapted to carry digital data as well as speech.

The overwhelming advantages of digital techniques have persuaded many PTTs, including British Telecom, to convert their networks from analogue to digital transmission.

This chapter reviews the development, construction and operation of transmission systems and pays special attention to voice telephony. The succeeding chapters examine the use of the telephone network for transmitting data.

The Telephone System

Although a telephone system is made up of many elements, the primary components are:

— the terminals (typically telephones);

— the transmission lines;

— the exchanges or switches.

These elements, when working together, enable a telephone associated with one exchange to be connected to any other telephone, either in its own country or on a system in another country.

The UK telephone system began towards the end of the 19th century. In 1880 the crown challenged the right of a private company to operate a telephone service on the grounds that a telephone message was encompassed by the Telegraph Act. The Act, which previously gave the Post Master General the monopoly of the public telegraph system, now required telephone companies to be licensed and to pay an annual royalty. In addition the companies could not operate outside an imposed five mile radius and could not construct lines between towns.

The first UK telephone service was set up in 1879 by the Telephone Company Ltd. This company installed private point-to-point speech circuits using telephones fitted with Graham Bell transmitters and receivers. In August 1879 they opened the first telephone exchange in Coleman Street in the City of London and thereby gave service to seven or eight telephone renters. By the end of 1879 two more exchanges had opened, and the number of Telephone Company Ltd subscribers had reached 200.

The licence restrictions of the Telegraph Act, which meant that it was not uncommon for subscribers to be associated with more than one exchange, were relaxed in 1884 to allow companies to operate anywhere in the UK and to construct lines between towns. However, some restrictions remained and private companies found it difficult to obtain permission for rights of way to lay underground circuits. These restrictions resulted in the development of an almost totally overhead external wiring system.

Despite the licence restrictions, many private telephone companies were formed and on 1st May 1889 the majority of them merged to form the National Telephone Company (NTC). However, the merger did not solve the problem of obtaining rights of way for underground circuits and so the trunk system was taken over by the Post Office in 1892.

In 1898 a House of Commons Select Committee recommended that the Post Office, and several local authorities, be allowed to provide telephone facilities in competition with the NTC whose licence expired on 31st December 1911. The combined result of the Select Committee's recommendation and the expiry of the licence was that, with the exception of Hull and Portsmouth, the whole of the UK telephone system came under the control of the Post Office. This state of affairs continued until 1914 when Portsmouth also came into Post Office control, leaving Hull, to this day, an independent operator.

In 1969 the Post Office became a Statutory Corporation with monopoly powers to provide telecommunications and mail services in the United Kingdom, as it previously did as a Department of State.

The 1981 British Telecommunications Act

The British Telecommunications Act of 1981 set up a new public corporation, British Telecom (BT), to run the telecommunications and data processing business of the Post Office. The Act also enabled the Secretary of State to appoint the British Approvals

Board for Telecommunications (BABT) as an approval authority for telecommunications apparatus.

In addition, the Act permitted the Secretary for Trade and Industry to license other organisations to provide telecommunications services, thus forming the basis for liberalisation.

In 1982 the *Mercury* Licence was issued allowing Mercury Communications to offer switched telecommunications services within UK. Two licences for cellular radio systems were issued a year later to Racal Vodac's *Vodafone* and Telecom Securicor Cellular Radio's *Cellnet*.

The 1984 Telecommunications Act

The Telecommunications Act of 1984 enabled licensing and approval procedures to be improved and also outlined the statutory foundations for the liberalisation of UK telecommunications. However, the main objective of the Act was to create the watchdog Office of Telecommunications (Oftel) and to privatise British Telecom, forming British Telecommunications plc.

By virtue of Section 9 of the Act certain telecommunications systems could be specified as Public Telecommunications Systems which would be run by Public Telecommunications Operators (PTOs). This enabled the Secretary of State to grant PTO status to Mercury Communications Limited and Hull City Council.

Although there are other licensed telecommunications services offered to the public, they are not legally treated as Public Telecommunications Systems and their operators are not regarded as PTOs. However, the 1981 and 1984 Acts have together formed an arena in which all licensed service operators can compete to provide services and apparatus throughout the UK.

The Branch Systems General Licence (BSGL)

The BSGL was created under Section 7 of the 1984 Telecommunications Act and brought the area of liberalisation into the subscriber's premises. Under the conditions of this licence, apparatus which are BABT approved for connection to a PTO network without the use of a tool, may be installed by the user. The licence also makes it clear that more complex devices must be installed and maintained by a designated maintainer. This means that instead of purchasing equipment and installation services from PTOs, users can choose suppliers to suit their technical and budgetary requirements.

The Extent of Telecommunications

The UK telephone system has evolved over the last 100 years, and together with other national networks forms one of the largest and most complex man made systems in the world. The magnitude of this achievement is perhaps not fully appreciated by the majority of users, although the statistics themselves are impressive. For example, in the early part of 1990, there were approximately 25 million exchange lines connecting users to the UK public telecommunications network. In America, there were nearly 200 million lines, and more than 500 million throughout the world.

Local Line Distribution

In the early days of the UK telephone system, the connections between the customers and the exchanges were provided by overhead wires carried on telegraph poles and connected to frames located on the roof of the exchange. The pair of connecting cables was, and still is, referred to as the *local loop*. In later years the telegraph poles acted as distribution points for a number of customers. A number of other distribution points, known as pillars and cabinets, could also be found between customer premises and exchanges.

Although the various distribution points still exist, the *local loop* has gradually been moved underground providing a more expensive but also much more error free connection. The *local loop* is the first stage of the connection between customers separated by local, national or international boundaries.

A modern exchange hierarchy has many levels (*see* Figure 2.1) and it is usual to incorporate additional cross-connections between, for example, two local exchanges. The cross-connections ensure that a call does not have to pass through every layer of the hierarchy, thus reducing the load on other exchanges. The control of a large switching system, such as a PTO network is a complex task which requires the speed and capacity of a computer. Exchanges controlled in this manner are referred to as Stored Program Control (SPC) exchanges.

The Transmission Line

The electrical signals used to transmit information through telecommunications networks take the form of electromagnetic waves travelling at speeds approaching the speed of light (3×10^8 m/s). The transmission lines which carry the signals exhibit four electrical characteristics which cause the electromagnetic waves to dissipate along the length of the line. Therefore, as the electrical information signals are transmitted through the line, they lose power or attenuate.

The four characteristics of a transmission line which cause this power attenuation are:

—   the series resistance R of the cables;

—   the leakage conductance G between the cables;

—   the inductance L of the cables; and

—   the capacitance C between the cables.

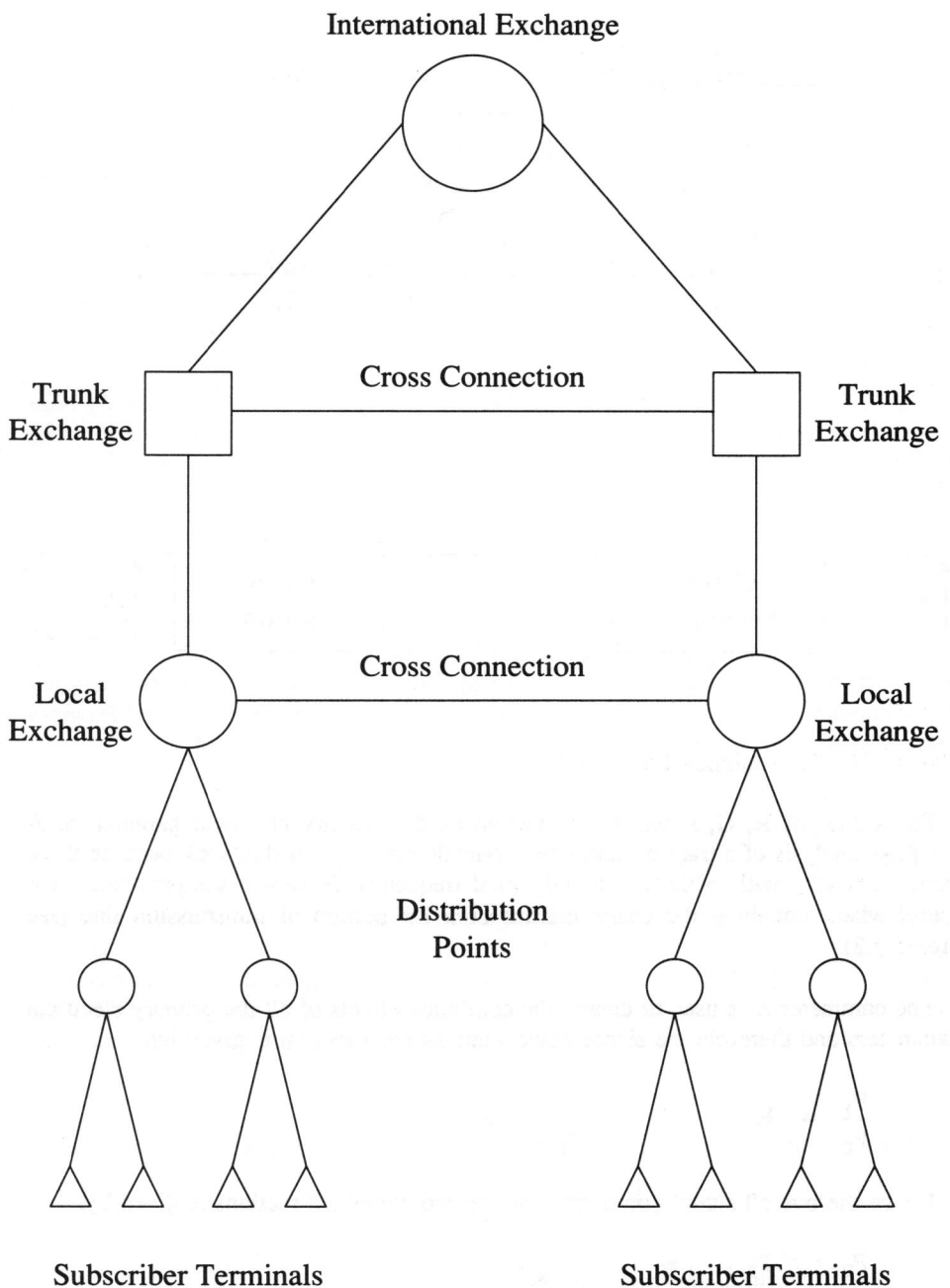

Figure 2.1 A modern exchange hierarchy

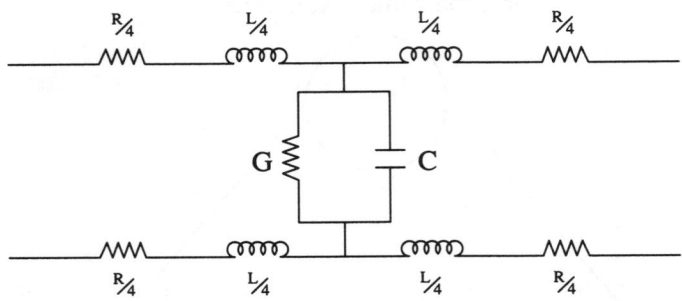

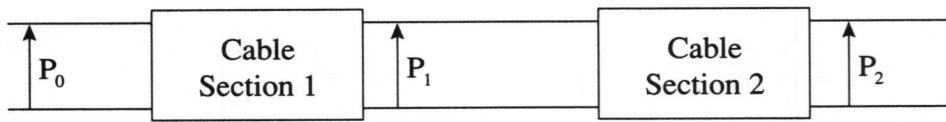

Figure 2.2 Transmission line model

The values of R, G, L and C are known as the primary electrical parameters. A complete analysis of a transmission line is outside the scope of this book because these parameters vary with cable length and signal frequency. However, the parameters are useful when modelling the characteristics of a 1m section of transmission line (*see* Figure 2.2).

The parameter K is used to denote the combined effects of all the primary electrical parameters and therefore the signal power ratio of each section is given by:

$$\frac{P_1}{P_0} = K \qquad \text{and} \qquad \frac{P_2}{P_1} = K$$

Hence the overall signal power ratio of the two combined sections is given by:

$$\frac{P_2}{P_0} = \left(\frac{P_2}{P_1}\right) \times \left(\frac{P_1}{P_0}\right) = K^2$$

In general, for a transmission line of length d, the signal power ratio is given by:

$$\frac{P_d}{P_0} = K^d$$

The decibel (dB) is a logarithmic measure of the ratio of two power levels, in this case P_d and P_0 and so on:

$$10 \ \log_{10} \left(\frac{P_d}{P_0} \right) = 10 \ \log_{10}(K^d) \ \ dB$$

Rearranging this equation gives the power loss of the combined section:

$$d \ \ 10 \ \log_{10}(K) = -\alpha \ d \ \ dB$$

In this equation:

$$\alpha = -10\log_{10}(K)$$

is a measure of the cable attenuation in dB/m. The introduction of the minus − sign gives a positive value for α given that $K < 1$ (ie attenuation).

For example, if the power at a distance of 1m is half of the original power, then the attenuation is given by:

$$\alpha = -10 \ \log(1/2) = 3 \ \ dB$$

If the power at a distance of 1m is twice the original power, ie the signal is amplified, then the attenuation is given by:

$$\alpha = -10 \ \log(2) = -3 \ \ dB$$

It is worthwhile noting that negative attenuation means the same as amplification or gain.

Table 2.1 shows the relationship between power ratios and decibel losses.

Table 2.1 Power ratios and decibels

dB	Power Ratios (power sent to power received)	dB	Power Ratios (power sent to power received
+3	2:1	+23	200:1
+6	4:1	+26	400:1
+9	8:1	+29	800:1
+10	10:1	+30	1000:1
+13	20:1	+33	2000:1
+16	40:1	+36	4000:1
+19	80:1	+39	8000:1
+20	100:1	+40	10000:1

Frequency Response

The conductive and inductive primary electrical parameters are dependent on the frequency of the signal as well as the distance it travels. The ability of a particular transmission line to carry or respond to different frequencies is referred to as its *frequency response*.

Due to the combined frequency response of G and L, high frequency signals are generally attenuated to a greater degree than low frequency signals. Figure 2.3 demonstrates this by showing the cable attenuation of a paper-covered 0.9mm copper cable, commonly used in the UK.

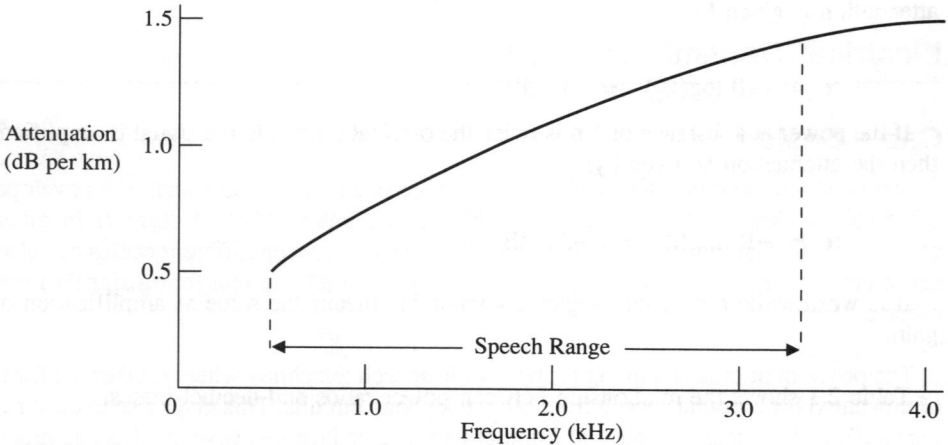

Figure 2.3 Attenuation of a paper covered 0.9mm copper cable

When a complex speech signal is transmitted for a long distance through the cable, the higher frequencies will be attenuated much more than the lower frequencies. If this effect is severe then the speech signal will become unintelligible to the listener at the end of the cable. Different diameter cables have different limiting distances for the satisfactory transmission of speech waveforms, and some examples using ordinary two-wire cables are given in Table 2.2.

Table 2.2   Limiting distances of cables

Diameter of Copper Conductor	Limiting Distance for Satisfactory Speech
0.63mm	9.5km
0.9mm	14.5km
1.27mm	20.0km

Loading

The power of a signal at any point in a circuit is given by the following expression.

Power = Voltage x Current

In an AC waveform the values of current and voltage are constantly varying and so it is not usual to use peak values. Instead, the Root Mean Square (RMS) value is used. This is defined as the square root of the average value of the squares of all the instantaneous currents or voltages taken over a complete cycle. The RMS value of a pure sinusoidal waveform is 0.707 times the peak value.

Electrical Transmission Media

Cable Design

Currents which flow in telephone cables produce magnetic fields which can envelope other cables close in proximity. This results in one cable inducing currents in other cables; an effect known as cross-coupling. Capacitance between different cables can also cause cross-coupling which impairs a wanted signal by adding interfering signals from other cables.

The problem of cross-coupling is apparent in speech telephony where it takes the form of unwanted background speech from other telephony circuits. This gives rise to the term *crosstalk,* and the main purpose of cable design is to reduce this adverse effect as much as possible.

This section concentrates on the two types of electrical transmission cable design which are shown in Figure 2.8.

Twisted Pair Cable

If a single wire was used to provide a connection between a telephone and an exchange, it would pick up capacitive and inductive interference from overhead cables as well as capacitive and inductive crosstalk from adjacent wires in the same cable.

To overcome these effects, each connection to the exchange is made with a pair of wires in close proximity to each other so that any interference which occurs is picked up by both wires. There is therefore no interference between them and the pair is said to be *balanced*. To maintain symmetry as far as possible, and to further reduce crosstalk, the pair is twisted together. Adjacent pairs have random positions along the length of a multi-pair cable so that a given pair receives interference uniformly from all the other pairs.

The attenuation of a typical twisted pair cable when carrying a 50 kHz signal is roughly 5 dB/km, making it very suitable for transmission of voice and digital data at up to 1 Mbits over short distances and without amplification.

Coaxial Cable

Coaxial cables consist of an inner signal-carrying conductor surrounded by an outer screen. The space between the two conductors is filled with dielectric (usually air) and spacers are used to stop the screen from collapsing and touching the inner conductor.

The arrangement of the conductors helps to reduce crosstalk by almost entirely containing the electric and magnetic fields within the outer screen. The screen also has the effect of reducing the amount of external interference which reaches the inner conductor. Coaxial cable for high bandwidth transmission uses an extruded aluminium screen, whereas braided copper is sufficient to shield low bandwidth or baseband data.

A pure sinusoidal signal has coincidental maximum voltage and current values (see Figure 2.4), and hence the power in the circuit is also at its maximum. In these circumstances the circuit is said to be *resonant*.

In many parts of the telephone network, the cables run for long distances at close proximity. The effect of this is that a circuit becomes capable of storing electrical energy and is said to have *capacitance* (primary electrical parameter C). Electrical capacitance in cables (or other electrical devices) enables them to store electrical energy for long periods and then discharge it very quickly. In some circumstances, such as flash photography, this effect is very useful, but in a telephone circuit the effect is highly undesirable. When telephone circuits exhibit capacitance, the moment of maximum current occurs before the moment of maximum voltage (see Figure 2.5). Because the signals become out of phase, the product of current and voltage is always less than maximum, resulting in a loss of power.

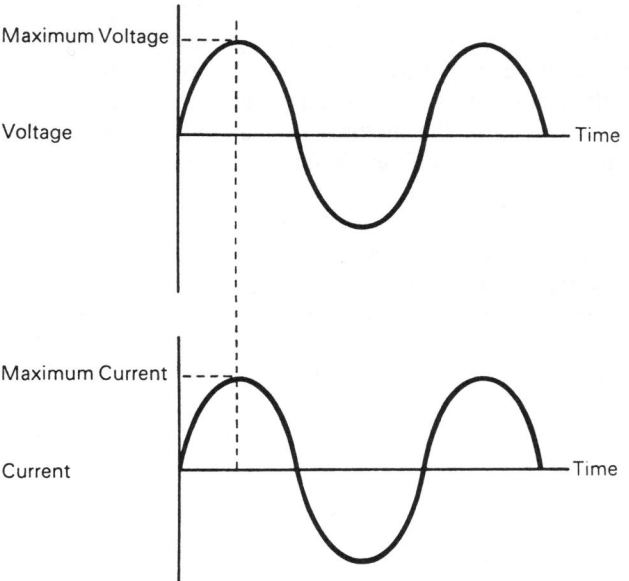

Figure 2.4   A resonant circuit

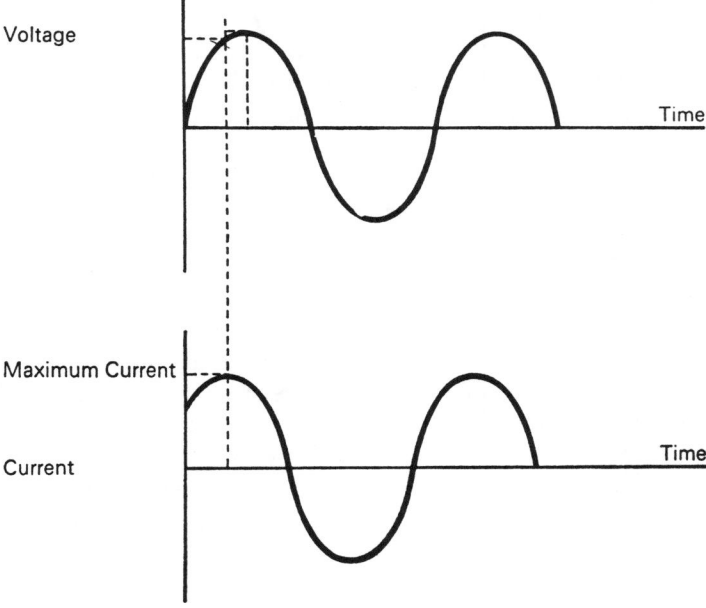

Figure 2.5   A circuit exhibiting capacitance

An opposite effect to capacitance is exhibited in circuits which have coils of wire. This effect is known as *inductance* (primary electrical parameter L) and results from a change in the current in a coil. When this happens, the force of the resulting inductance opposes the current flow. In a telephone circuit the current is constantly fluctuating and the inductance in the circuit curbs the build up of current so that it falls behind the build up of voltage. This results in the moment of maximum current occurring after the moment of maximum voltage (see Figure 2.6) and therefore the maximum power is lost.

The inductance of a coil is useful because when it is added to a telephone circuit it counteracts the effects of capacitance and reduces the power loss. This process is known as *loading*. In practice the loading coils are housed in iron cases which are buried underground or located in manholes at appropriate distances along the length of the cable. Loading a telephone circuit to maximum effect dramatically increases the limiting distance for satisfactory speech (see Table 2.3).

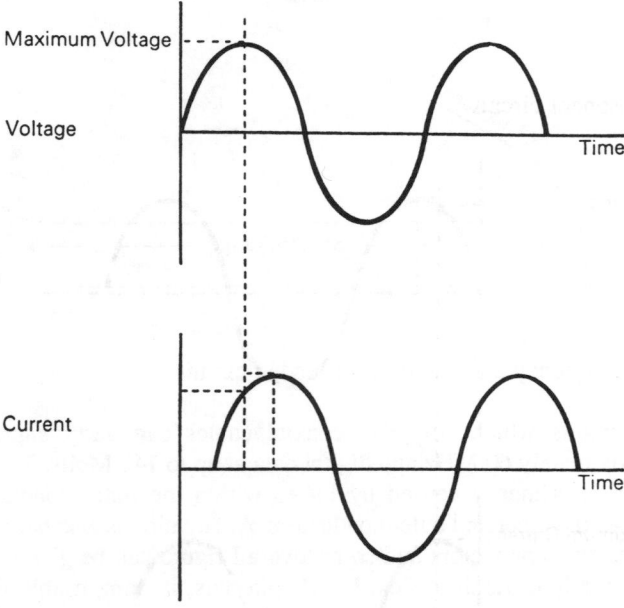

Figure 2.6   A circuit exhibiting inductance

Table 2.3   Limiting distance and loading

	Limiting Distance for Satisfactory Speech	
	Unloaded	*Loaded*
0.63mm conductors	9.5km	35km
0.9mm conductors	14.5km	67km
1.27mm conductors	20.0km	112km

Although loading serves to counteract the power losses caused by unwanted capacitance in a telephone circuit, it also has an adverse effect on the *bandwidth*, or range of frequencies which the circuit can effectively carry. The frequency response of a loaded circuit, shown in Figure 2.7, has a sharp cut off point beyond which frequencies cannot be effectively carried.

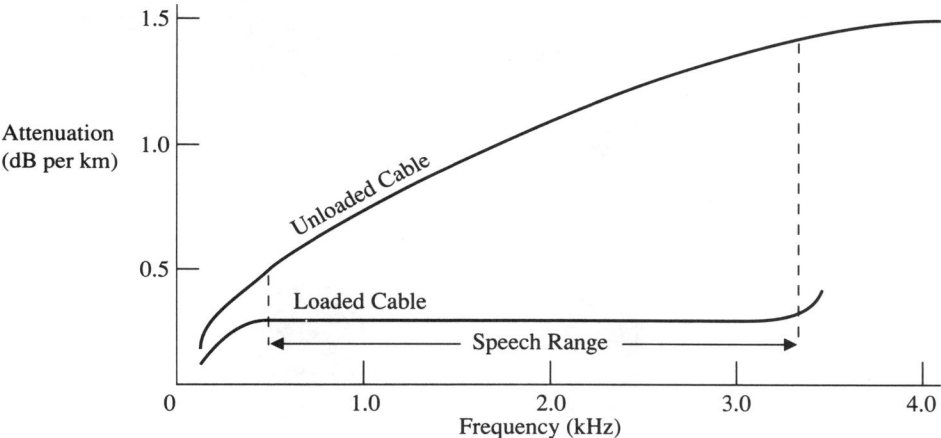

Figure 2.7   The frequency attenuation of a loaded circuit

Transmission circuits which comprise coaxial cables can carry signals up to a frequency of approximately 60 MHz and digital data at up to 140 Mbits. The attenuation of a coaxial cable is primarily caused by losses within the inner conductor and the capacitance between the inner and outer conductors. Attenuation is also dependent on the relative radii of the two conductors and so no overall figure can be given. The coaxial cable itself is physically difficult to flex, but despite this, its long usable life span and ease of application make it a popular choice for high frequency transmission.

Twisted Pair Cable

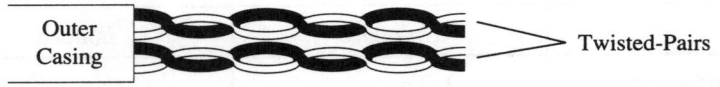

Coaxial Cable

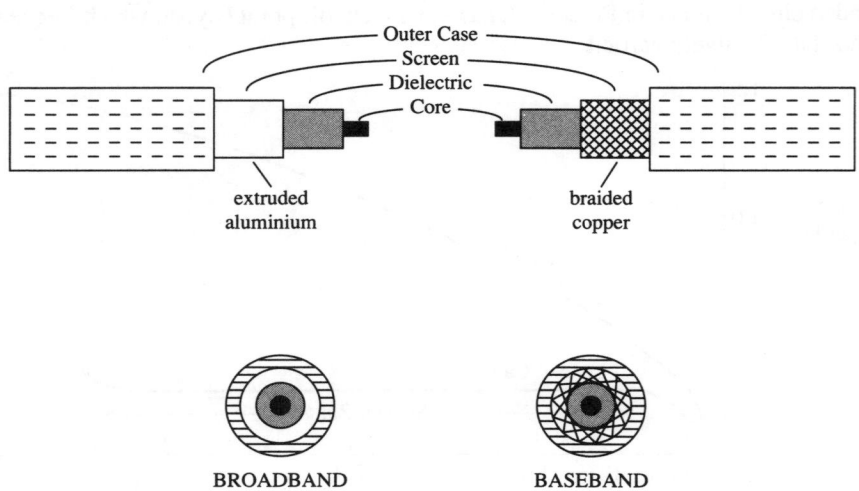

Figure 2.8 **Cable designs for twisted pair and coaxial cables**

Filters

The interference which affects a telecommunications circuit can be further reduced if the frequency of the interference is outside the bandwidth of the wanted signal. In the simplest form of this process, a specially designed arrangement of inductors and capacitors filters out the high frequency interference (such as clicks and squeals) and the low frequency interference (such as hum).

These electronic filter circuits can be designed to have a cut off at any desired frequency and depending on the arrangement of the filter, frequencies above or below this point will be rapidly attenuated. Exact values of the cut off are calculated mathematically and as design techniques become more complex the cut off becomes sharper and the filters become more accurate.

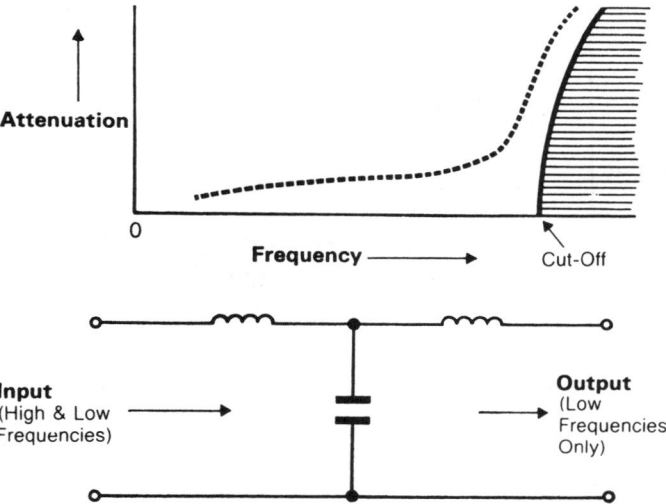

Figure 2.9 A low-pass filter

Figure 2.9 shows how capacitors (C) and inductors (L) can be arranged to produce a *low-pass* filter. This type of filter allows all frequencies below the cut off to pass almost without attenuation, but blocks all frequencies above the cut off. The solid line shows the frequency response of an ideal low-pass filter whilst the dotted line shows the frequency response of a practical filter.

From Figure 2.7 it can be seen that the attenuation is independent of frequency over the range 300 to 3400 Hz (the range of speech frequencies) and above the cut off point at 3400 Hz, the loss in the cable increases rapidly. This shows that the effect of loading the cable has made each wire pair behave like a low-pass filter.

The circuit in Figure 2.10 represents a *high-pass* filter which allows higher frequencies to be passed whilst greatly attenuating the frequencies below the cut off point. A *band-pass* filter, as illustrated in Figure 2.11, allows a band of frequencies to be passed whilst blocking of the frequencies above and below the higher and lower cut off points.

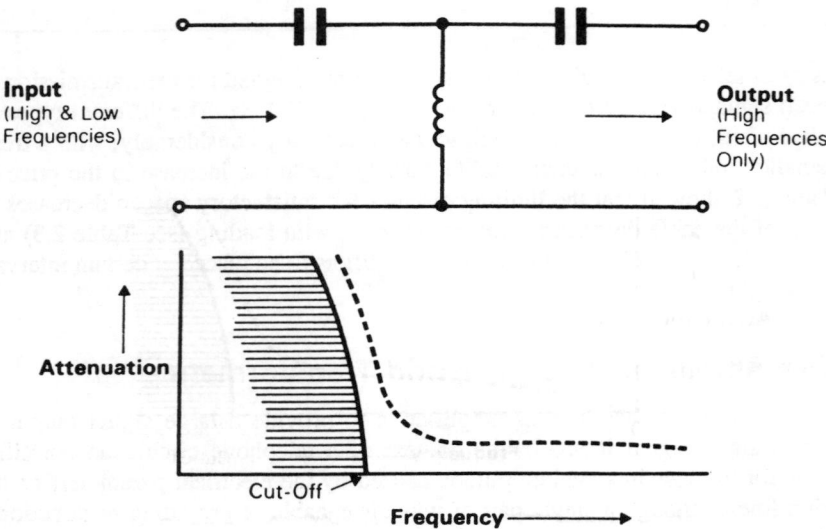

Figure 2.10   A high-pass filter

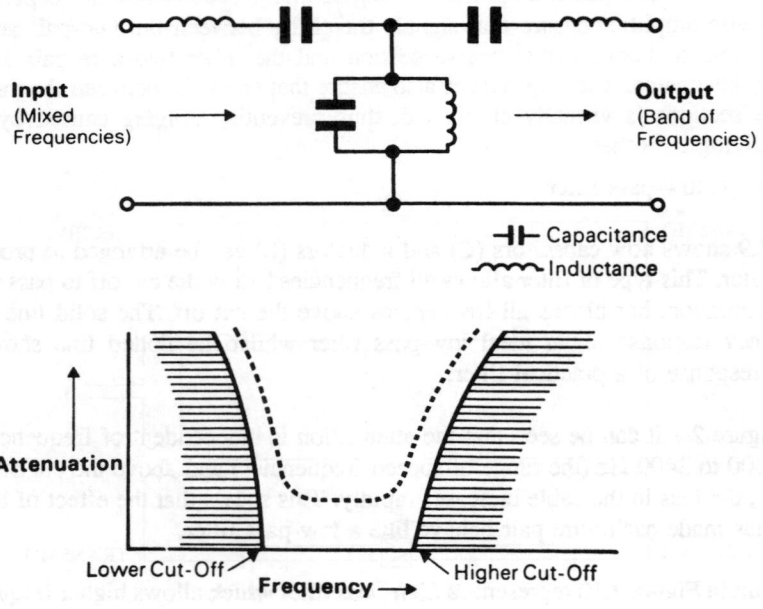

Figure 2.11   A band-pass filter

Amplified Circuits

The majority of telephone circuits within a network are designed for the transmission of audio speech signals with a frequency range of 300 to 3400 Hz. The different types of cable, or *line plant,* over which the circuit is routed can vary considerably, with a trend towards smaller and smaller diameter cable, largely due to the increase in the price of copper. Table 2.2 showed that the limiting distance for satisfactory speech decreases as the diameter of the cable decreases. This occurs even with loading (see Table 2.3) and therefore many telephone circuits now require amplifiers to be placed at certain intervals along their length.

Two-Wire Amplification and Hybrid Transformers

The primary function of any sort of amplifier is to provide a large signal output in response to a small signal input. In the context of a telephone circuit, an amplifier compensates for the loss in signal amplitude caused by the electrical parameters of the transmission line. Although a single pair of wires is capable of providing bi-directional and simultaneous transmission, amplifiers are traditionally one way.

Inserting a single amplifier onto a pair of wires will result in speech being transmitted in one direction only. This is obviously unsatisfactory for speech conversation and so two *hybrid transformers* separate the transmission into two directions, each of which is passed to an individual repeater amplifier (see Figure 2.12). The *line balance impedances* of the two-wire amplifier ensure that signals travelling between the two-pair and the transmit section, or between the receive section and the other two-wire pair are not dramatically attenuated. The impedances also ensure that crosstalk between the transmit and receive sections is virtually eliminated, thus preventing *singing* caused by each amplifier driving the other.

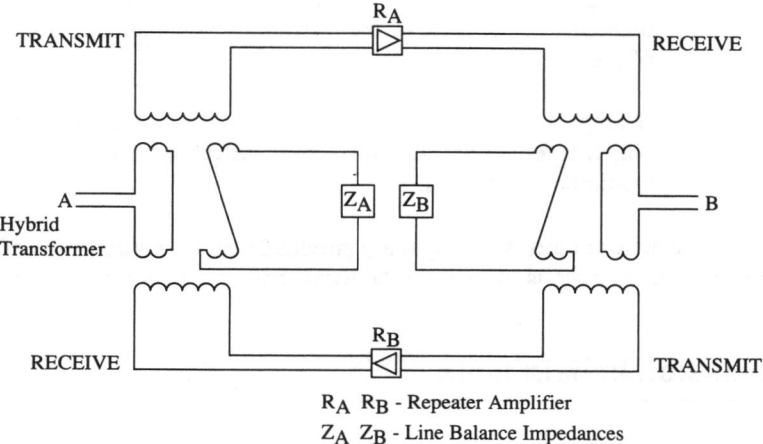

R_A R_B - Repeater Amplifier
Z_A Z_B - Line Balance Impedances

Figure 2.12 A two-wire amplifier

Four-Wire Transmission

On long distance circuits the effectiveness of two-wire transmission is limited because of problems maintaining the stability of the transformers. The solution to this problem is effectively to split the two-wire amplifier in half, and provide four-wire transmission (see Figure 2.13). This method uses one-wire pair for transmitting and the other for receiving.

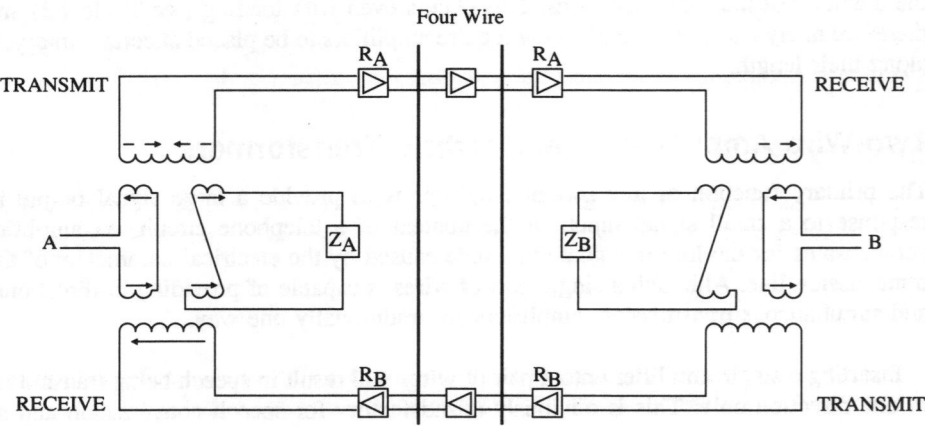

Figure 2.13   Four-wire transmission

Amplification is provided by repeater stations at regular intervals along the four-wire transmission circuit because:

— if amplification was only provided at the receiver, the signal would be unintelligible because of the interference accumulated over the length of the circuit; and

— if amplification was only provided at the transmitter, the required output level needed for satisfactory reception would be so high that it would be impractical.

Although four-wire circuits were originally provided over the entire length of the audio line plant, modern practice is to provide the local loop as a two-wire circuit.

Transmission Impairments

The elements of a telephone circuit which adversely affect a signal can take many forms, but in all cases they reduce the intelligibility of the signal received. This section reviews the main types of transmission impairments which are found in telephone circuits.

Sidetone

Sidetone is defined as the reproduction in a telephone receiver of sounds picked up by the associated transmitter. A specific level of sidetone creates a live sounding connection and allows telephone users to adopt a normal volume level of conversation.

The effect of sidetone on a conversation is subjective but it is generally thought that too much sidetone causes a speaker to lower the voice, thus diminishing the volume of the received signal. Too little sidetone, on the other hand, causes a speaker to raise the voice, thus giving abnormally high received signal levels. Large knock-on effects of sidetone can make understandable conversation very difficult.

Insertion Loss

Insertion loss is defined as the loss in decibels between one end of a circuit and the other, relative to a specific frequency of 800 Hz, and within the range 30 to 3400 Hz.

$$\text{Insertion loss} = 10\log_{10} \left(\frac{P_{sent}}{P_{received}} \right)$$

In other words, if an 800 Hz signal was transmitted on a line with a +3dB insertion loss, the received signal would have half the power of the original signal.

The losses which occur in a transmission line are generally dependent on frequency, and so insertion loss is sometimes referred to as *frequency distortion*. In order to make the distortion independent of frequency over a certain bandwidth, an electrical network known as an *equaliser* is introduced into the transmission line. The equaliser has the inverse attenuation characteristic (with respect to frequency) to that of the line. Figure 2.14 shows how the inclusion of an equaliser results in the combined result of line and equaliser loss being constant across the required bandwidth.

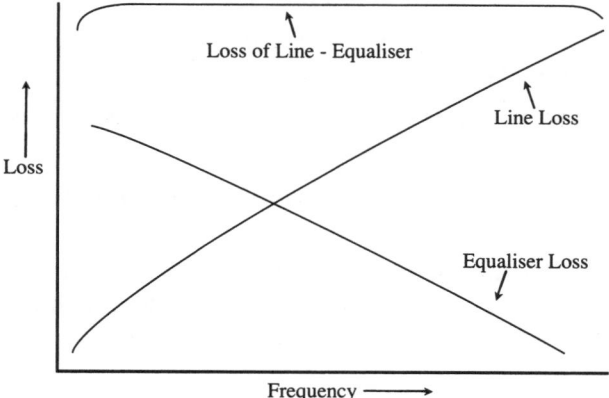

Figure 2.14 Line equalisation

As an alternative to the equaliser, an amplifier can be introduced with a gain/frequency characteristic which is a replica of the loss/frequency characteristic of the transmission line (see Figure 2.15).

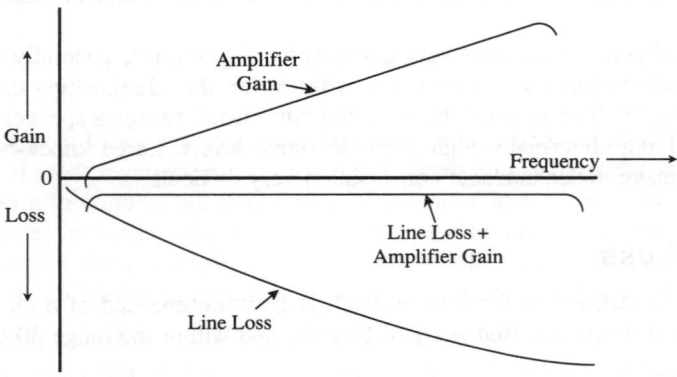

Figure 2.15   Amplifier equalisation

Noise

Noise can be considered as anything added to an information signal which makes it more difficult to extract the information from that signal. The electronic components found in telephone systems exhibit four inherent types of noise. These are:

— thermal noise, caused by the random movement of electrons within the line plant and circuit components;

— shot noise, caused by electrical currents crossing the boundaries between different types of conductor;

— flicker noise, caused by imperfections in the materials which make up the electronic circuit components; and

— burst noise, caused by contamination of the materials which make up the electronic circuit components.

Noise may also be introduced by outside sources such as power lines, lightning or heavy machinery.

Each noise signal which is introduced into the telephone circuit is amplified along with the speech signals. The repeater stations must, therefore, be located at intervals which ensure that the noise power (N) at the input of the amplifier is small compared to the signal power (S).

The *signal to noise ratio (SNR)* is often the limiting factor in determining the degree of amplification which can be obtained and is extremely important when considering data transmission. The SNR is usually expressed in decibels and is given by:

$$SNR_{dB} = 10 \log_{10} \left(\frac{S}{N} \right)$$

The larger this figure, the greater the strength of the signal compared to that of the noise.

Echo

Difficulty can arise on long circuits when reflections or echoes of words are heard a short time after the speaker said them. The echoes occur wherever there is a change in the electrical characteristics of a circuit, for example at the junction of a two-wire and four-wire circuit. The echo effect is rarely a problem within a relatively small geographical area, but when the connections span thousands of kilometres it can seriously inhibit discernible speech, or, more importantly, data transfer.

To combat this problem *echo suppressors* are usually fitted onto intercontinental circuits, between the transmit and receive pairs, permitting only one pair to transmit or receive at a time. However, echo suppressors also prevent the simultaneous bi-directional (true duplex) transmission of data and therefore have to be removed if this kind of transmission is required.

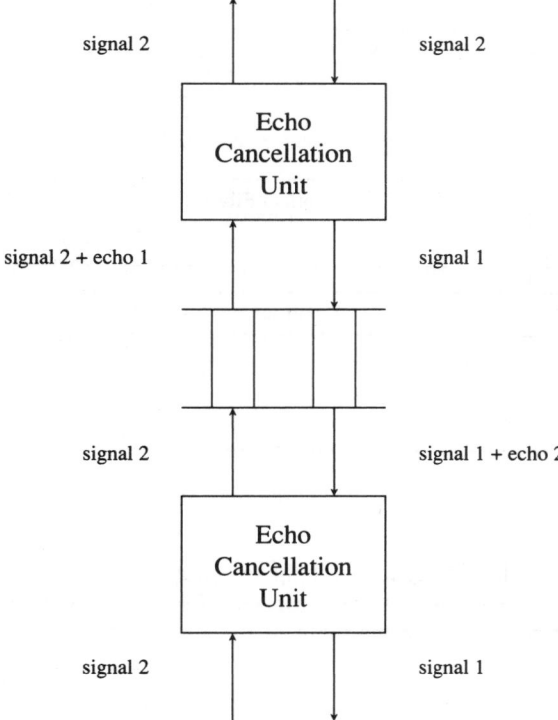

Figure 2.16   Echo cancellation

Echo cancellation is another method of reducing the level of the echo signal, thus making it possible to use the line for simultaneous two-way transmission. The echo cancelling device isolates the received signal by subtracting from it any echo interference from its own transmit signal as well as any other low-level interference signals which are present. Figure 2.16 shows how an echo cancellation system eliminates the echoes received from the junction of a two-wire and four-wire circuit.

Optical Fibre Transmission

The previous sections of this chapter have all been concerned with the clerical transmission lines which have been in use for more than a century. More recent developments in transmission media became possible when, in 1966, glass fibres were first proposed for use in optical communication systems. This section discusses the main elements of optical fibre transmission systems, outlining the principles and technologies involved.

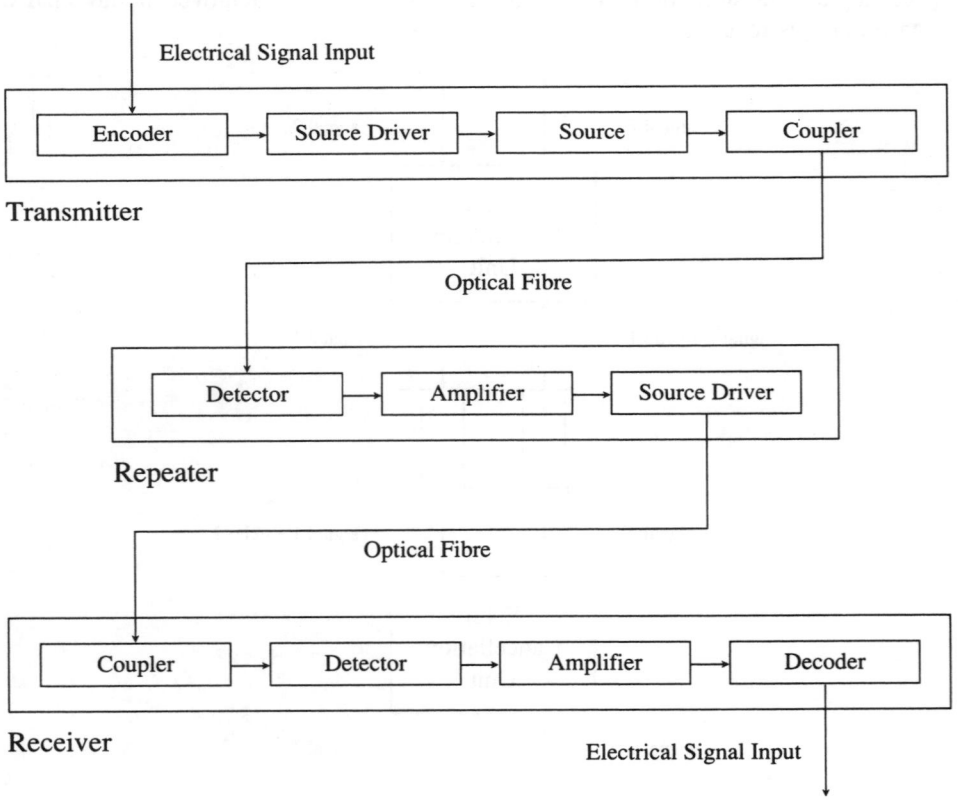

Figure 2.17 An optical fibre system

The Optical Fibre System

A generalised diagram of an optical fibre system used for analogue and digital transmission is shown in Figure 2.17. The signal to be transmitted from one end of the system to another traces a path through a number of separate sections.

In an analogue system, the *signal encoder* predistorts the information signal in order to compensate for the distortion inherent in later stages. This section in a digital system provides the analogue to digital conversion of the information signal and hence produces the correct digital signal.

The *source driver* takes the output signal from the encoder stage and modulates it to provide an electrical signal which is suitable for driving the source. The source then changes the driving signal into an optical signal. The different types of source are discussed later in this section.

The function of the *coupler* is to introduce the optical signal into the optical fibre. To do this properly, the coupler must have a low coupling power loss and be exactly the same diameter as the *optical fibre* which transmits the optical signal from transmitter to receiver. The principles of optical transmission are discussed in a later section.

Depending on the source and the length of the fibre, it may be necessary to use a *repeater* to regenerate the optical signal. This section is specifically designed to boost the level of the optical signal and, if necessary, reshape it ready for retransmission onto the next section of fibre.

The *detector* stage contains a photo-detector whose purpose is to detect efficiently the optical signals passed to it from the receiving coupler and to turn them into electrical signals. Some of the different types of detectors are examined in a later section.

The *amplifier* which takes the output from the detector must be free from distortion so that it can enhance the electrical signal from the detector and reshape it for proper further use. The *decoder* then converts the raw electrical signal from the amplifier into a replica of the original signal. This section may also shape the signal into a format which will be determined by its intended application.

Optical Sources

In general an optical source must be reliable, economically viable and compatible with the type of fibre used. There are three types of source which exhibit these properties:

- the light emitting diode;
- the injection laser diode; and
- the Nd:YAG solid state laser.

Light Emitting Diode (LED)

The two principle LED structures found in optical fibre systems (see Figure 2.18) are distinguished from each other by the direction of the light which is emitted. Each structure uses two types of material known as p-type gallium arsenide (p-GaAs) and n-type gallium arsenide (n-GaAs). The two materials differ in the degree of impurity (usually aluminium) present in the gallium arsenide, and the way in which it is introduced.

Edge Emitting LED

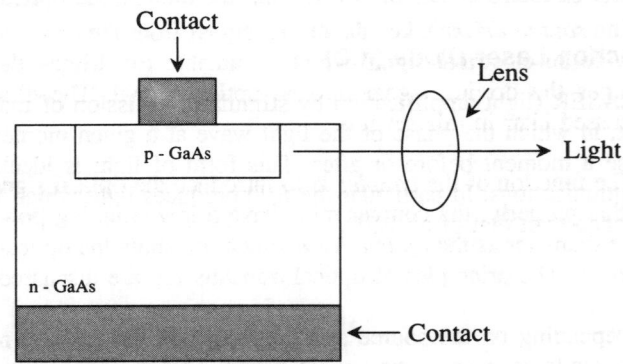

Side Emitting (Burrus) LED

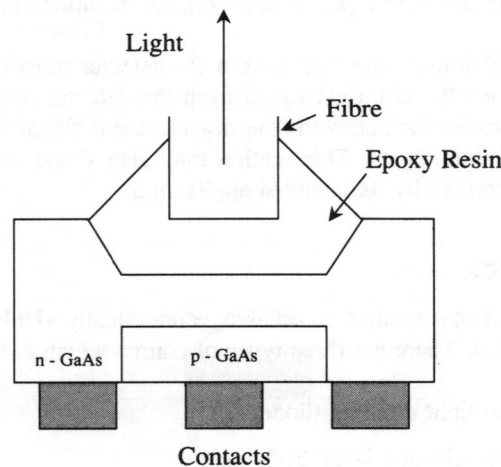

Figure 2.18 LED structure

As its name suggests, the *surface emitting* or *Burrus* LED produces light at its surface and is suitable for direct connection to the fibre. The light emitted at the edge of the *edge emitting* LED, however, has to pass through a concentrating lens before being transferred to the fibre. The maximum data rate of both LEDs is approximately 100 MHz.

The efficiency, power output and emission area of the Burrus LED, make it the most widely used LED structure. This is despite its unsuitability to drive bundles of fibres and its complex fabrication process. However, because both structures produce spontaneous or *incoherent* light in which the photons, or light units move in random directions, neither structure is suitable for driving the monomode optical fibre.

Injection Laser Diode (ILD)

A LASER (light amplification by stimulated emission of radiation) produces *coherent* light, in which the phase of the light wave at a given moment in time is related to its phase a moment before or after. This form of light is ideally suited for optical fibre communications. In particular the ILD produces light which has greater directivity and radiance than that of an LED.

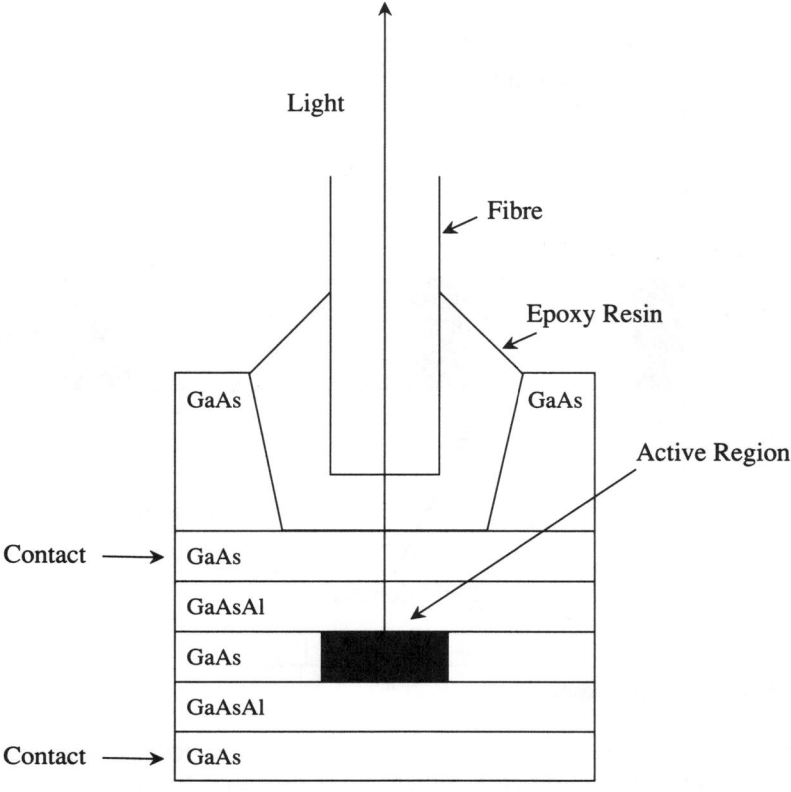

Figure 2.19 Injection laser diode (ILD)

The most promising type of ILD is the Gallium Arsenide Aluminium (GaAs AI) striped double-heterostructure design (see Figure 2.19) The stripe confines the light to a small cross section and hence it is possible to couple the emitting area directly with the core of the optical fibre. Although direct coupling is expensive to achieve and reduces lifetime, it does provide a high coupling efficiency and makes the ILD ideal for long-haul fibre systems. A great deal of research is being conducted in this respect, to try to increase the lifetime of an ILD.

The lifetime of an ILD is also governed by the current which is driving it, and so it is much better to use pulse modulated light (in excess of 1 GHz) than continuous light. Despite the fact that pulsed light increases the optical power and the efficiency of the ILD, it does tend to reduce its speed and its upper operating frequencies.

Nd:YAG Laser

The light from neodymium-doped yttrium-aluminium-garnet (Nd:YAG) laser is produced with a solid state rod (see Figure 2.20) driven by a semiconductor diode. The length of the rod allows in-phase feedback from the mirrored and semi-mirrored end surfaces. This ensures the emission of coherent light waves with identical phases and frequency and gives the laser a very high coupling efficiency.

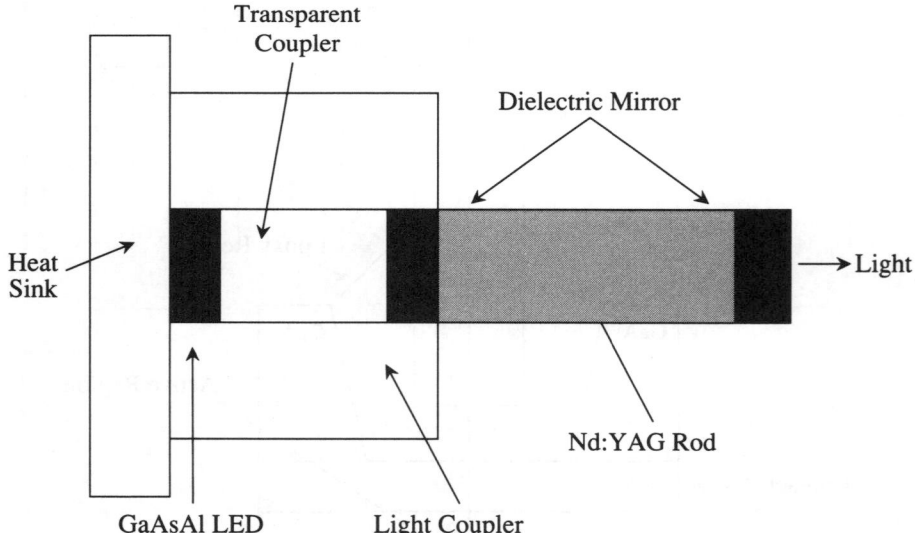

Figure 2.20 Nd:YAG laser

In-phase feedback makes the Nd:YAG laser ideally suited for high bandwidth applications over long-haul systems. Although it is expensive to produce, its lifetime is greater than that of the LED or ILD, making it the most widely used type of solid state laser.

Optical Fibres

There are three types of optical fibre structures currently in common use:

—   the step-index multimode fibre;

—   the graded-index multimode fibre; and

—   the step-index monomode fibre (see Figure 2.21).

Each structure has a set of individual characteristics which make it suitable for a particular type of transmission and application.

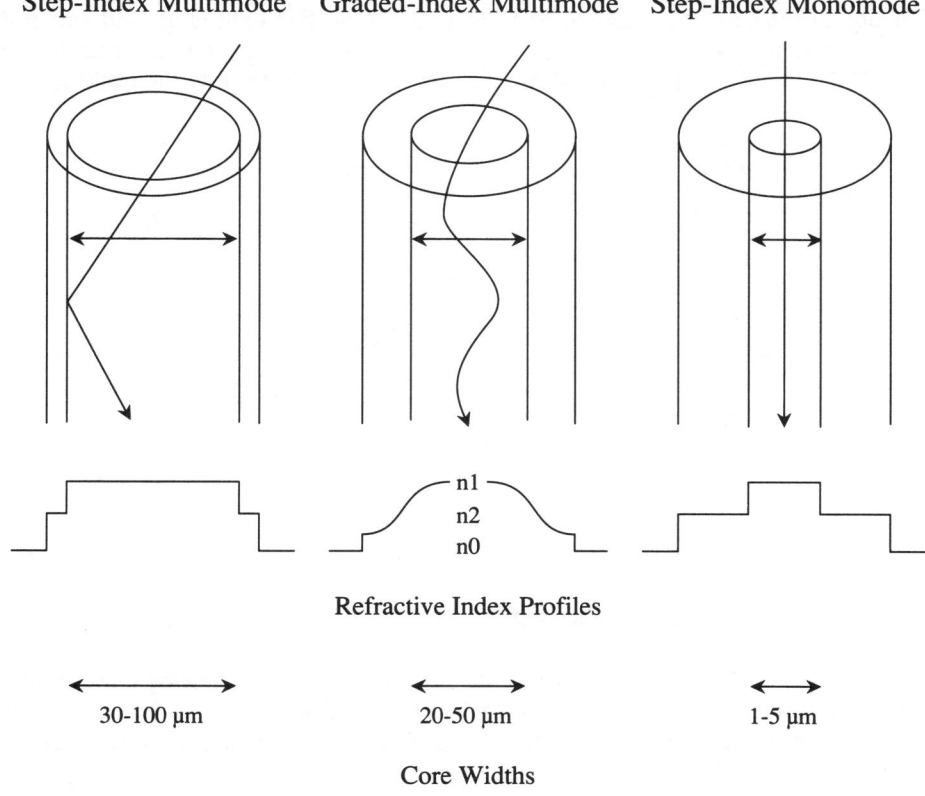

Figure 2.21   Optical fibre structures

One such distinguishing characteristic is the *refractive index* profile. The refractive index is defined as the ratio of the sine of the angle of incidence to that of the angle of reflection when a light ray passes from one medium to another.

Step-Index Multimode Fibre

This is the simplest form of optical fibre and consists of an inner glass core with refractive index n_1 surrounded by a lower refractive index cladding n_2. The relative index difference is given by:

$$\Delta n = \frac{(n_1 - n_2)}{n_1}$$

and is usually of the order of 0.01 (1%) for low-loss fibres.

Light rays are guided through the fibre by a series of *total internal reflections* at the junction of the core and cladding (see Figure 2.22). So that light is trapped within the glass core in this way, the light rays must enter at a shallow angle relative to the fibre axis. If the incident angle is too steep, then the light ray is lost into the cladding.

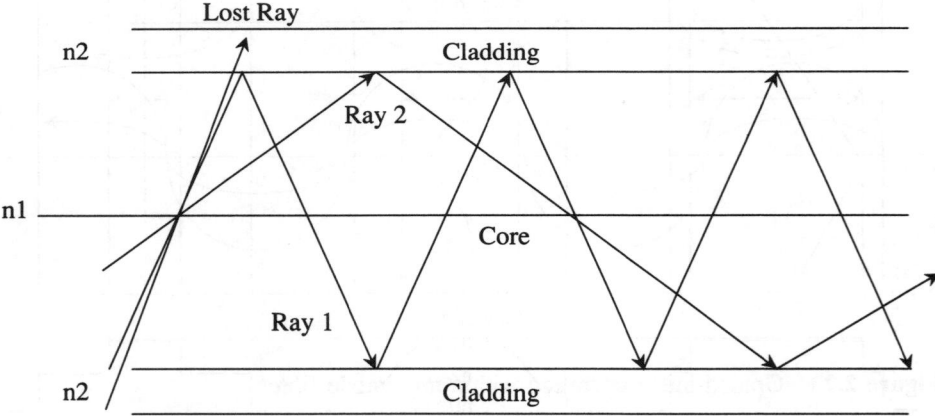

Figure 2.22 Total internal reflection

The multimode nature of the fibre results in each light ray propagating throughout the fibre at a different angle and therefore each ray has a different transit time. This means that if a short pulse of light is launched into a fibre of length L, and the rays enter at all angles, then the pulse which exits in the fibre is spread out in time by an amount given by:

$$\Delta t = \frac{n_1 L \Delta n}{c}$$

where c is the speed of light in the medium.

The effect of *pulse spreading* or *pulse dispersion* restricts the rate at which pulses may be transmitted through a fibre of length *L*, and is therefore an important design consideration. For short-haul transmission systems such as data links within buildings, ships and aircraft, this type of fibre is quite suitable. However for long-haul or higher bandwidth applications, other types of fibre must be used.

Graded-Index Multimode Fibre

This type of fibre has an inner core where the refractive index decreases from a maximum at the centre to a minimum at the cladding. Rays which enter at shallow angles are confined to the centre of the core, whereas steeper angled rays penetrate the low refractive index region where their velocity becomes higher.

Ideally, the refractive index profile is approximately parabolic. In this situation, the time taken for each ray to reach a common point is the same (see Figure 2.23) hence there is no pulse spreading. Since there is no practical profile which results in all ray paths having the same delay, some degree of pulse spreading is inevitable in a graded-index fibre, again this can be the limiting factor in the design of long-haul systems carrying high bandwidth data.

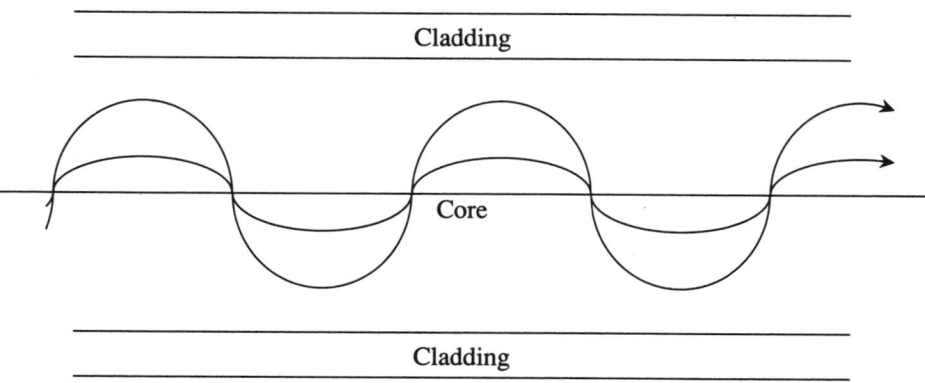

Figure 2.23   Graded-index operation within multimode fibres

Step-Index Monomode Fibre

The two structures described so far are both multimode fibres in which there are many different propagation angles (modes). The main reason for this multitude of modes are large core widths and large refractive index differences.

However, if the core and the index difference is made small enough the fibre will support essentially one mode, or angle. This type of fibre is known as a monomode fibre (see Figure 2.24), and because it exhibits no pulse dispersion it can support very high frequencies.

Optical Fibre Losses

The effect of pulse dispersion is just one of the limitations in using an optical fibre. In addition, there are effects known as:

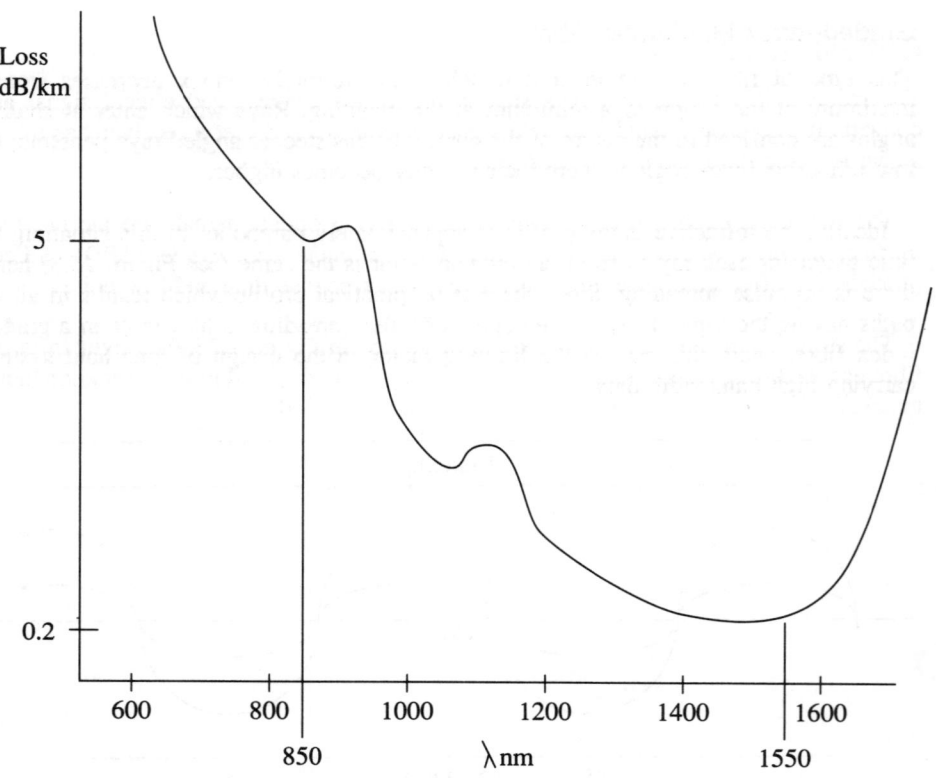

Figure 2.24   Optical fibre losses

- *chromatic dispersion* which results from light rays travelling at different velocities;

- *waveguide scattering* due to irregularities at the core-cladding interface; and

- *impurity absorption* which is primarily due to the presence of water ions and metal impurities in the core.

The combined effect of all these losses is to attenuate the power of the optical signal as it travels along the length of the cable. Figure 2.24 illustrates the attenuation characteristic of typical optical fibres and shows how scattering and absorption reach a minimum at a wavelength of 1550 nm. Contemporary optical fibre systems operate at this frequency and at frequencies of 850 nm and 1300 nm.

Optical Receivers

The receiver in an optical fibre system demodulates the received optical signal and converts it into proportional electrical signal. It is important that a receiver has a high signal-to-noise ratio and a low error rate.

The two most important devices used for receiving optical signals are the *positive-intrinsic-negative photodiode (PIN)* and the *avalanche photodiode (APD)*. Silicon photodiodes offer small size, high sensitivity, fast response time and low noise.

The structure of the PIN photodiode is illustrated in Figure 2.25. The light from the fibre passes through a non-reflective, transparent glass casing and into the semiconductor material where it causes a current to flow through the diode.

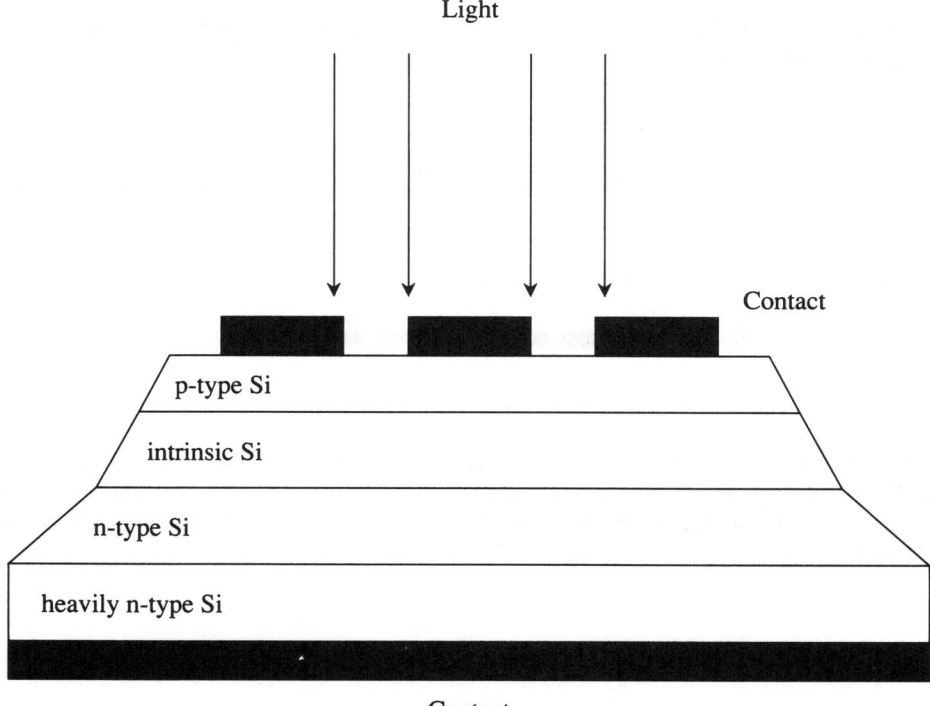

Figure 2.25 The structure of the PIN photodiode

By adding more impurity, the number of electrons in the n-type part of the PIN photodiode can be increased and an avalanche photodiode (APD) is created. Despite the inherent noise associated with this type of receiver, it is far more sensitive than the PIN photodiode and is therefore used in long-haul systems.

System Considerations

Since optical fibres enable long distance transmission of analogue and digital data with high reliability and a low cost bandwidth ratio, fibre systems are capable of replacing any other transmission medium in nearly all applications.

The optical fibre medium offers many other features which are not available from other transmission media. These include:

—   protection from electrical interference;

—   lack of detectable radiated signal;

—   very high bandwidth; and

—   reduction in weight and volume.

These, and other features make optical fibre systems ideally suited for a wide range of applications, including:

—   inter-exchange trunk circuits;

—   data transmission in aircraft where lightweight components are essential;

—   cable television systems;

—   industrial process control systems; and

—   military communications.

Many of these applications have operated for many years. However, one of the most recent innovations in optical fibre systems is the *fibre distributed data interface (FDDI)*. This is a rapidly emerging technology which uses optical fibre to carry data between networked devices at speeds of up to 100 Mbit/s.

3

Modulation and multiplexing

Introduction

Communication, in any sense of the word, is simply the transference of information from one place to another. Thus, in a typical model of a communications system, it is always possible to identify an *information source* and an *information sink*. The communications path between these two points is given the general term of *channel* regardless of the technology involved. The final elements in this basic communications system model ensure that the information passes over the channel in a suitable manner. This process requires the conversion of the information from the source into a form appropriate to a given channel, and its reconversion at the other end for the sink to understand.

The above description of a communications system can be used to fit any scenario. Take writing a letter as a communications process. The information source is the brain of the correspondent and the information contained in it is converted into a suitable form for transmission by hand movements and an implement such as a pen. The letter is conveyed across the postal system, which constitutes the channel. The recipient's brain is the information sink and the reverse conversion from written letter to reader comprehension takes place via the eyes.

The above example is trivial, but it underlines a most important point, which is that information cannot be conveyed without a transmission process (in this case the pen movements). Furthermore, the form of transmission is dictated exclusively by the characteristics of the channel and the nature of the information. This is especially true in an electronic communications system. In this chapter we will look at the transmission techniques that have evolved to cater for the various channels encountered in such systems, in particular, the two most common transmission processes, modulation and multiplexing.

Modulation

Modulation is the process whereby a high frequency wave is made to carry a lower frequency wave. There are many instances where a transmission medium will convey high frequency signals, and this is when modulation is required. Radio provides a good example. High frequency electromagnetic waves propagate well through space, but low frequency speech and music signals do not. In radio, therefore, the speech or music signals are allowed to modulate a high frequency carrier of several hundred kilohertz (kHz), and the modulated high frequency signal is then broadcast. At the radio receiver this modulated high frequency wave is demodulated to retrieve the original speech or music signal.

Another example of the need to modulate occurs in sending data over standard telephone circuits. As discussed in the previous chapter, frequencies below 300Hz are increasingly attenuated on such circuits, and since a digital data signal has a spectrum extending right down to 0Hz, it would experience an intolerable amount of distortion. Modulation is the answer.

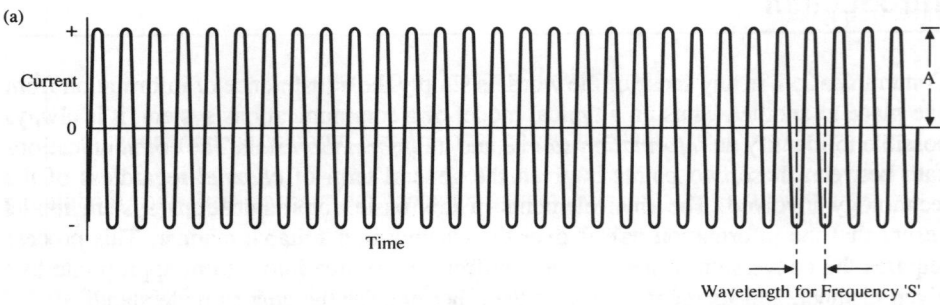

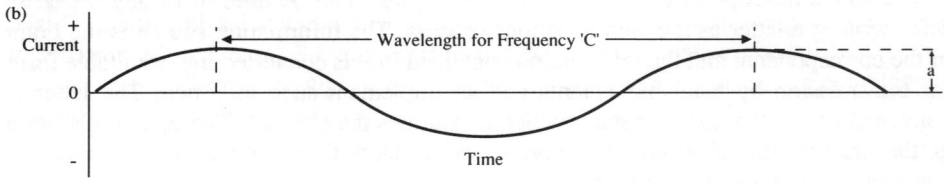

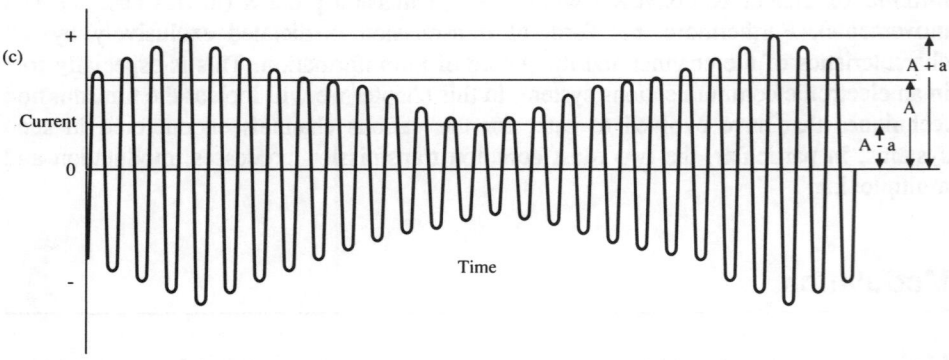

Figure 3.1 The principle of amplitude modulation

Various modulation techniques are available and these can be classified by the characteristic feature of the carrier which is modified. The three main categories are amplitude, frequency and phase modulation. These techniques and some of their variants will be described in this chapter.

Amplitude Modulation

Figure 3.1 illustrates the effect of a speech wave modulating a carrier where amplitude modulation is employed.

The typical 'envelope' shown in the lower part of the figure is in fact a compound of three frequencies:

-     c   the original 'carrier' frequency;

-     c + s   the sum of the 'carrier' and modulating frequency;

-     c − s   the difference between the carrier and the modulating frequency.

It is usually much more convenient to express signals and thus demonstrate the effects of modulation in terms of frequency. Using Fourier theory as discussed in Chapter 1 it is possible to do this, and in the example of Figure 3.1 such an approach would confirm that the signal consists of only the three component frequencies identified. This is represented in the frequency domain as shown in Figure 3.2 (note that the presence of a single component frequency is simply denoted by an upward arrow).

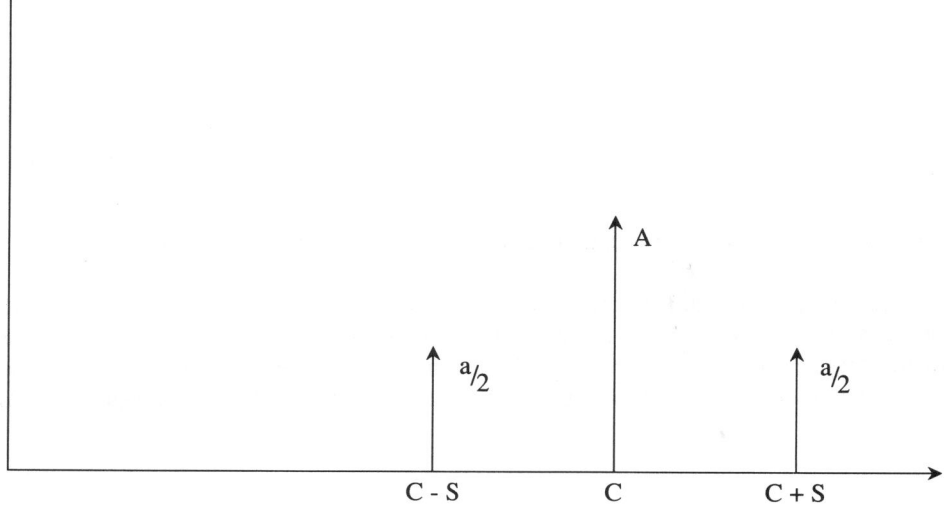

Figure 3.2 Frequency domain: single component

However, the single frequency scenario that these diagrams use is a great simplification. A cursory glance at a typical speech signal, for instance, will reveal that it is far from being a perfect sine wave. In fact it is a combination of many sine waves of different frequencies, and so if Fourier analysis were to be performed on such a signal, the result would indicate a spectrum such as the one shown in Figure 3.3. Thus, amplitude modulating this signal produces the frequency plot shown in Figure 3.4.

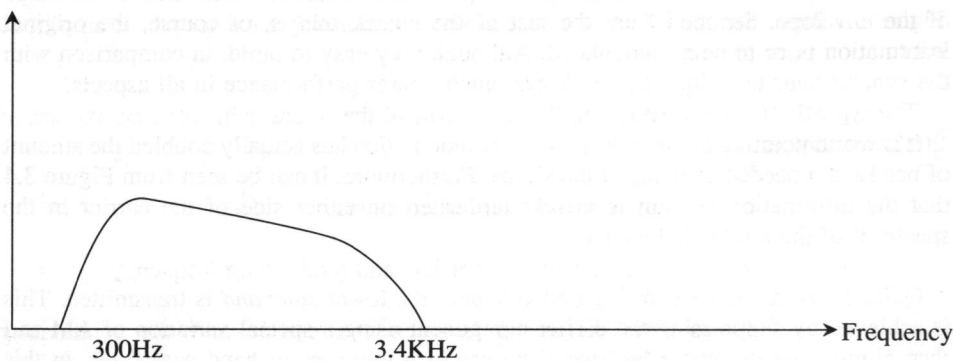

Figure 3.3 A typical speech spectrum

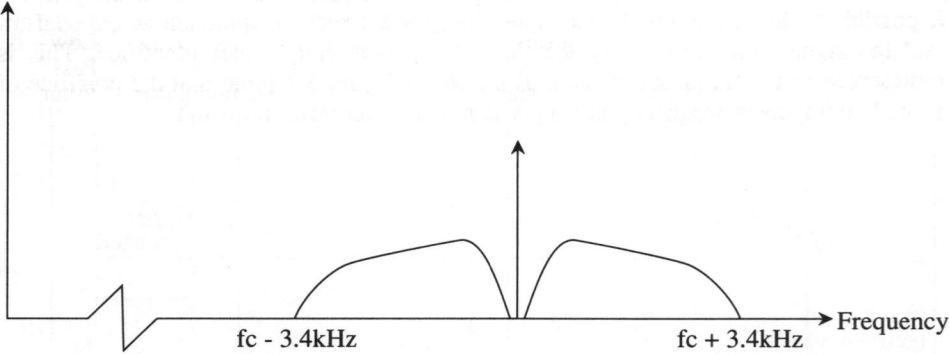

Figure 3.4   Amplitude modulation of a speech signal onto a high frequency carrier

At the receiving end, a locally generated carrier, of the same frequency and in the same phase relationship as the original carrier, is introduced (to restore the conversation to the original speech frequencies). This technique is known as 'demodulation'. The demodulator works on the same basis as the modulator in that three frequencies will be produced:

$$c = c$$

$$c + (c - s) = 2c - s$$

$$c - (c - s) = \text{(the original speech frequencies)}$$

The third frequency is the original speech frequency (or band of frequencies); it may be filtered off by means of a low-pass filter and amplified as required. Since it requires an accurate version of the original carrier frequency (ie, a precise clock) this form of demodulation is described as synchronous or coherent.

Another method of demodulating a simple AM signal is to use an envelope detector circuit. This is an extremely straightforward piece of electronics which 'tracks' the shape of the envelope. Stripped from the rest of the signal, this is, of course, the original information prior to being modulated. Although very easy to build, in comparison with the synchronous technique above, it has much poorer performance in all aspects.

It is worth pointing out that the process of modulation has actually doubled the amount of bandwidth needed to transmit the signal. Furthermore, it can be seen from Figure 3.4 that the information content is merely replicated on either side of the carrier in the spectrum of the modulated signal.

Quite often in speech carrying systems, only the lower *sideband* is transmitted. This is achieved by suppressing the carrier component using a special variation of AM and then eliminating the upper 'sideband' by use of a low-pass or band-pass filter. In this way power requirements are reduced and the number of channels which can be accommodated in a given frequency range is increased. Figure 3.5 shows how this process might be organised.

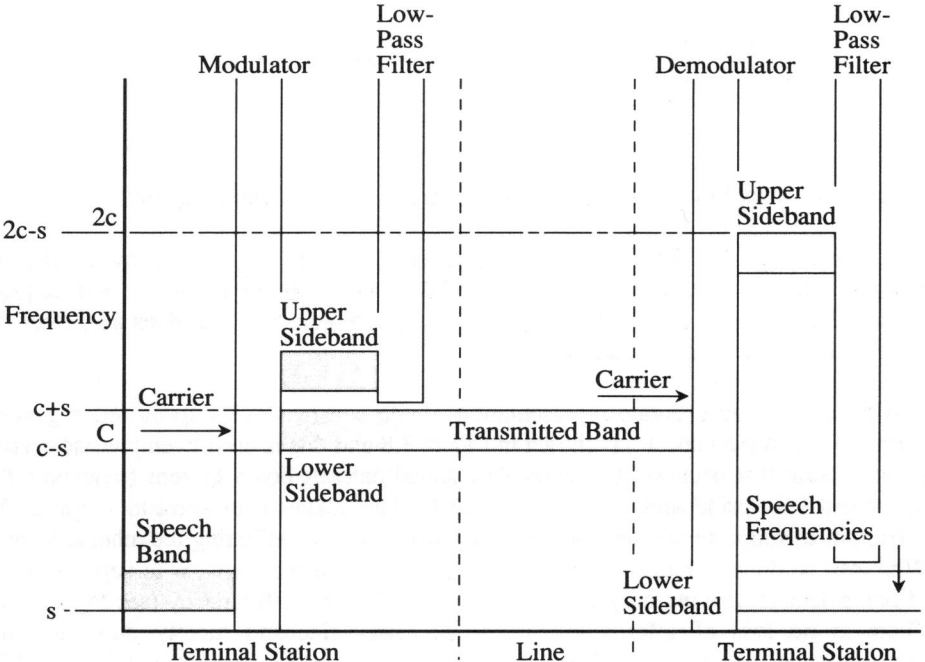

Figure 3.5 Principle of lower sideband suppressed carrier working

Amplitude Modulation using Digital Signals

Figure 3.6(b) illustrates a carrier wave modulated in amplitude by the binary data shown in Figure 3.6(a). A special case of amplitude modulation is when the lower of the two amplitude levels is reduced to zero; the modulation process then reduces to switching the carrier on and off as illustrated in Figure 3.6(c). However, the variation in transmitted energy makes this technique unsuitable for data transmission over telecommunications networks except using optical systems where it is widely employed.

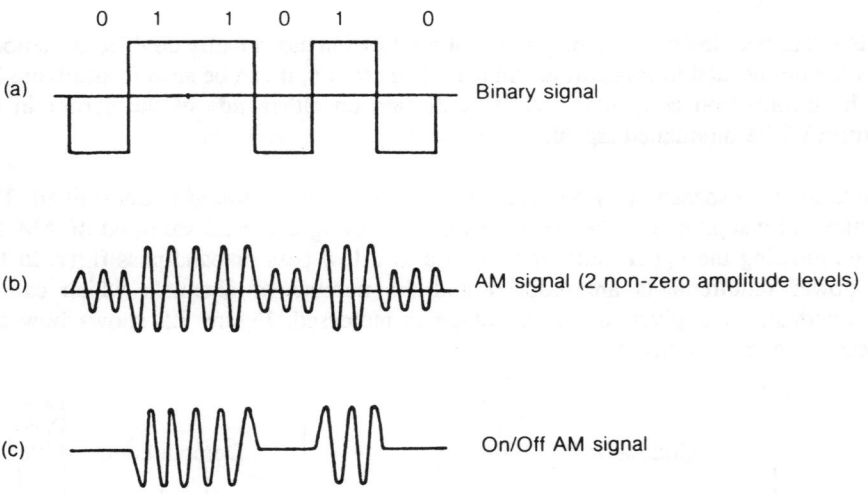

Figure 3.6 Amplitude modulation with binary digital modulating signal

A square wave like the one in Figure 3.6(a) contains high frequency components, and in a practical AM system the data signal would be passed through a low-pass filter prior to modulation. This rounds off the square wave (see Figure 3.7), but does not affect the information content of the data signal.

Although not an accurate representation of the square wave's spectrum, a general method for denoting spectra is shown in Figure 3.8 and this is used to emphasise certain points. Note that because the binary data signal extends down to zero frequency, the upper and lower sidebands actually meet at f_c. This makes it difficult to suppress the carrier, or to suppress one sideband and the carrier, without affecting the other sideband. What can be done to reduce the bandwidth of the modulated signal is to suppress most of one sideband, leaving only a vestige of it near the carrier frequency (see Figure 3.9). There is no loss of information since the lower sideband merely duplicates the information in the upper sideband. The technique is called vestigial sideband (VSB) modulation.

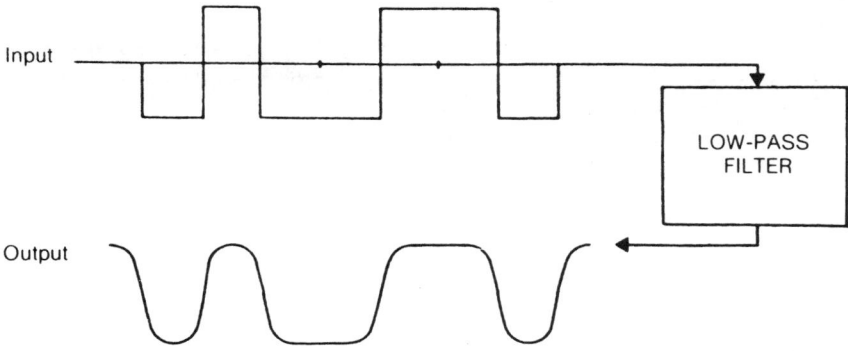

Figure 3.7   Rounded square wave produced by a low-pass filter

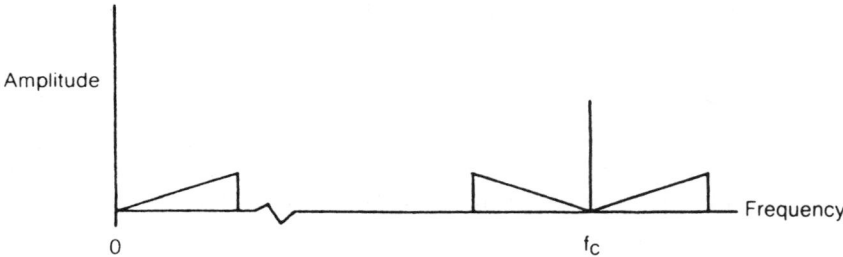

Figure 3.8   Sidebands in an AM signal with baseband modulating waveform

With clever filter design it is possible to suppress the carrier in VSB systems. It leads to part of the upper sideband being suppressed, but the vestige of the lower sideband which remains supplies the missing frequencies.

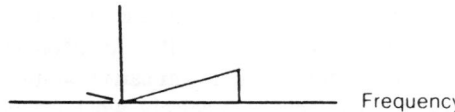

Figure 3.9   Vestigial sideband AM

True single-sideband amplitude modulation with a digital modulating signal can only be achieved by scrambling the original data (ie randomising the bit-stream) in order to remove low frequency components caused by long strings of 1s or 0s. This has the effect of separating the sidebands from the carrier (see Figure 3.10), thus making it feasible to filter out one sideband and the carrier.

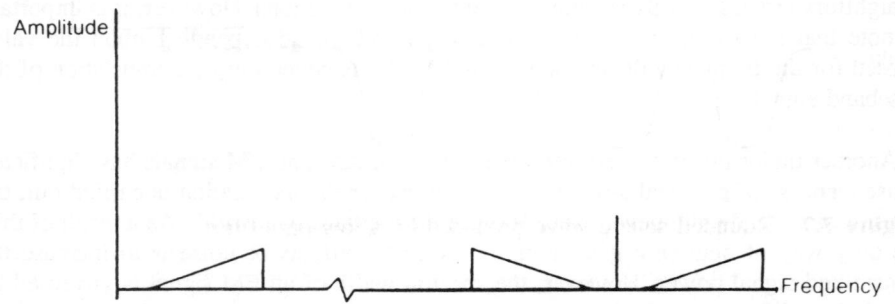

Figure 3.10   Sidebands in an AM signal with scrambled baseband modulating signal

Pulse Amplitude Modulation (PAM)

Pulse Amplitude Modulation (PAM) when used with a digital modulating waveform, provides a means of coding more than one bit per baud, by encoding the binary data signal as a signal with more than two levels (sometimes called an m-ary signal).

For example, the bits of a binary data signal could be sampled in pairs. There are four possible combinations of a pair of bits, and thus each pair could be encoded as one of four amplitude levels. The encoded four-level signal has half the baud rate of the original data signal, and can be used to amplitude modulate a carrier in the usual way.

Frequency Modulation

Since many of the traditional channels used for telecommunications transmission are subject to fluctuating signal losses, pure AM has not often been used. Another disadvantage of AM which makes it less suitable when digital messages are to be modulated is that a different amount of transmitter power is required for each of the two bit values which means that one value is more susceptible to noise than the other. Clearly, both these limitations of AM are as a direct result of its inconstant envelope and thus there is strong motivation for maintaining a uniform carrier amplitude and finding another carrier attribute to vary instead. This is precisely what frequency and phase modulation both achieve.

In frequency modulation, the idea of a carrier frequency is somewhat more notional than in AM. The 'instantaneous' frequency of the carrier is continuously altered to reflect the signal level of the modulating message. In the simple case of one cycle of a sine wave, the two amplitude peaks would be represented in the final FM wave by the two greatest departures either side of the carrier frequency. This introduces the concept of *frequency deviation* which is the ratio of the maximum departure and the carrier frequency itself.

The spectrum of FM signals is exceedingly difficult to derive, and even in the most straightforward case involves complex mathematical functions. However, it is important to note that the spectrum varies in both shape and spread depending upon the value chosen for the frequency deviation and unlike AM, is never simply a translation of the baseband signal.

Another major difference between the spectra of AM and FM signals has significant consequences for practical systems. In AM, however the modulation is carried out, the transmission bandwidth never exceeds twice the message bandwidth. As a result of this, the only way of obtaining greater immunity to the effects of noise is to increase the transmitted signal power. However, the spectral width of an FM signal is governed by the choice of frequency deviation, and is in many cases greater than twice the message bandwidth. By judicious alteration of the frequency deviation it is actually possible to trade bandwidth for noise performance.

Frequency Modulation of Digital Signals

Figure 3.11 illustrates a binary data signal and the corresponding carrier when frequency modulation techniques are employed. Note that the carrier consists of two frequencies (sometimes referred to as tones) and that it switches abruptly between these two frequencies in sympathy with the binary value of the data. This method of carrying binary information is termed frequency shift keying (FSK).

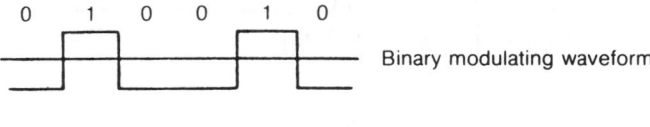

Binary modulating waveform

FM signal

Figure 3.11   Frequency shift keying (FSK)

In pure FSK transmission there are no initial restrictions placed on the tones used to represent the binary states, although the further apart they are, the greater the frequency deviation and this in turn means a larger bandwidth requirement. In the case of analogue signals it was pointed out that spectral analysis of FM is very complicated. However, in the digital case there is a neat simplification that can be made, and this is demonstrated in Figure 3.12. The signal can be regarded as the sum of two amplitude shift keying (ASK) components and thus it is not surprising to find that the bandwidth required for FSK signalling is greater than for ASK. However, it is more resilient to error and so tends to be used for low data rates in situations where the bandwidth available is not a serious restriction.

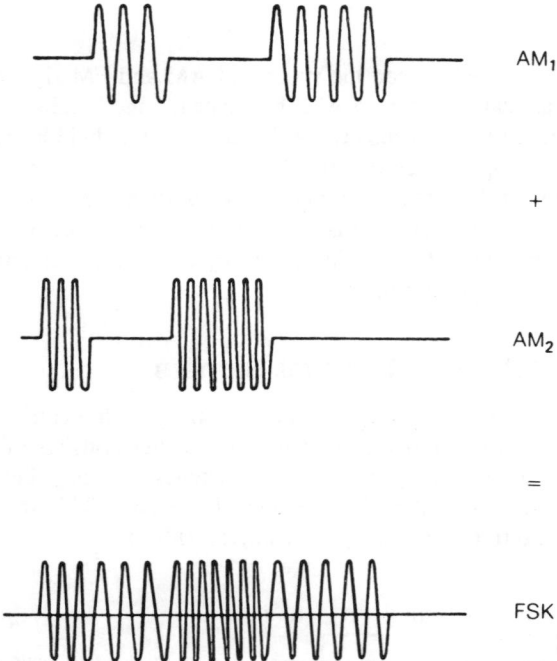

Figure 3.12   FSK as the sum of two AM signals

Consider the simple low speed FSK system described in the V21 recommendation. With this system binary 1 is represented as 980 Hz and binary 0 as 1180 Hz. The data rate is 300 bit/s. True duplex, or two-way, is provided by means of a second channel operated in the reverse direction on the same link using two different frequencies, 1650 Hz for 1 and 1850 Hz for 0.

Phase Modulation
Analogue phase modulation is not often used but its digital counterpart, phase shift keying (PSK), has a very prominent role in data transmission. Imagine two oscillators both generating a sine wave at the carrier frequency, but 180° (half a cycle) out of phase.

One oscillator is connected to line whenever there is a 0 in the data signal, and the other whenever there is a 1. The waveforms are shown in Figure 3.13.

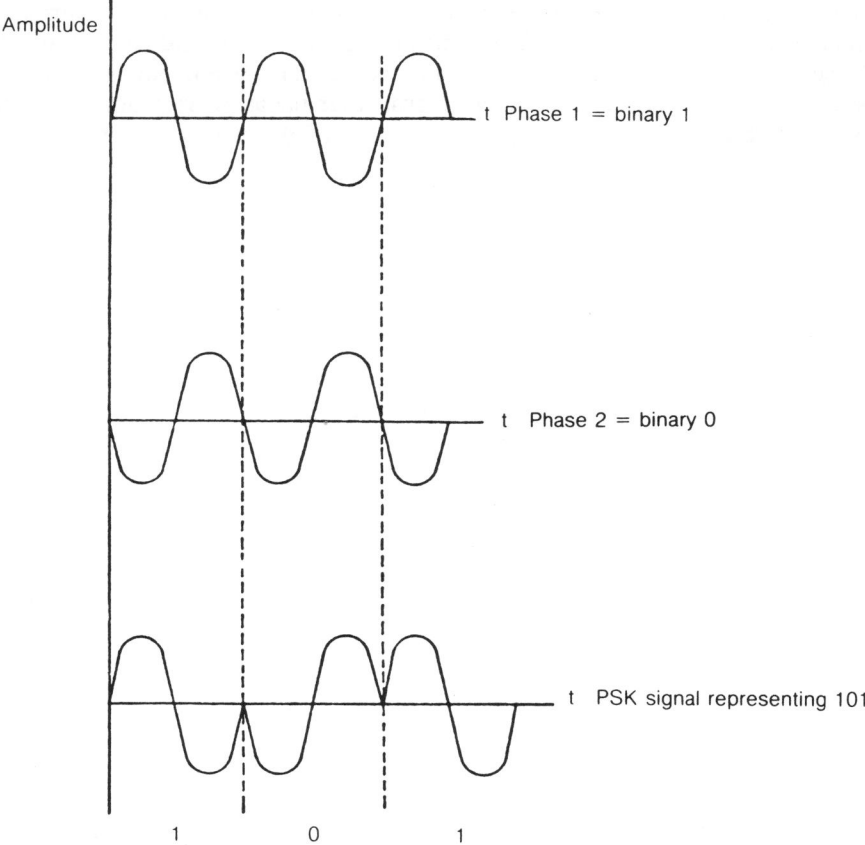

Figure 3.13   Phase shift keying (PSK)

If the receiver has a reference carrier against which to compare the phase of the incoming signal, it will be able to demodulate the signal. However, the receiver must be given some indication at the start of transmission as to which phase represents a 0 and which a 1.

A modulation system in which a carrier is generated locally at the receiver and used to demodulate the incoming signal is known as a *coherent* system. The phase modulation system just described is a coherent system. It is also known as fixed reference phase modulation.

One of the problems encountered in practical implementations of PSK systems is the accurate referencing of the carrier phase in the detector. If no other information is given, the receiver may regenerate the carrier 180° out of phase, which would result in every bit being detected as its inverse. To avoid this, a technique known as differential encoding is employed, which is demonstrated in Figure 3.14. The principle behind this scheme is that the information is carried by the changes in phase and not in the absolute phases as for ordinary PSK. From Figure 3.14 it can be seen that a binary 1 is transmitted as a change in phase from the previous bit-period (regardless of what it was) and a binary 0 is represented as a continuation of the signal phase. No matter what the phase of the reference signal at the receiver, changes in phase and retentions of previous phase are detected as such, and so the 1s and 0s of the original bit-stream can be identified.

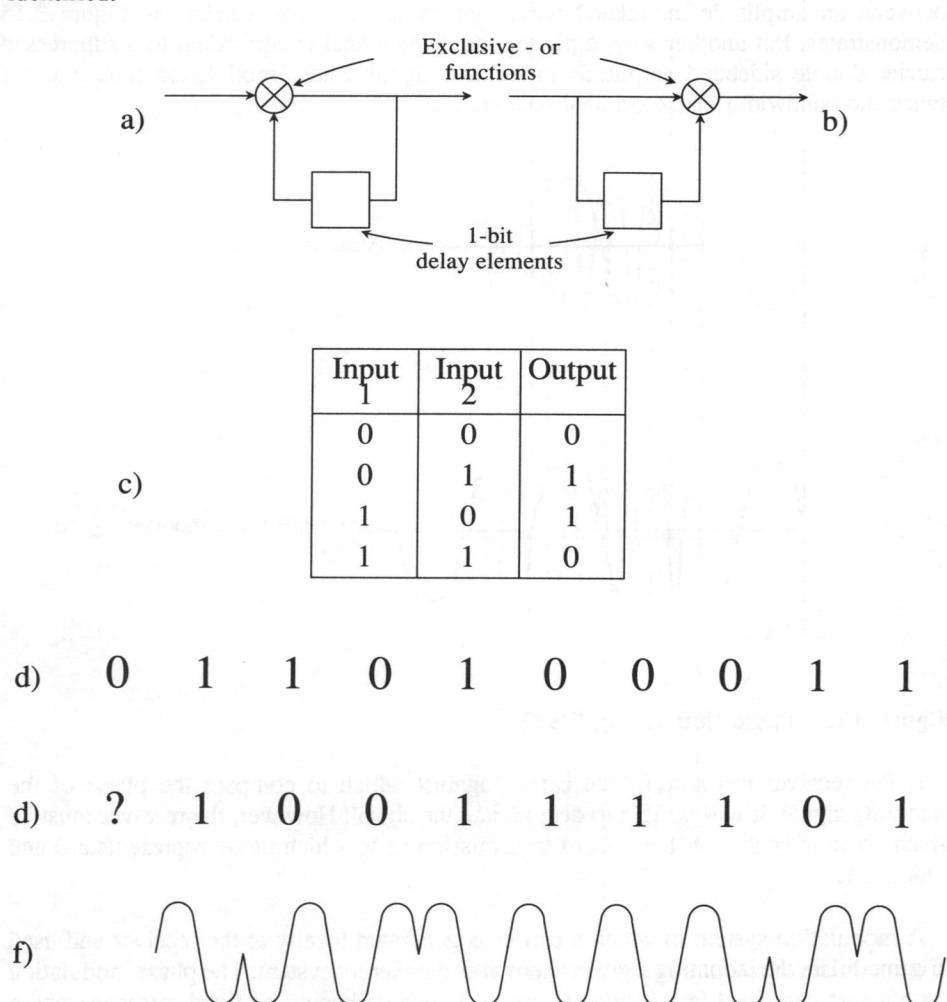

a)

Exclusive - or functions

b)

1-bit delay elements

c)

Input 1	Input 2	Output
0	0	0
0	1	1
1	0	1
1	1	0

d)   0   1   1   0   1   0   0   0   1   1

d)   ?   1   0   0   1   1   1   1   0   1

f)

Figure 3.14 Differential encoding and PSK (a) differential encoder, (b) differential decoder, (c) exclusive-OR operation

Unlike FSK described above, which could be detected coherently and non-coherently, PSK can only be demodulated with the aid of carrier regeneration. Unfortunately, this requires extra complexity in receiver design to obtain a synchronised reference signal, one attraction of FSK is that a simplified, if non-optimal, receiver can be used. However, the theoretical performance of PSK is better than that of any FSK technique (even coherent detection). Additionally, PSK is more spectrally efficient, in that it consumes less bandwidth for a given bit rate.

Phase modulation using a binary modulating signal can be considered as the difference between an amplitude modulated wave and an unmodulated carrier, as Figure 3.15 demonstrates. Put another way, a phase modulated signal is equivalent to a suppressed carrier double sideband amplitude modulated signal. Phase modulation thus requires twice the bandwidth of the original data signal.

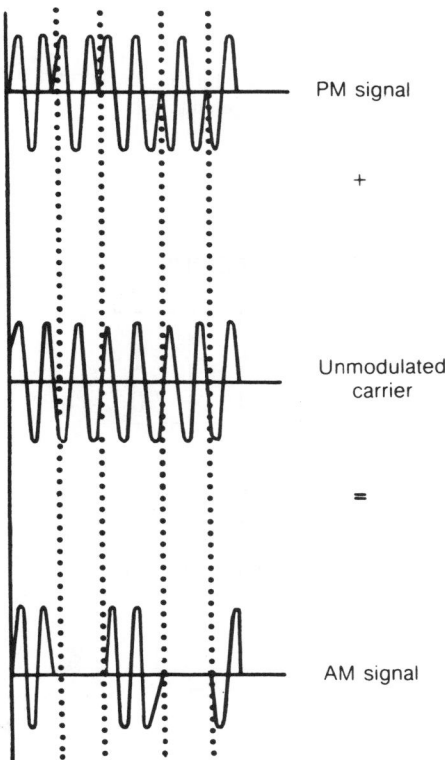

Figure 3.15 Relationship between PM and AM

Multi-Phase Modulation

Two-phase modulation as described above codes one bit of data per phase change. With more than two phases, it would be possible to code more than one bit per phase change, and thus to increase the bit rate without altering the modulation rate. To code two bits per phase change, for example, we would need four possible phase changes to represent the four combinations of two bits (00, 01, 10, 11).

An attraction of this is a useful property of sinusoidal waves, known as orthogonality. It can be mathematically proven that both a sine wave and cosine wave carrier can be used simultaneously and detected quite separately using coherent demodulation on the two respective carriers. Thus a modulation and demodulation scheme such as shown in Figure 3.16 can be followed. As long as the two recovered carriers are correctly phased, there should be no crosstalk between the information in streams A and B. The cosine and sine processes in this modulation are sometimes referred to as *in-phase* and *quadrature* or *real* and *imaginary*.

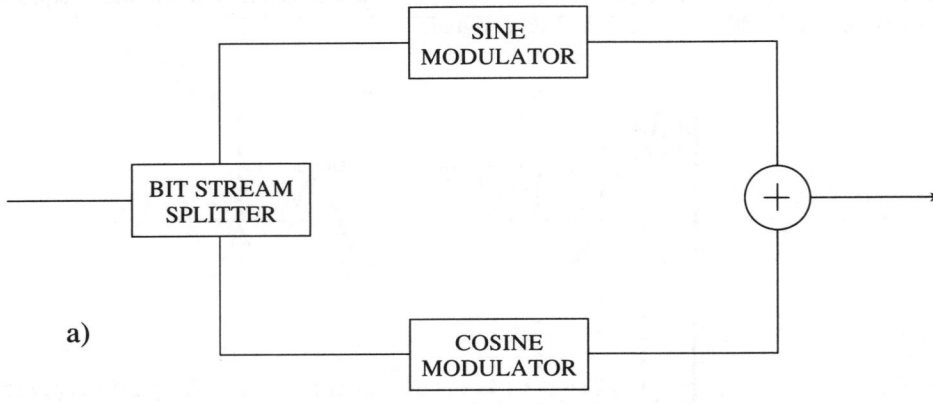

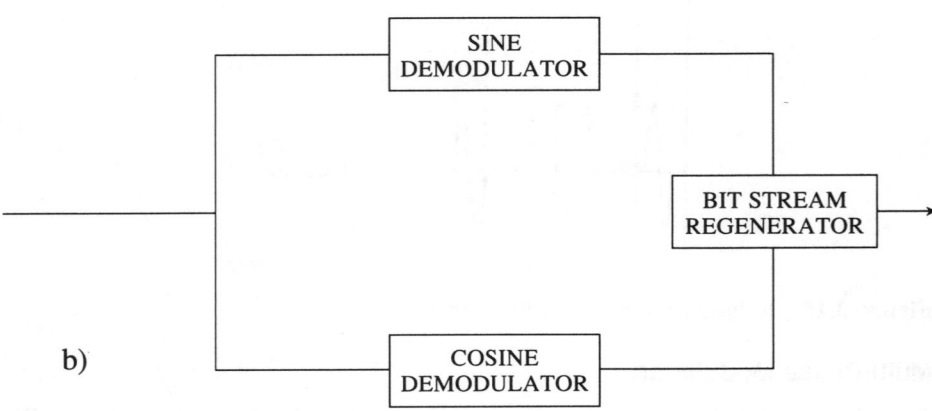

Figure 3.16 Modulation and demodulation scheme

Using differential modulation, one technique would be to divide the data signal into pairs of bits (*dibits*), and to shift the phase of the carrier signal in one of four ways, according to the dibit combination. Table 3.1 shows one possibility.

Table 3.1 Coding of dibits

Dibit	Phase change
00	0°
01	90°
11	180°
10	270°

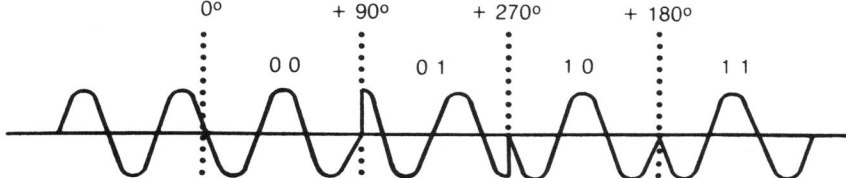

Figure 3.17 4-Phase modulated signal

The phase changes shown in Table 3.1 are phase shifts relative to the previous phase of the carrier. Figure 3.17 gives an example of a carrier modulated in this way. An alternative method of presenting the information given in Table 3.1 is to use a type of phase diagram showing possible phase shifts (see Figure 3.18). Such a diagram provides a very convenient method of illustrating the states that a transmitted carrier can occupy, especially for the more complicated phase and amplitude modulation techniques.

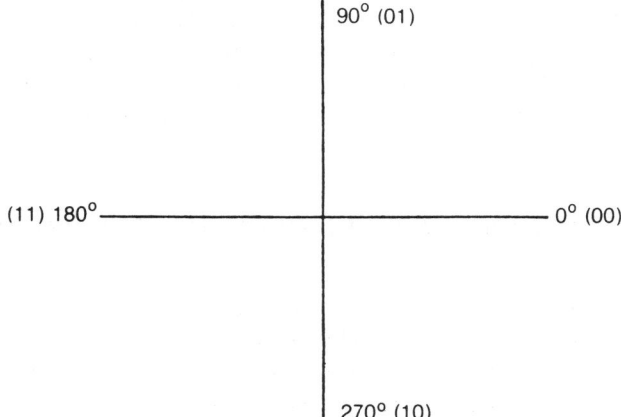

Figure 3.18 Phase diagram showing phases in 4-phase modulation

Notice that in using the dibit-phase associations from Table 3.1, only one of the two carriers is active during the transmission period of any particular dibit. This can be avoided by transmitting a final set of phases which are always the result of a balanced combination of equal amplitude, real and imaginary components (following more faithfully the modulation scheme shown in Figure 3.16). Although the phase diagram of this approach (see Figure 3.19) only shows a 45° rotation of phases from that in Figure 3.18, the net effect is to permit more efficient usage of the two carriers available.

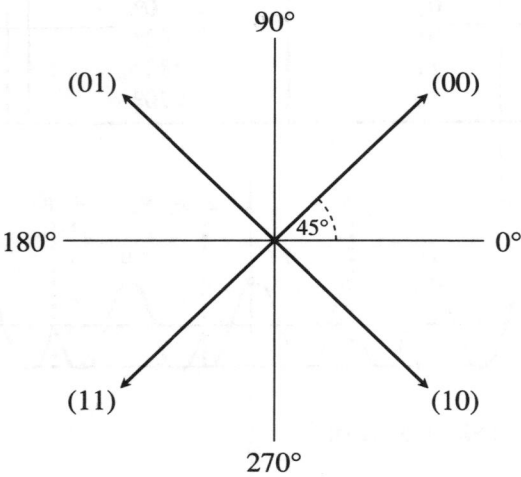

Figure 3.19 Phase diagram showing phases in 4-phase modulation when in-phase and quadrative bit-streams act in tandem

We have already seen that, for a binary modulating signal, two-phase modulation is equivalent to double sideband suppressed carrier (DSB-SC) amplitude modulation. Four-phase modulation is equivalent to two DSB-SC waves with carriers 90° out of phase with each other, being transmitted simultaneously. The two corollaries of this are first that quadrature phase modulation consumes no more bandwidth than binary phase modulation but has twice the bit rate capacity and that like all forms of phase modulation it is a synchronous modulation method requiring a clock in the modem and coherent detection at the receiver.

Phase and amplitude modulation (quadrature amplitude modulation)

The quadrature phase modulation techniques described above can also be considered a special case of a superset of modulation schemes known as quadrature amplitude modulation (QAM). It is possible to combine phase modulation and amplitude modulation to give a further increase in the number of bits per baud. Figures 3.18 and 3.19 show the possible phase shifts in one of two possible levels for each of these phase shifts, there would be eight possible states which the carrier could adopt in each baud period (see Figure 3.20). This would allow three bits per baud to be carried.

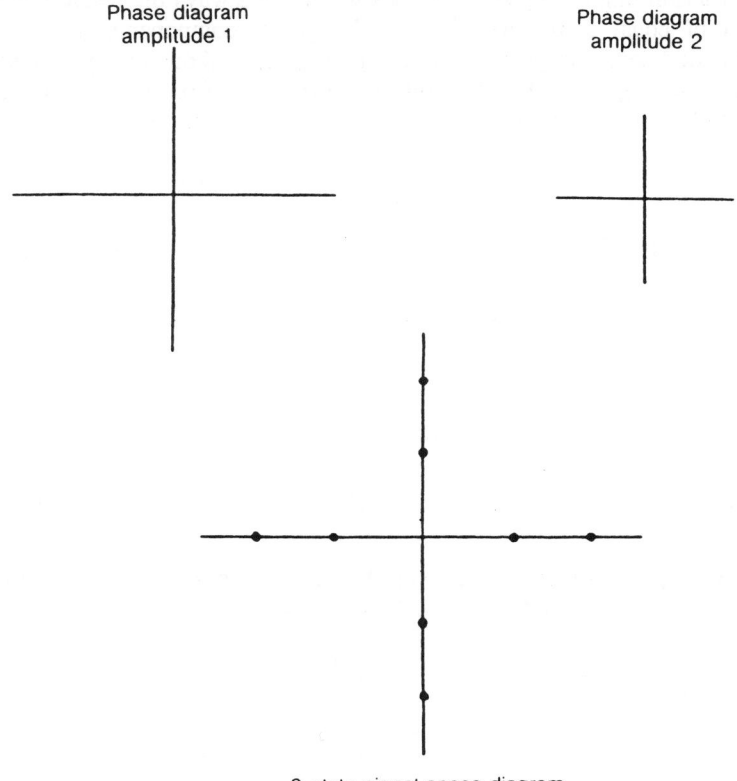

8-state signal space diagram

Figure 3.20   Phase-amplitude-modulation (4 phases, 2 amplitudes)

Modulation technology has witnessed the evolution of intricate phase-amplitude combinations. The number of combinations in these schemes is always 2^n so that each of them represents a unique n-bit symbol. The phase diagrams are usually drawn as series of points, and called *signal state diagrams*. Examples of such representations for 16 and 32 point QAM (n=4 and 5 respectively) are shown in Figure 3.21.

One practical implementation of these two constellations is as follows. Notice that for 16-point QAM the state diagram shows four *quadrants* each with four points in them. In Figures 3.21a and 3.21b these have been annotated as A, B, C and D for convenient reference. Here we have a scheme whereby two of the four data bits in each symbol are used to identify one of these quadrants, and the remaining two identify a point within this quadrant. Since the usual phase ambiguity problems will affect only the ability of the receiver to determine from which quadrant the transmitted data was, differential encoding is only required on these two quadrant bits.

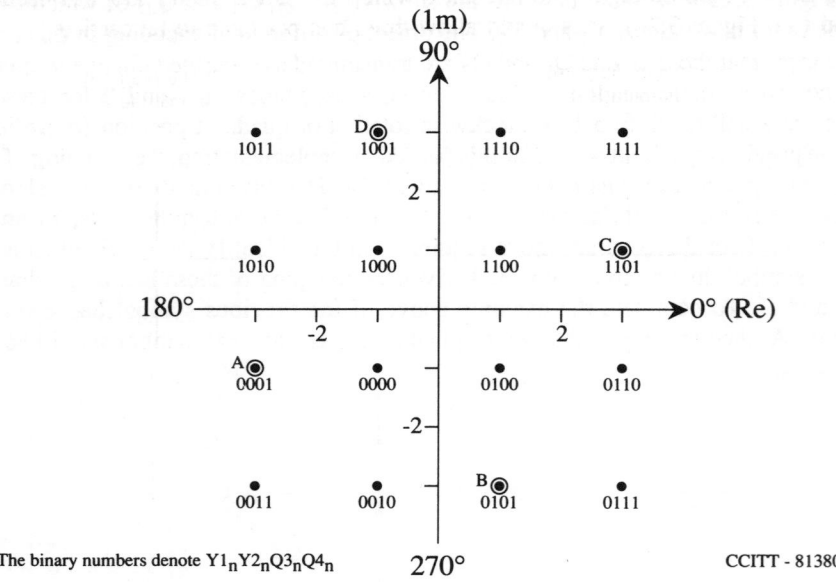

Figure 3.21a 16-point structure with non-redundant coding for 9600 bit/s and subset A, B, C and D states used at 4800 bit/s and for training

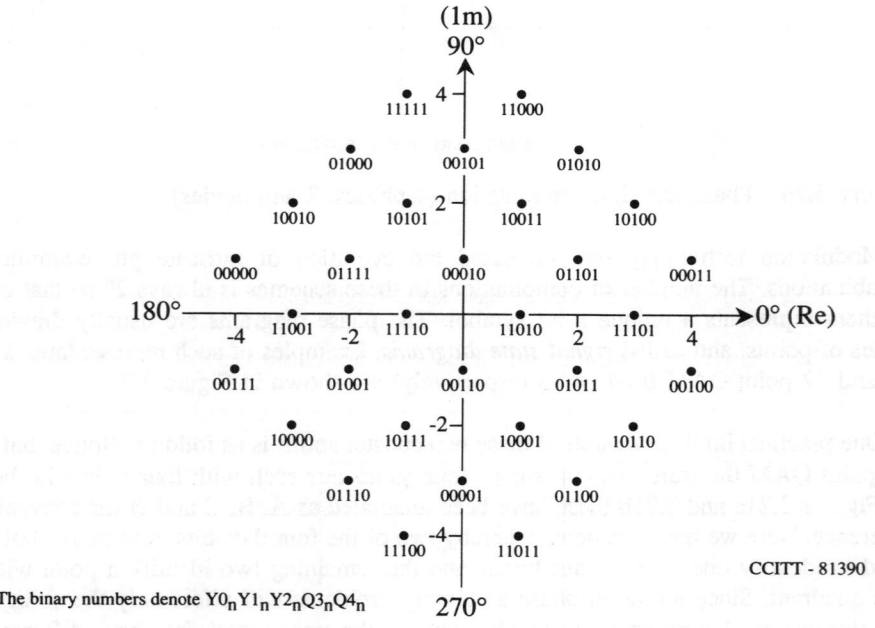

Figure 3.21b 32-point structure with trellis coding for 9600 bit/s and states A, B, C, D used at 4800 bit/s and for training

In Figure 3.22 four initial data bits are shown (Q_1, Q_2, Q_3 and Q_4) of which two are encoded into new values, Y_1 and Y_2. The object of differential encoding in this context is to ensure that the data bits Q_1 and Q_2 are transmitted as a defined change in quadrant (and not an absolute quadrant value). For instance, values of 1 and 0 for these bits respectively will result in a 180° clockwise rotation of quadrant position (regardless of what he previous quadrant was). For this particular implementation, the mapping of these two bits to quadrant change is given in Table 3.2a. The differential encoding algorithm in this case is not straightforward and must ensure that the transmitted bits, Y_1 and Y_2, change to reflect the quadrant move required and thus identify the quadrant to be sent for this symbol time period. Table 3.2b gives the mapping of these bits to quadrants A, B, C and D. To complete the example above, if the previous symbol had been from quadrant A, then to convey the two original data bits, the next symbol should be from quadrant C.

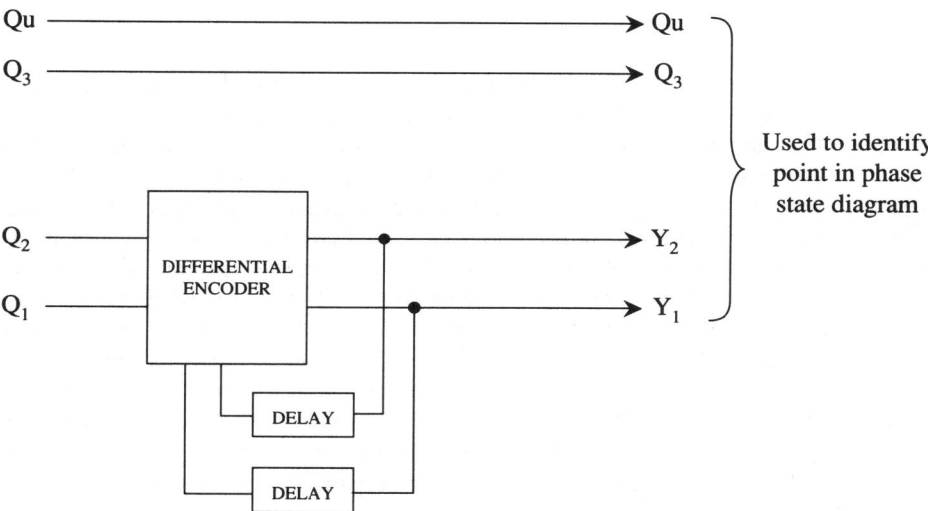

Figure 3.22 Differential encoding phase state diagram

Table 3.2a Coding of two data bits intoa quadrant phase change

Data Bits		Phase Change
Q_1	Q_2	
0	0	90°
0	1	0°
1	0	180°
1	1	270°

Table 3.2b Encoded bits necessary for actual transmitted quadrant phase

Encoded Bits		Actual Quadrant
Y_1	Y_2	
0	0	A
0	1	B
1	0	D
1	1	C

In the same implementation, use is made of the 32-point QAM constellation as a suggested alternative for achieving a four bits per symbol throughput. A fifth bit is calculated from a complex mathematical operation (which includes delay elements) on two of the four symbol bits and the five bits are then used to identify a point in the state diagram. The points in 32-point QAM are closer together than 16-point QAM, which normally indicates greater susceptibility to noise (and therefore errors). However, in this case, since one of the five bits has a mathematical dependency on two others (spread over a number of symbol periods by the delay elements), there are limits on the ways in which the state diagram points may traverse from one quadrant to another. This technique actually provides extra resilience against errors and is known as *trellis coding*. There are a number of trellis codes which are applicable to different constellations. In mathematical terms these are all forms of convolutional error correcting codes (see Chapter 8). In this instance, four bits carry information (thus achieving the same overall throughput as for 16-point QAM) and the fifth is responsible for error correction. (See also Table 7.7 in Chapter 7).

Multicarrier Amplitude Phase Modulation

Another of the amplitude phase modulation techniques relies on the simultaneous transmission of multiple carriers. One particular technique employs 48 carriers, separated by 45 Hz spacings. By a combination of phase and amplitude modulation, each carrier can occupy one of 32 discrete states in each baud period, permitting five bits per baud to be carried. Thus the 48 carriers can carry 240 bits per baud. For operation at 9600 bit/s the modulation rate need only be 40 bauds; such a slow rate is very tolerant of the phase and amplitude hits so common in the telephone network. The actual bandwidth used is 2240 Hz. Modulation and demodulation are all performed digitally by a microprocessor.

Multiplexing

The modulation techniques described above all have a common purpose in that they permit the use of a transmission channel otherwise unsuitable for the transportation of raw data. Once the necessary conversions have been made by the modulators, it may well be the case that the spectrum occupied by the modulated signal will be only a fraction of that available on the transmission channel, leading to inefficient usage of what may be an expensive resource. This is particularly common when low data rate devices are involved (since a low data rate leads to a low signal spectrum), and thus techniques have been developed to reduce this inefficiency.

Multiplexers and Concentrators

The terms multiplexer and concentrator are often used interchangeably, particularly by suppliers. In general, a concentrator is a more sophisticated device than a multiplexer; it does not offer a transparent route for the data on each channel, but rather manages the terminals connected to it and communicates on their behalf with a host computer. In many cases, these concentrators are special-purpose devices tailored to a single user's requirements.

A distinction sometimes drawn between the two devices is that a multiplexer only has the capacity to handle the sum of the transmission rates of the input circuits, that is the sum of the outputs cannot exceed the sum of the inputs. In contrast a concentrator has buffering capability to queue inputs which exceed capacity. However, as multiplexers become increasingly intelligent, this distinction is being eroded.

The purpose of both multiplexers and concentrators is to enable a number of devices, usually connected locally by relatively low speed circuits, to share a more expensive, higher bandwidth circuit.

Most time-division multiplexers employ sophisticated error detection and recovery techniques in transmitting data across a line. This very low level of residual errors has made the use of cheap asynchronous terminals a viable alternative to more expensive synchronous devices.

The rest of this chapter examines the techniques which multiplexers have employed. These started with frequency division multiplexing and then time division multiplexing to provide channels at high speeds.

Frequency-Division Multiplexing (FDM)

Frequency-division multiplexing (FDM) divides the available bandwidth of a communications channel into a number of independent channels, each having an assigned portion of the total frequency spectrum. Figure 3.23 shows the frequency bands of the first four channels in a typical FDM system. The *bearer* circuit, as the multiplexed circuit is called, has a nominal bandwidth of 3000 Hz giving 12 channels each of 240 Hz bandwidth. One of the limitations of an FDM system will be seen from Figure 3.23; that is guard bands or safety zones are needed to prevent overlapping of the electrical signals, resulting in under utilisation of the available bandwidth.

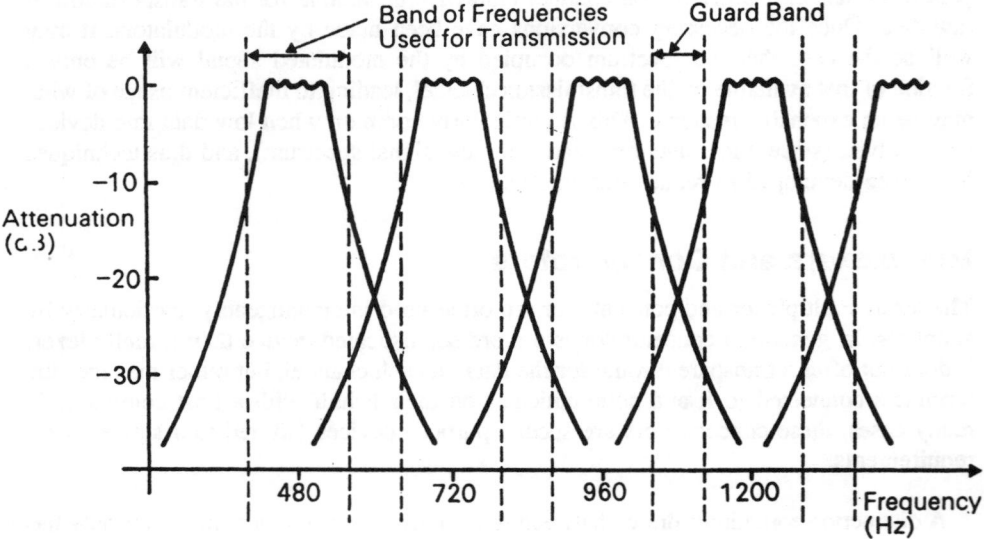

Figure 3.23   Frequency bands of a FDM system

The data signalling rate of each channel on the example system is 110 bit/s to cater for the numerous 10 character/s data-printers in use. If all the channels were in use at once, a total of only 1320 bit/s would be transmitted which is far less than the data signalling rate capability of the bearer circuit (up to 19200 bit/s or more using suitable high-speed modems). More expensive filtering equipment can improve the aggregate bit rate on FDM systems but this would not be likely to exceed much above 2000 bit/s in the example case.

A further disadvantage, inherent to FDM systems, is that the error performance of some channels may be poorer than others due to the loss/frequency ratio and group delay characteristics of the bearer circuit. For example, signals in a channel in the higher part of the frequency spectrum will be more severely attenuated than the signals in a lower channel; this problem can be largely overcome by equalisation but at added cost.

The main advantage of FDM for a user is to be found in those applications where the low aggregate bit rate is not a problem. However, due to the low speed constraint and its relative inefficiency, FDM is little used now.

Time-Division Multiplexing (TDM)

A time-division multiplexer works on the principle of taking data from a number of sources and allocating each of these sources a period of time or *time slots*. The individual time slots are assembled into *frames* to form a single high speed digital data stream. Pulse code modulation (discussed in Chapter 4) systems use time-division multiplexing to interleave a number of speech conversations which are first converted from analogue to analogue to digital form. The output from data terminals is already digital and this simplifies the time-division multiplexing process. The interleaving takes two different forms, *bit-interleaving* and *character- or byte-interleaving*.

Bit-Interleaved TDM

Figure 3.24 shows a simplified representation of the bit-interleaving process. It will be seen from the diagram that, if synchronisation between the multiplexing and de-multiplexing functions is lost, there is a danger that bits may be delivered to the wrong outputs.

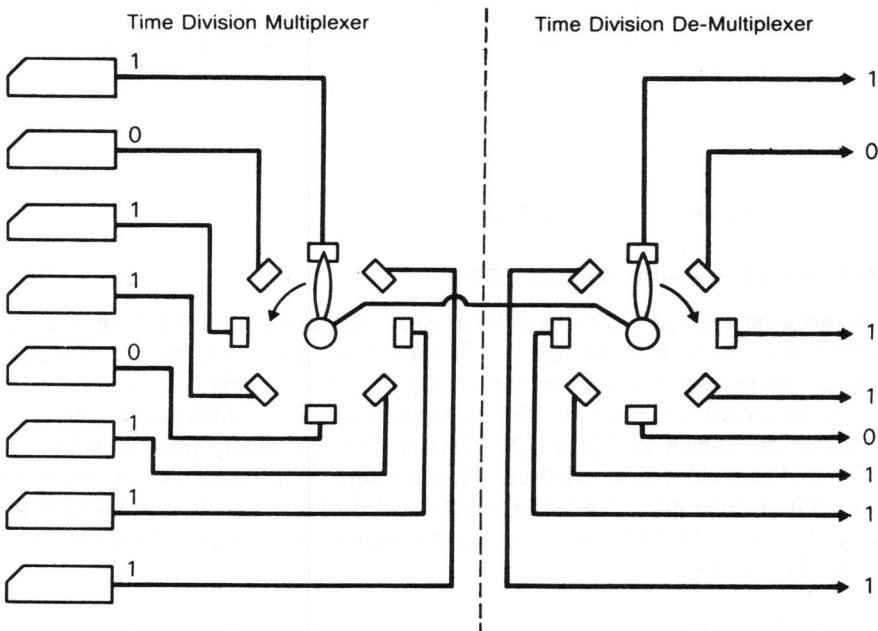

Figure 3.24 Principles of bit-interleaved time-division multiplexing using simple commutators

Synchronisation is maintained by assembling the individual bits sampled from each terminal into a frame so that nominally one frame consists of one bit from each of the terminal inputs to the system.

A unique bit system termed a *framing pattern* is included within each frame (or can be distributed throughout a number of frames) and must be detected a predetermined number of times by the de-multiplexing stage before synchronisation is assumed. Similarly, the loss of frame synchronisation is not assumed until the framing pattern has failed to be detected a predetermined number of times. If synchronisation is lost due to any factor such as line disturbance, re-synchronisation is not assumed until this has been detected several times.

In practice, the digital output from any time-division multiplexer must be converted into analogue form for transmission over telephone circuits and back from analogue to digital at the receiving end. The modems used may be an integral part of the multiplexer but are more often separate as shown in Figure 3.25. Synchronous modems are used so that bit synchronisation can be maintained throughout the system.

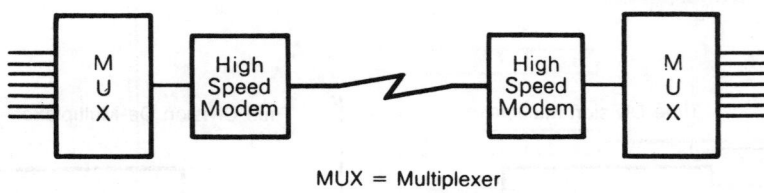

MUX = Multiplexer

Figure 3.25 Modems required in TDM

Character- or Byte-Interleaved TDM

A character-interleaved multiplexer interleaves characters received from low speed channels on to the high speed bearer circuit. Each interleaved character occupies eight bits in the frame, but this may vary depending on the code level being transmitted and the need for a status bit. The simplified elements of a character-interleaved time-division multiplexer are shown in Figure 3.26. Data is received serially from each low speed channel and assembled in a serialiser. When a complete character has been received, it is shifted in parallel form to a buffer register and then transmitted in serial form to the high speed line at the high speed line rate during the time slot period allocated to that low speed channel.

The scanner operates at a speed related to each low speed character period. For a 10 character/s terminal, scanning for transmission to the high speed line would take place every tenth of a second. If a buffer is empty, the related byte in the transmitted frame will be padded. Each frame consists of typically nine bits per low speed channel (this will vary in accordance with the terminal code being used) plus a synchronisation sequence and frame control information, the nature and length of which vary between manufacturers. The frame structure used in a typical character-interleaved system is shown in Figure 3.27.

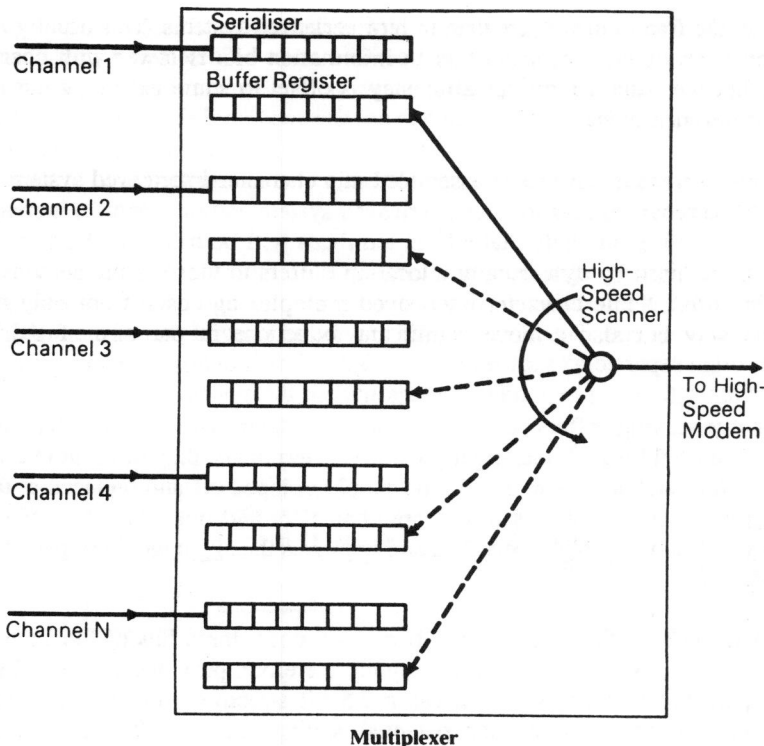

Figure 3.26 Simplified character-interleaved time-division multiplexer

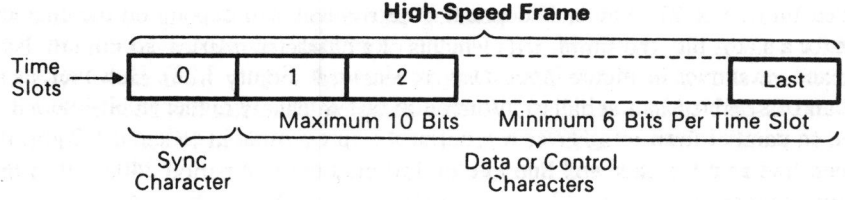

Figure 3.27 Frame structure in a character-interleaved system

The effect of lost synchronisation with bit-interleaved TDM was mentioned earlier and similar schemes have been developed in order to maintain frame synchronisation on character-interleaved TDM systems. Again this involves the use of a unique framing pattern within each frame which must be detected a predetermined number of times before frame synchronisation is assumed. Since characters rather than bits are interleaved in each frame, the frames are longer than in bit-interleaved systems. This usually means that the proportion of synchronising bits to information bits is lower with character-interleaving but the resulting greater efficiency is offset to some extent by the longer time taken to resynchronise.

Since whole characters have to be assembled in a character-interleaved system, more storage has to be provided than in a bit-interleaved system and this tends to increase the costs. However, as it is unlikely that all the terminals will be in use at the same time, storage can be reduced by dynamically allocating buffers to those terminals which are actually being used. Some character-interleaved multiplexing designs not only reduce costs in this way but also achieve significant reductions in the size of the TDM equipment.

Character-interleaving offers a number of advantages over bit-interleaving. In particular, when used for multiplexing asynchronous terminals, they offer more efficient transmission. Start and stop elements can be simply stripped off and replaced at the de-multiplexing stage, as the units sampled are characters and not bits. This effectively compresses the data on the high speed line and improves the aggregate low speed bit rate of the TDM link.

On 10 character/s machines using ASCII or IA 5 code, three bits of the 11 bits per character are used for start/stop, and bit stripping, therefore, provides a potential saving of over 27 percent of line time. IA 2 has a 7.5 bit structure with 2.5 bits used for start/stop and provides even greater potential savings of over 33 percent. Thirty character/s machines use one start and only one stop element in a 10 bit character structure but bit stripping still offers an attractive 20 percent improvement factor. In practice, the need for synchronisation and control bytes within frames reduces the low speed bit rate of character-interleaved multiplexers to some extent, but this is still usually between 10 percent and 20 percent higher than the rate of the bearer circuit; that is a 2400 bit/s circuit will typically bear some twenty five 110 bit/s terminals — an aggregate low speed bit rate of 2750 bit/s. The actual improvement will depend on the character structure used by the terminals and whether terminals of mixed speed are being multiplexed (examples of mixed input rates are shown in Figure 3.28). However, as the proportion of synchronisation bits to information bits is usually higher in bit-interleaved systems, character-interleaving gives a significant improvement in efficiency. Typically, a bit-interleaved multiplexer will support 21 110 bit/s terminals on a 2400 bit/s bearer circuit (an appropriate low speed bit rate of only 2310 bit/s). Bit stripping is possible on bit-interleaved systems, but the added logic necessary reduces their cost advantage. A major advantage of character-interleaving over bit-interleaving is that noise bursts affect fewer terminal users. A 10 bit burst would affect a maximum of two terminals whereas up to 10 terminals would be affected in a bit-interleaved system (although note that this is used to advantage in bit-interleaved systems when error protection is applied — as discussed in Chapter 8.). The relatively better raw error performance of character-interleaved systems has to be weighed against the added length of time they take to resynchronise.

	Low-speed inputs	Data signalling rate of the bearer circuit	Aggregate low-speed bit rate
Example 1	8 x 110 bit/s 12 x 134.5 bit/s 8 x 300 bit/s	4800	4894
Example 2	11 x 110 bit/s 1 x 1200 bit/s	2400	2410
Example 3	11 x 110 bit/s 12 x 134.5 bit/s 2 x 300 bit/s 2 x 600 bit/s	4800	4624

Figure 3.28 Typical examples of mixed input rates possible on a character-interleaved multiplexer

Most TDM systems can accept input from terminals using different speeds and codes but because the multiplexer system operates on a synchronous basis it is necessary to know in advance the speeds and codes to be handled and to program for them. TDM systems can also handle data from synchronous data terminals where the terminal transmission is in the block mode rather than character by character. For this type of transmission, bit interleaving is usually employed as this provides a degree of bit sequence independence not possible with character-interleaved systems and in any case is simpler and cheaper. It should be remembered that data characters within a block do not usually include start and stop bits and thus the facility of bit stripping which is a significant advantage of character-interleaving would not be applicable.

Where a TDM system is required to handle a mixture of input speeds and codes, it is usual for groups of channels and their related computer ports to be dedicated for particular speed and code operation. Some types of multiplexer, however, can provide a facility which enables some or all channels of the system to handle terminals operating at any speed or code within a predetermined range — this facility is variously known as *adaptive speed control* or *automatic bit rate* (ABR) and *code level selection*. To provide the facility, all messages from the terminals connected to the multiplexer must be prefixed by an agreed character which, when compared with internally programmed set characters at the multiplexer, enables the multiplexer to determine the speed and code level being employed and to adapt the channel conditions to handle that type of terminal. This facility has a particular advantage for computer operators who use the PSTN to provide users input to the multiplexer. For example, if users in Birmingham were to be connected to a London computer centre over a TDM multiplexed link between Birmingham and London, access to the TDM multiplexer would be gained by dialling a local call in the Birmingham area. If the terminals users in Birmingham were all one speed and used the same code, they could all be given the same number to dial for access to any one of the TDM ports (see Figure 3.29). The concentration function of the PSTN can be seen in this example and, typically, a contention ratio of four terminals to one input port is used, giving high utilisation of the TDM channels. The effect of having terminals of two different speeds, 110 bit/s and 300 bit/s, can be seen from Figure 3.30.

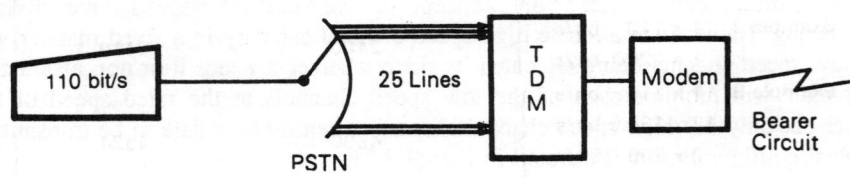

Figure 3.29   Access into a 'local' TDM using the PSTN as a concentrator

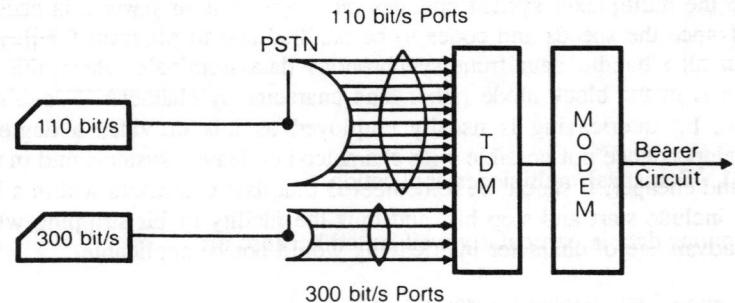

Figure 3.30   Access into a 'local' TDM from different speed terminals using the PSTN for concentration

At one time, different parts would be used for the two types of terminal and separate telephone numbers provide two distinct access routes. Since 110 bit/s ports may be free while 300 bit/s are overloaded the efficiency of the whole system is reduced.

Although adaptive speed control overcomes this problem, we cannot obtain something for nothing. The penalty, apart from the slightly increased cost of the multiplexer equipment, is that the channel capacity of the multiplexer system will be restricted and must be determined on the basis that all channels may need to operate at the highest speed required, that is 300 bit/s in the example Figure 3.30; time slots are not allocated dynamically.

When using adaptive speed control, it is necessary for the ports that serve the host system also to be able to adapt to the required terminal speed and this can be arranged by sampling a prefix character in a similar manner to that described for the multiplexer. This would be the normal way of configuring the host system.

Statistical (Time/Division) Multiplexers

Statistical multiplexers extend the concept of character-interleaved time division multiplexing. Rather than allocate high speed channel capacity in a fixed manner with each low speed channel being allocated its share whether it needs it or not, a statistical multiplexer will monitor each of the low speed channels at the rated speed of that channel but only use high speed channel capacity when there is data to be transmitted. A typical configuration is illustrated in Figure 3.31.

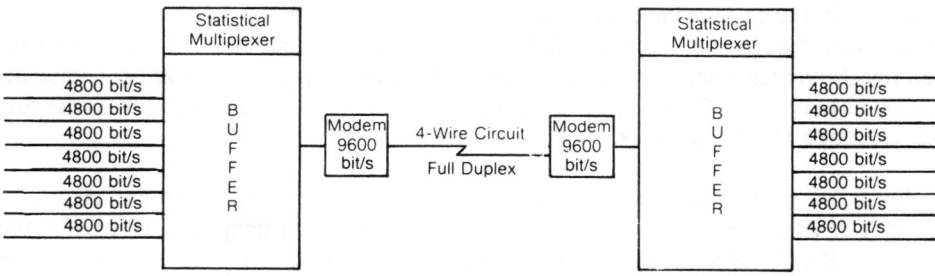

Figure 3.31 Statistical multiplexer application

Asynchronous data is particularly well suited because of:

— low channel utilisation;

— start/stop bit stripping;

— statistical multiplexing providing error correction;

— use being made of data compression.

Where the low speed channels are lightly utilised, it is possible for the multiplexed link to handle data from low speed channels with aggregate data rates far in excess of the high speed channel speed. It should be noted that the maximum throughput of a statistical multiplexer is the same as that of a basic time-division multiplexer; however, the time-division multiplexer will only achieve that throughput when each of the low speed channels is operating at maximum capacity — a very rare occurrence in practice. A statistical multiplexer, on the other hand, serves more, but less heavily utilised, channels to achieve a higher average throughput across the multiplexed link.

As the combined data rates of the low speed channel exceed that of the high speed channel, it is possible for the data flow from the low speed channel to exceed the throughput capability of the multiplexed links.

To cope with this eventuality, statistical multiplexers contain large buffers which are capable of absorbing short-term peaks in demand. However, it is important that there is a mechanism capable of suspending data flow into the multiplexer should there be any danger of buffer overflow. One of the two methods is usually used to achieve this flow control. Either the multiplexer outputs the *x-off characters* to the terminals via the *receive data circuit* in the terminal interface, or the *clear to send* signal in the terminal interface is dropped (when it is safe for data transfer to resume, either the character *x-on* is input, or the *clear to send* signal is raised) (see Chapter 7 for an explanation of interchange circuits).

Most multiplexers will implement the *hold off* signal when a given buffer reserve is reached. It is therefore important that the terminal or computer should stop outputting data before the reserve is exhausted. To this end, the input/output characteristics of the terminal/computer should be matched carefully. For example, since some computers output as many as 256 bytes after receiving *x-off*, it is crucial that the multiplexer should have enough reserve capacity.

For some types of terminals, notably printers, the flow into the terminal should be capable of being suspended if data loss is to be avoided; for example, if the printer runs out of paper. Those terminals use the same flow control mechanism as described above; in this case the multiplexers should suspend the data flow into the terminal and then avoid data loss resulting from an overflow of the multiplexer's buffers.

A typical configuration for a TDM system is shown in Figure 3.32.

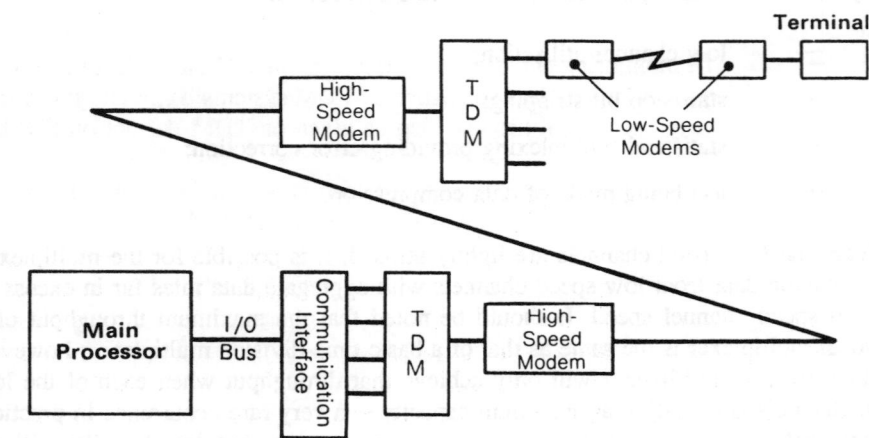

Figure 3.32   A typical TDM configuration

FDM versus TDM

Many of the arguments for and against these two fundamentally different forms of multiplexing have been discussed; in conclusion they can usefully be summarised as follows:

-   The simplicity of FDM brings costs advantages when only a few low speed channels are required. At present FDM is economic when fewer than about 10 or 12 channels are needed, the actual breakeven figure depending on which particular manufacturers' equipment is being compared. TDM is progressively cheaper per channel than FDM as the number required rises beyond 12 as higher speeds are required.

-   Due to the need for guard bands on FDM systems, the aggregate low speed bit rate achievable is less than on TDM systems on which no guard channels are required and some data compression can be applied. This not only affects the costs per channel but also limits the ability of FDM systems to expand. For example, a TDM system over a typical leased circuit could expand to accommodate up to twenty five 100 bit/s terminals. The more modern FDM system would be unlikely to handle more than eighteen 110 bit/s terminals. Some FDM channels may give a worse performance than others due to the line characters of the bearer circuit. TDM systems give a constant performance on all channels of the same speed.

-   TDM systems are more flexible than FDM systems. Time-division multiplexers can intermix terminals of different speeds and with different synchronisation methods. System and channel configuration changes can also be effected more easily.

-   The monitoring and diagnostic facilities on TDM are much better than on the relatively less sophisticated FDM systems. High and low speed parity check facilities can also be given on TDM but not on FDM.

-   STDMs have inherent error protection which is an advantage for asynchronous data which is normally unprotected.

4

Analogue to digital conversion

Until the middle of the 20th century, the information which was sent across the public telephone networks was in an *analogue* form. This was mainly due to the fact that the only kind of information being passed over the telephone networks was speech.

As binary computers with digital storage and retrieval capabilities began to appear, the complexity of electronic systems grew. It soon became apparent that computer systems needed to communicate and that great advantages could be derived from the timely provision of information to users at their workplaces. Since the systems used binary operating methods, communication took place through digital transmission systems and equipment, using signals made up of discrete *bits* of information. Even with some very clever modulation techniques the bandwidth for data and voice from a single analogue link is relatively low. Digital techniques can provide at least 2 megabits/sec to a user which can alternatively be used to transmit 32 simultaneous voice calls.

The digital networks which evolved for these, and other systems, offered many additional benefits over analogue networks:

-      they were cheaper to install and maintain;
-      they reduced the amount of noise which accumulated;
-      they enabled fast and efficient signal processing; and
-      they allowed integration of different traffic types.

As the number of computers grew, so did their reliance upon digital networks. It soon became obvious that this type of network could be further utilised by carrying digitally encoded forms of analogue signals such as speech.

This chapter examines some of the methods of digitally encoding an analogue signal for transmission through a digital network.

A Comparison of Signal Types

Analogue Signals

An analogue signal has an amplitude which varies continuously between upper and lower limits with respect to time (see Figure 4.1). These signals are most commonly derived from variable physical quantities such as speech, light, heat or pressure.

Analogue

Upper

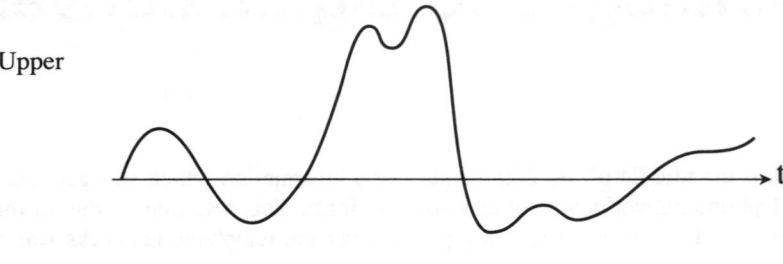

t

Lower

Digital

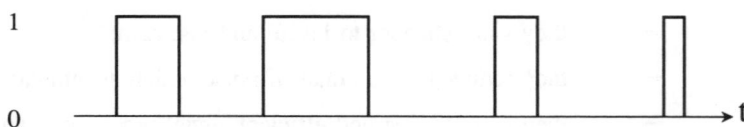

1

0

t

Figure 4.1   Analogue to digital signals

The upper and lower limits of an analogue signal are determined by the physical quantity it is representing. In the case of speech, the amplitude of the signal represents the loudness of the voice and varies between the limits of absolute silence and loud shouting.

Digital Signals

In contrast, digital signals as shown in Figure 4.1, have discrete levels with respect to time. The levels of a *binary* digital signal represented by a presence or absence of a pulse electricity or light (in optical fibre systems) and are also referred to as 1 or 0. Signals of this type are produced by digital devices, such as computers, which operate using binary digits.

Sampling

The first stage in any process which converts or encodes a signal from its analogue form into a digital form is known as *sampling*. This involves measuring a characteristic of the signal, such as amplitude or frequency, at frequent and regular intervals.

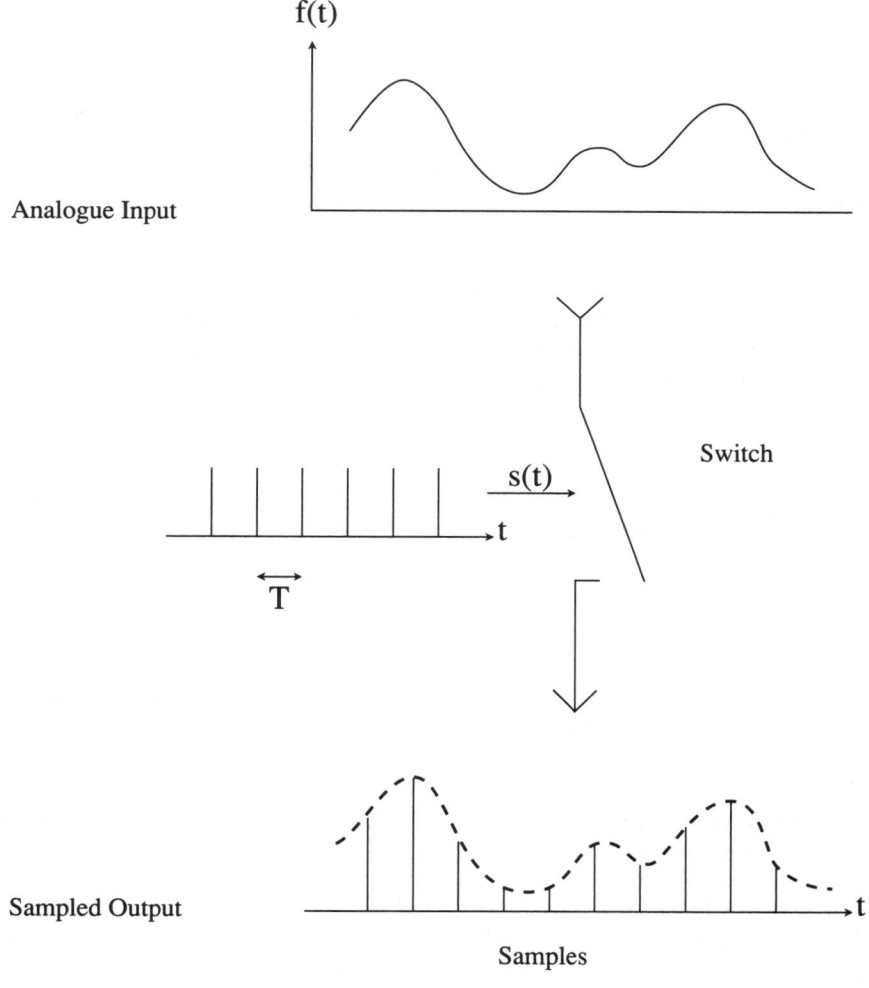

Figure 4.2 Sampling an analogue signal

Figure 4.2 shows a basic sampling device in which the switch is opened and closed momentarily every T seconds, thus allowing the instantaneous amplitude of the signal to appear at the output. The output signal is known as a *pulse amplitude modulated* signal.

Nyquist Sampling Theorem

The sampling process described earlier is basically the product of two mathematical functions: the input signal f(t), which has an upper and lower frequency, and the train of narrow sampling pulses s(t).

The input signal f(t), and any other analogue signal can be mathematically expressed as a series of superimposed waves, each with a different frequency and amplitude. When sampling, it is vital to know the frequency of the highest frequency component contained in the original signal.

When the frequency of the sampling pulses (f_s) is high enough then the sampled output contains all the information which was present in the input signal. The frequency at which this occurs is known as the *Nyquist rate* and is derived from the Nyquist Sampling Theorem which states that:

> for the complete reconstruction of the original signal from the sampling train and the sampled output signal, the sampling rate (f_s) must be at least twice the highest analogue frequency (f_{max}) present in the original signal. That is:

$$f_s \geq 2 \ f_{max}$$

If sampling occurs at a frequency below the Nyquist rate then the output signal is distorted and is not a true representation of the input.

Alias Distortion

When samples are taken at a frequency which is less than the Nyquist rate then an effect known as *alias distortion* or *aliasing* occurs. It is this distortion which makes it impossible to reconstruct the original signal accurately.

Although aliasing can be explained using complex mathematics which are outside the scope of this book, it is useful to give a visual example of the aliasing effect and how it prevents information recovery.

When an old fashioned cine camera films the spoked wheel of a moving bicycle, the wheel can often appear to be stationary or moving backwards. This effect is caused by the speed, or sampling frequency of the camera falling behind the frequency at which the wheel is turning, that is:

$$f_{shutter} \ < \ 2 \ f_{wheel}$$

By looking at just the wheel and ignoring the background, it is impossible to determine the wheel's true speed and direction.

Pulse Modulation

Pulse modulation (also known as analogue modulation) is the process by which an analogue signal is converted into a series of pulses. There are three types of pulse modulations.

— *pulse width modulation*, which changes the width of the pulses in sympathy with the original signal;

— *pulse position modulation*, which changes the position in time of the pulses; and

— *pulse amplitude modulation*, which changes the individual height of the pulses.

Pulse Width Modulation (PWM)

The PWM technique uses the amplitude of an analogue signal to vary the duration or width of the pulse (see Figure 4.3). The maximum duration of the pulse is governed by the maximum amplitude of the original signal. This technique is analogue because the duration of the pulse can vary between zero and the maximum.

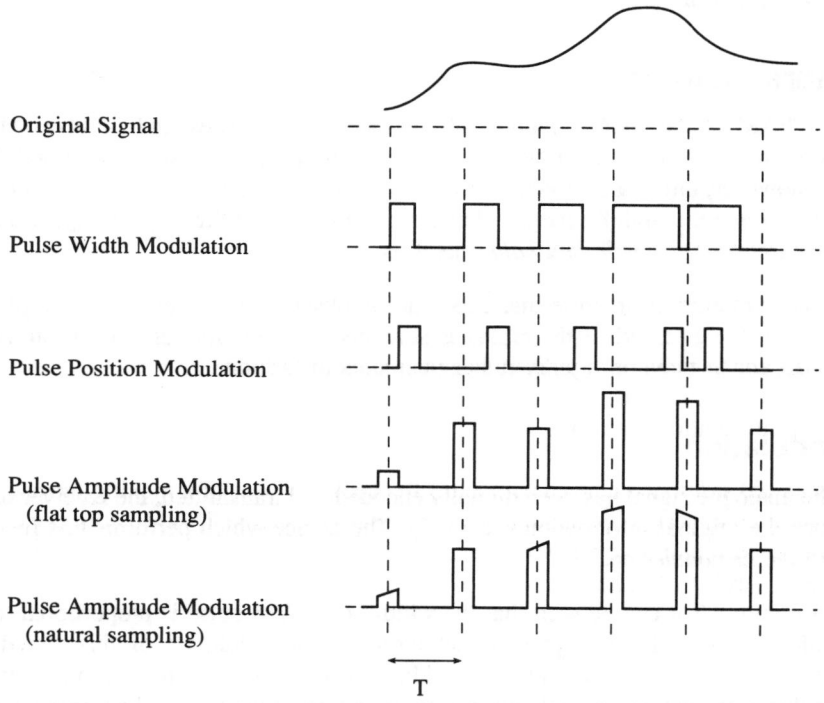

Figure 4.3 Pulse modulated signals

The advantage of using PWM is that the samples have constant amplitude and are therefore reasonably immune to any noise which may be apparent in the system. However, because the duration of the pulse is so variable, the technique is very inefficient in its usage of the available bandwidth. It is also impossible to time-division multiplex a PWM signal.

Pulse Position Modulation (PPM)

An analogue signal can also be represented by varying the time between a constant width pulse and a reference time (see Figure 4.3). In PPM, the maximum difference between the pulse and reference time is related directly to the maximum amplitude of the original signal. Once again this is a type of analogue modulation because of the infinitely variable position of the pulse.

The advantages and disadvantages of PPM are much the same as those associated with PWM and so PPM is rarely used in communication systems.

Pulse Amplitude Modulation (PAM)

As described in a previous section, the PAM technique involves varying the height of constant duration and position pulses. The sampled output in Figure 4.3 shows PAM using *flat top* and *natural* sampling.

Aperture Distortion

Flat top PAM is the more common of the two sample types and involves taking extremely short samples which are *held* for the duration of the sampling period. This process, however, introduces *aperture distortion* which is caused by the almost complete failure to sample the high frequency changes of amplitude in the original signal during the lower frequency *sample-and-hold* operation.

A visual example of aperture distortion can be observed in a slow motion replay of a tennis ball being served. If the replay is slow enough then aperture distortion makes the ball appear to move in a jerky rather than smooth fashion.

Demodulation

Once the analogue signal has been digitally encoded and transmitted, the receiver has to reproduce the original information accurately. The device which performs this function is called the *demodulation*.

From Figure 4.3 it can be seen that the width of a PWM pulse is proportional to the amplitude of the analogue signal at that moment. Demodulation of this signal can therefore be done by time-averaging the width of each pulse over the sample period T. This produces an analogue signal which is approximately the same as the original signal. By increasing the sampling frequency or reducing the sample period, the approximation becomes more accurate.

To demodulate a PWM signal is it necessary for the receiver to have a pulse generator which replicates the sampling pulse train. By combining this signal with the PPM signal in an electronic device called a *Reset-Set (RS) Flipflop*, a PWM signal can be obtained which can be demodulated as described above.

The transmission process is completed by passing the analogue signal, produced by the demodulation process, to a telephone.

Distortion Filtering

Two types of distortion have been mentioned so far: alias distortion and aperture distortion.

The effects of both of these can be reduced with the use of filters.

Anti-aliasing Filter

The effects of aliasing can be reduced to virtually nothing by using a low-pass anti-aliasing filter. The filter is designed to pass only those frequency components of the original signal which are less than half the sampling frequency. This ensures that the sampling frequency is always at least twice the maximum frequency of the original signal. In turn, the Nyquist sampling theorem is satisfied and no aliasing occurs.

In the bicycle and cine camera scenario, the anti-aliasing filter would ensure that the bicycle wheel is rotating at a frequency which is less than half of the frequency of the camera shutter. This would guarantee that the film showed the wheel moving at its true speed in the right direction.

Reconstruction Filter

The aperture effect is basically caused by the rapid attenuation or loss of high frequency components in the sampled output signal. The purpose of the reconstruction filter is to allow for the aperture effect by effectively boosting or equalising the attenuated high frequencies.

In the slow motion replay scenario, the reconstruction filter would insert stills portraying the intermediate steps of the ball into the film. This would make the tennis ball appear to move smoothly in the slow motion replay.

Pulse Code Modulation

None of the PWM, PPM and PAM techniques are digital because they do not use time-discrete pulses of constant amplitude and duration. The information contained in digitally modulated signals is represented by the presence or absence of a pulse. As a result, digitally modulated signals are much less susceptible to the noise which affects pulse modulated signals.

Pulse Code Modulation (PCM) is a truly digital technique, based on PAM, in which the amplitude of an analogue signal is sampled before being coded into a sequence of digital 1s and 0s, or on and off pulses. The PCM process is widely used in modern telecommunications networks and has become standardised throughout the world as CCITT Recommendation G.711.

In keeping with the analogue pulse modulation methods, the rate at which the samples are taken must conform to the Nyquist sampling theorem by being at least twice the highest frequency component of the original analogue signal. The samples are taken at regular and frequent intervals and are immediately *quantised*.

Quantisation

The process of linear quantisation involves making an approximation of the original signal using discrete levels. Each level represents a different signal amplitude and has an associated binary code as shown in Figure 4.4.

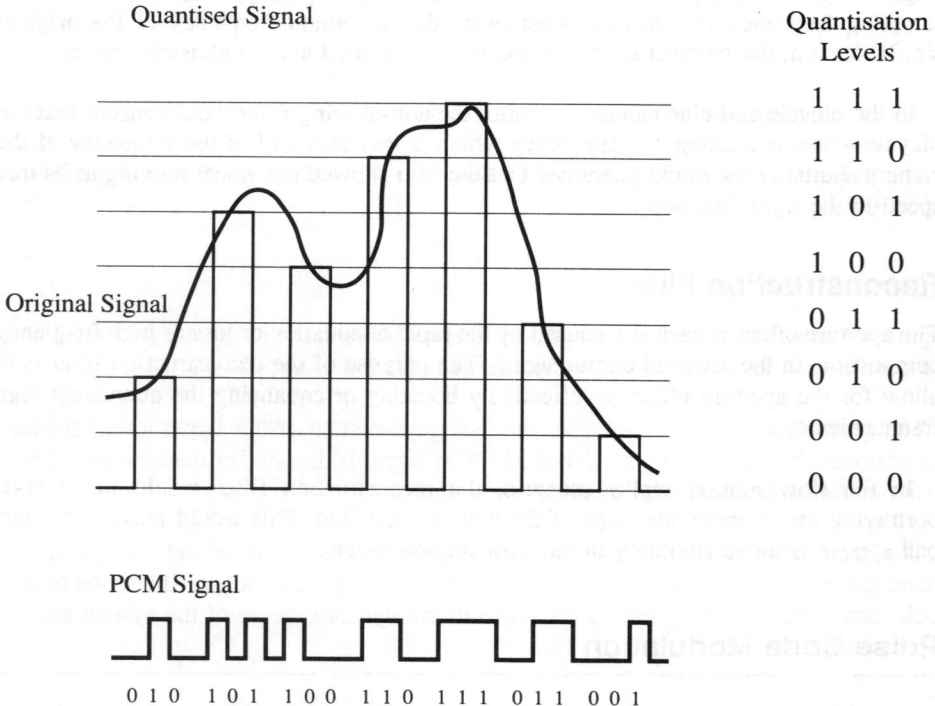

Quantised Signal

Quantisation Levels

1 1 1
1 1 0
1 0 1
1 0 0
0 1 1
0 1 0
0 0 1
0 0 0

Original Signal

PCM Signal

0 1 0   1 0 1   1 0 0   1 1 0   1 1 1   0 1 1   0 0 1

Figure 4.4 Quantisation

In the example illustrated in Figure 4.4, there are 8 discrete levels represented by 8 individual codes containing 3 bits. For more accurate quantisation of the signal, more discrete levels would be used.

For the transmission of speech through a modern digital system, each sample is usually represented by 8 bits (7 bits in North America), and therefore:

$$\text{number of quantisation levels} \quad = \quad 2^N$$
$$= \quad 2^8$$
$$= \quad 256$$

where N is the number of bits per sample.

Dynamic Range

The dynamic range of a PCM system is given by the ratio of the amplitudes of the smallest to the largest signal which can be transmitted. The dynamic range is expressed in decibels and is often represented as:

$$DR \quad = \quad 20 \log_{10} (\text{number of quantisation levels})$$
$$= \quad 20 \log_{10} (2^N) \text{ dB}$$
$$= \quad 6.02 \text{ x } N \text{ dB}$$

where N is the number of bits per sample.

The dynamic range of the system is an important factor since it affects the quality of transmission.

Quantisation Noise

Figure 4.4 shows that there is always a difference between the original signal level and the quantisation level and this gives rise to *quantisation error*.

The quantisation error results from the misrepresentation of the signal levels outside the centre of the sampling pulse. When the PCM signal is decoded or demodulated these errors manifest themselves as quantisation noise.

The measurement of the level of quantisation noise is known as the signal to quantisation noise ratio (SNR_Q). The derivation of this figure is outside the scope of the book; however, the SNR_Q makes reference to the dynamic range of the system and is expressed in decibels:

$$SNR_Q \quad = \quad 4.77 \; - \; \alpha_{dB} \; + \quad 6N \quad \text{dB}$$
$$= \quad 4.77 \; - \; \alpha_{dB} \; + \quad DR \quad \text{dB}$$

where α is the peak-to-mean power ratio of the original signal, N is the number of bits per sample and DR is the system's dynamic range.

For speech, the peak-to-mean power ratio is 10 dB, and a modern digital transmission system uses 8 bits per sample. Therefore:

$$\text{SNR}_Q \quad = \quad 4.77 \quad - \quad 10 \quad + \quad (6 \times 8) \quad \text{dB}$$
$$= \quad 42.77 \quad \text{dB}$$

By using a technique which makes the discrete quantisation levels non-linear, the level of signal power to quantisation noise power can be reduced.

Companding

Linear quantisation, in which the discrete levels are a uniform distance apart, gives weak (low amplitude) analogue signals a large SNR_Q. In addition, systems which require a large dynamic range require many bits per sample and consequently a greater SNR_Q.

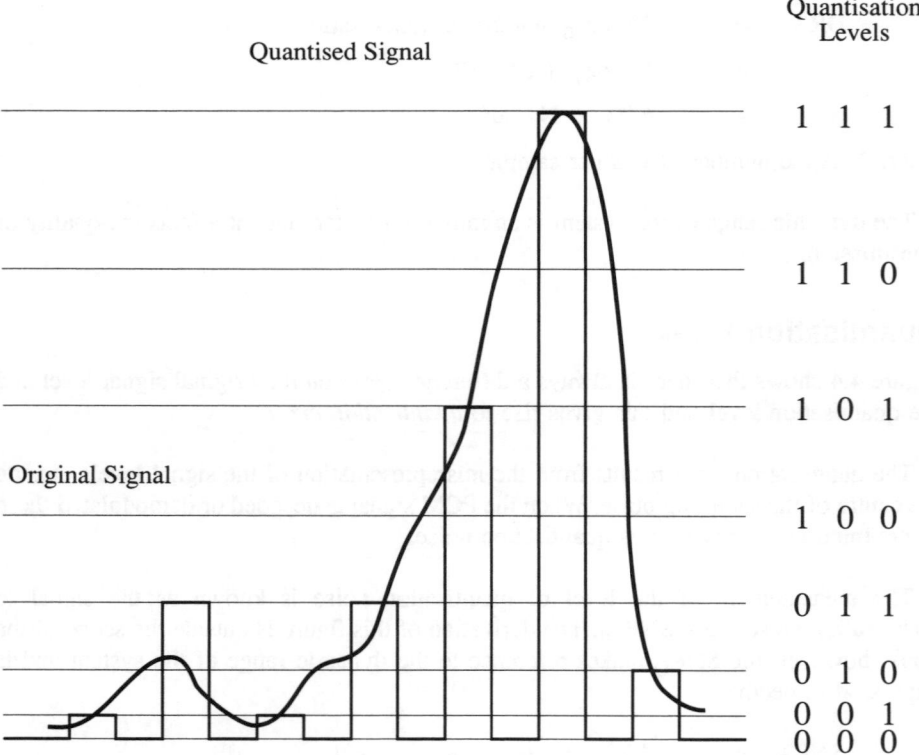

Figure 4.5 A companded signal

In order to improve the SNR_Q for weak signals, the spaces between discrete levels can be made non-linear so that a greater number of levels appear for lower amplitude signals (see Figure 4.5). This process is known as *companding* (a term which combines *compressing* and *expanding*).

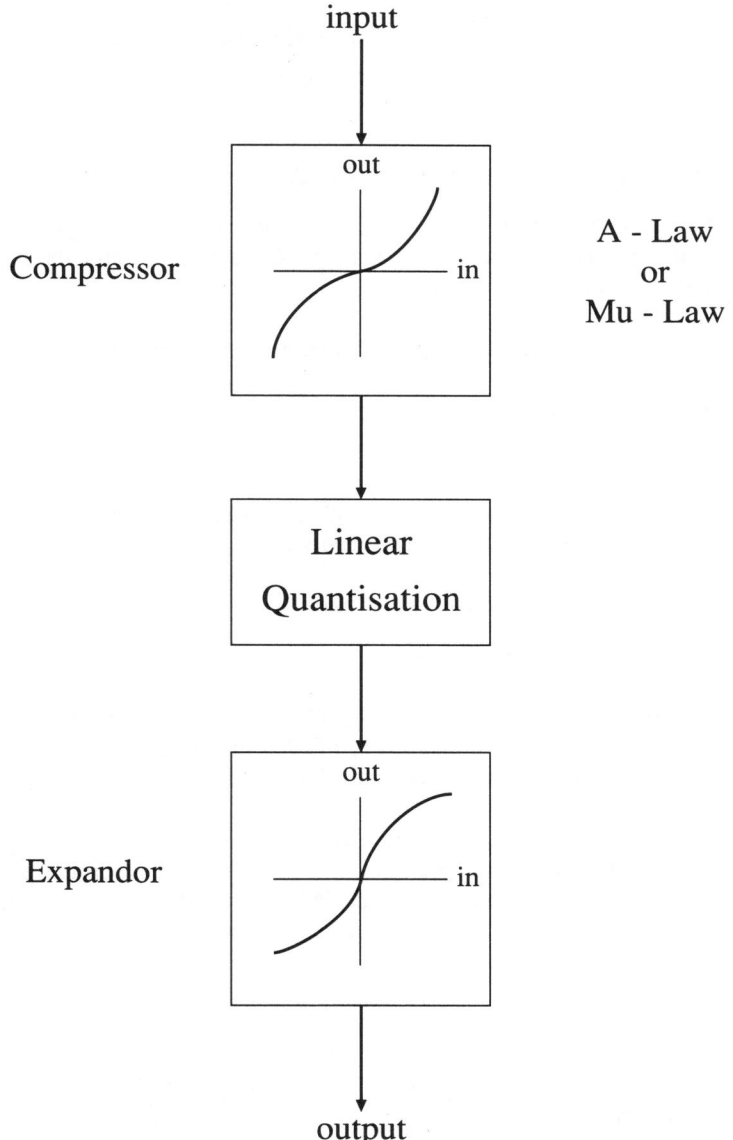

Figure 4.6 The companding process

The companding process illustrated in Figure 4.6, shows how a compressor, with a specified compression curve, gives a uniform output signal for a non-uniform input. This signal can then be linearly quantised. The signal is decoded at the end of transmission by the application of an expander which has the inverse characteristic of the compression curve.

The curve follows a path which is governed by the *companding law*. In Europe the companding law is known as the *A-law* and results in a different compression curve than the Northern American *Mu-law*.

Transmission of voice using PCM

Although voice conversation occurs over a frequency range of approximately 300 to 3400Hz, the anti-aliasing filters in normal telephone circuits limit the bandwidth to a slightly higher frequency of 4 kHz.

To comply with the Nyquist sampling theorem, the minimum sample rate is given by:

$$f_s \quad = \quad 2 \times 4 \text{ kHz}$$
$$= \quad 8 \text{ kHz}$$

Each of the 8000 samples taken in a single second is represented digitally by 8 bits, and therefore a PCM voice signal has a data rate of:

$$\text{rate} \quad = \quad 8 \text{ kHz} \quad \times \quad 8 \text{ bits}$$
$$= \quad 64000 \text{ bit/s}$$

The path down which this 64 Kbit/s signal, or bit-stream is sent is referred to as a channel. Using *time-division multiplexing* (TDM) techniques, a number of individual channels, each carrying a separate conversation, can be sent down a single line.

PCM, together with TDM, provides a cost effective and error-resilient method for using digital links to their full potential. The proliferation of digital equipment in a modern communications system has made PCM easier to generate and it is therefore used for digital transmission of speech through most of the world's digital telephone networks.

Demodulation

Pulse code demodulation involves decoding each group of amplitude representing bits as they arrive. This is the most difficult part of the process as factors such as companding must be taken into consideration.

The decoding procedure results in a PAM signal which, if the sampling frequency was high enough, can be a reasonably accurate reconstruction of the original signal.

Differential Pulse Code Modulation (DPCM)

The DPCM process involves sampling a signal and using a fixed predictor device to estimate, from a discrete set of values, the value of the next sample. The original estimate signals are then combined to form a difference signal which is quantised into 16 levels and encoded to a digital signal consisting of 4-bit codes. Figure 4.7 shows the principles of DPCM with quantisation into 4 levels for clarity.

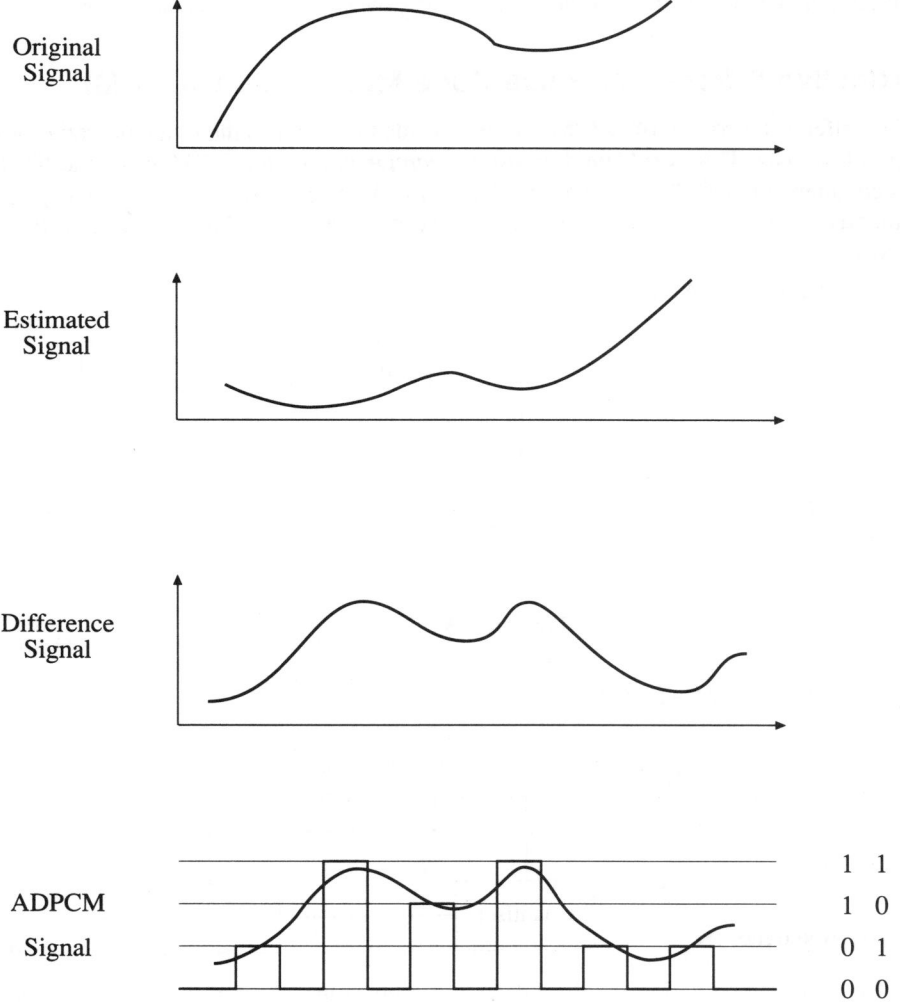

Figure 4.7   Differential PCM

In the European system (E1), 32 8-bit channels are multiplexed to give 256 bits per frame. Thirty user channels are available for voice, data or a mixture, the TDM using a drop and insert technique for the mixture. The other two channels are used for signalling and synchronisation as described in CCITT Recommendation G732. In E1 there is a total of 2048 Kbps. The North American system uses 1544 Kbps (24 channels) and is described in Recommendation G733. BT's Megastream service is based on the 2048 Kbps system. There are now CCITT Recommendations for several orders of higher bandwidth.

By using *differential PCM*, transmission rate can be reduced to 32 kbit/s and the fixed predictors can considerably improve the SNR_Q over direct quantisation PCM.

Adaptive Differential Pulse Code Modulation (ADPCM)

The differential coding of DPCM can be extended to adapt to the dynamic variations in speech signals. This is achieved by using *adaptive differential PCM* technique (CCITT Recommendation G.721) which introduce variable or adaptive quantisation step sizes. Adaptive differential coding improves the dynamic range of the system as well as its SNR_Q.

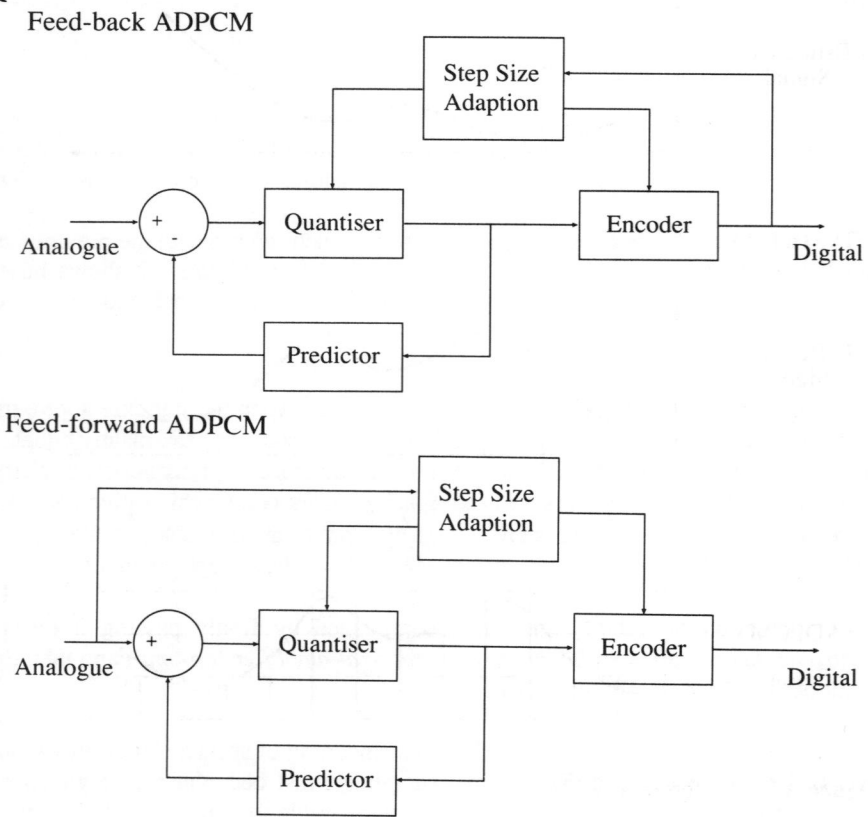

Figure 4.8 The ADPCM processes

Figure 4.8 shows an ADPCM system with feed-back and feed-forward adaptive quantisation. Feed-forward systems require the current step size to be transmitted along with the digital code, whereas feed-back systems use receivers which can determine the required step size from the incoming signal. Feed-back systems, however, are prone to transmission errors which can distort the reconstructed speech signal.

ADPCM systems provide good quality speech transmission at a rate of 32 kbit/s. The encoding algorithms being standardised by CCITT make the encoder signal independent, and therefore capable to handle music and tones as well as voice.

Decoding of received ADPCM signals involves passing them through a stage which has the inverse function of the feed-back or feed-forward stage of the encoder. The output from this stage is then passed through a digital to analogue convertor which reconstructs the original signal.

Delta Modulation

Linear Delta Modulation (DM)

Delta modulation is a digital technique which makes it possible to represent a single bit of information, thus reducing the number of bits required for coding. This saving can be extremely important for rapidly varying signals such as full audio music or video.

The technique involves making a comparison of each sample with the previous sample and transmitting a 1 if it is higher and 0 if it is lower. Figure 4.9 shows how delta modulation results in a staircase approximation of the original signal, and how errors can be introduced.

Startup errors occur at the beginning of quantisation when the staircase approximation differs dramatically from the input signal until they become approximately equal. When the input signal remains relatively constant but the quantised signal consists of alternating positive and negative steps (1s and 0s) *hunting* errors occur which give rise to *idling noise*. The *slope overload* errors are caused by a rapid rate of change in the input signal which exceeds the maximum rate of change of the staircase approximation.

Demodulation of a DM signal is accomplished by firstly passing it through an electronic device called an *integrating operational amplifier (op-amp)* and then through a smoothing filter (see Figure 4.10).

The main application of DM is the low cost encoding of speech signals in which slope overload must be avoided. To do this, sampling must occur at a rate which is high enough to make enough to make the SNR_Q comparable to that of PCM. The minimum pulse rate of delta modulation is usually therefore 32 kbit/s and anything less than this, such as 16 kbit/s produces an intelligible but noisy signal.

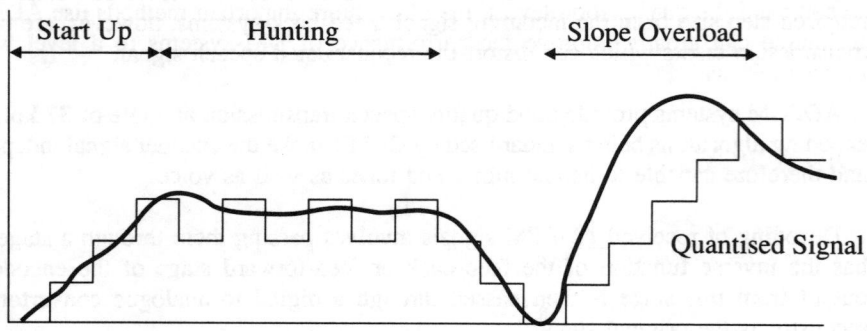

DM Signal

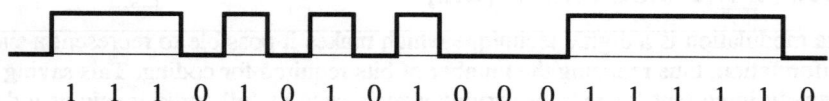

$$1 \; 1 \; 1 \; 0 \; 1 \; 0 \; 1 \; 0 \; 1 \; 0 \; 0 \; 0 \; 1 \; 1 \; 1 \; 1 \; 1 \; 0$$

Figure 4.9 Linear delta modulation

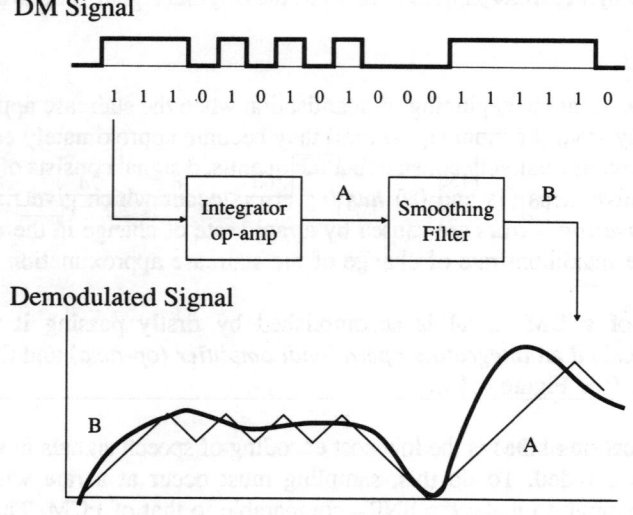

Figure 4.10 Integrating operational amplifier

Adaptive Delta Modulation (ADM)

There are a number of methods by which the pulse rate can be further reduced. One is to compand the quantisation levels; the other, more important methods use ADM. This technique can greatly improve the performance of DM systems to a level which is comparable with PCM.

Discrete ADM

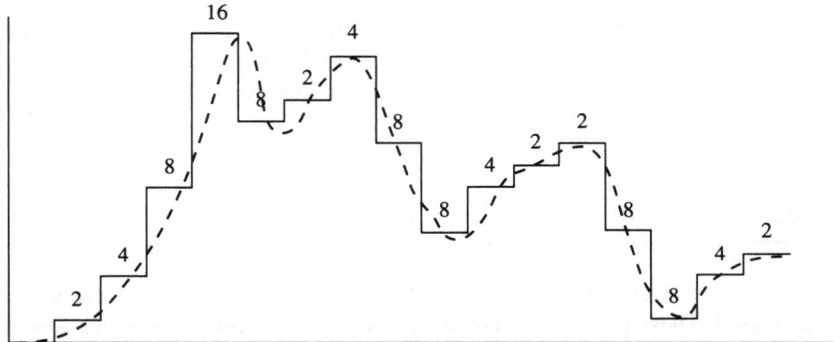

Variable ADM (CVSD)

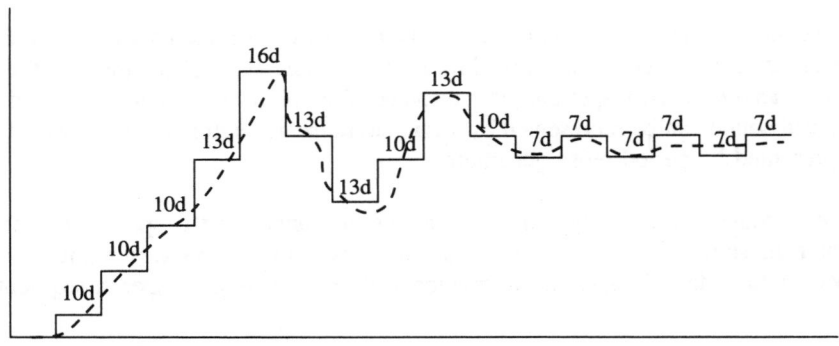

d = minimum step size

Figure 4.11 Adaptive delta modulation

In general, there are *discrete* and *variable* methods of adapting the DM system to the slope of the input signal (see Figure 4.11). Discrete ADM uses a logic device to analyse the number of 1s or 0s which appear in sequence. When, for example, slope overload occurs, the quantising step size is discretely incremented. Conversely, a long string of 0s causes the step size to be reduced. Because the step size changes at the same rate as sampling, this process is known as *instantaneous companding*.

Continuously Variable Slope Delta Modulation (CVSD)

Further reduction of the bit rate can be achieved by using variable ADM. This process varies the quantising step size in an almost continuous fashion, and is therefore often referred to as *continuously variable slope delta modulation (CVSD)*. The dynamic range of CVSD systems is between 30-40 dB which represents a 10 dB improvement on discrete ADM.

Predictive Coding

The methods of speech code which have been discussed so far, treat the voice as an analogue signal. These processes, known as *waveform encoding* could therefore be used to code any analogue signal.

However, a more effective method of transmitting voice involves the analysis of speech parameters and attempts to preserve the actual word information. This is known as *source encoding* and is done by breaking up the speech into its basic components and waveform characteristics.

Vocoders

The vocoder is a device which models speech as it leaves a person's vocal tract. Speech is assumed to be *voiced* (corresponding to the periodic flow of air through the vocal cords) or *unvoiced* (corresponding to a turbulent flow of air through a constricted cord). Unvoiced sound can be represented by a random noise generator, whereas voiced sound is represented by period pulse generator.

The vocoder works on the same principles and analyses the speech to determine whether the sound is voiced or unvoiced. Since speech is always varying, it is vital for a vocoder to update the speech information with each successive speech segment.

Linear Predictive Coding (LPC)

The LPC vocoder is based on a predictive model of the speech source, and reduces the bit rate which is necessary for speech transmission to as little as 2.4 kbit/s. The LPC system which is shown in Figure 4.12 consists of an analyser which calculates and predicts information about the speech from a number of fixed or discrete choices. In addition, a *pitch extractor* is used to extract all the essential frequency information from the speech. All this data is then passed to the digital encoder and onto the transmission link.

Encoder

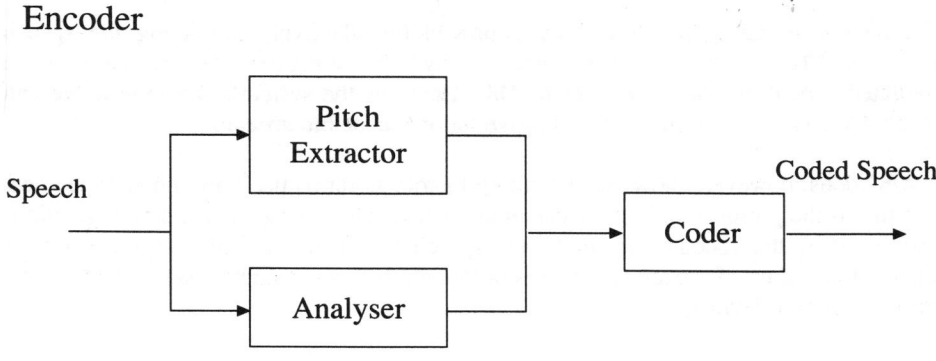

Synthesizer

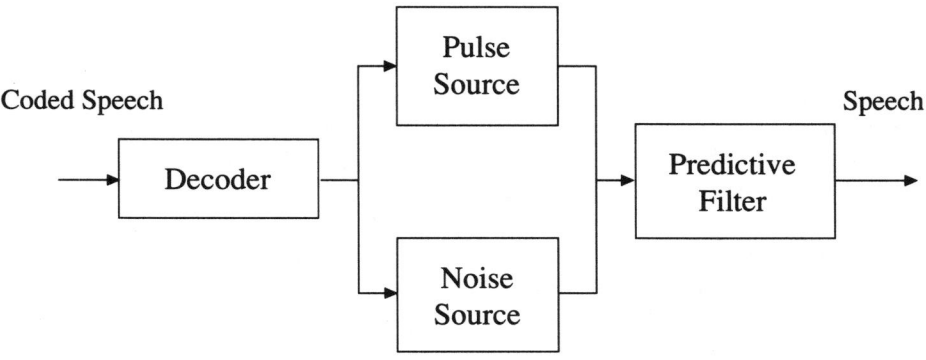

Figure 4.12 Linear predictive coding

Reconstruction, or synthesis of the speech waveform is done by using the decoder to select between a pulsed source or a *white noise* (depending on whether sound is voiced or unvoiced respectively). The resulting signal drives a predictive filter which reconstructs intelligible but synthetic sounding speech.

Adaptive Predictive Coding (APC)

The quality of LPC speech can be improved by adaptively predicting the speech waveform. The APC vocoder system does this by transmitting the difference between the predicted signal and the actual signal. This decreases the system's dynamic range and hence decreases quantisation noise to give more acceptable speech.

APC does, however, show some limiting factors as far as performance is concerned. The first is the presence of listener perception which can distinguish the small variations introduced by the vocoder. Secondly, the speech can be dramatically degraded by the incorrect decoding of speech parameters (although this can be compensated for by using error correction devices).

5

Telephone networks

The telephone network in every country is primarily a switched network and is, therefore, generally referred to as the Public Switched Telephone Network (PSTN). This network is still used for a high proportion of data transmission activities and is of vital interest to those who design and operate data communication systems. Since there are now many personal computer that are equipped with modems it is likely that more work stations use the PSTN for direct data transmission and for access to packet-switched networks than the specialised data communication networks. The PSTN is universally available and, in many cases, provides the least cost solution to data communication needs. Those users who are connected to the PSTN are, in most cases, the light duty users of data transmission but they are the majority of users and are generally less experienced with data systems than those who make continual use of their data work stations. Private leased (ie non-switched) lines are available in most industrialised countries and are rented to form corporate networks for all types of telecommunications traffic. Until ten years ago most private networks were based on analogue lines, but in recent years there has been a tremendous increase in the use of multi-channel digital links for both voice and data traffic. Some of the services that are based on digital links, such as Kilostream in Britain and Digital Data Services (DDS) in the USA, are available only for data communications.

Elements of the Telephone Network

Most of the components of the public telephone network are illustrated in Figure 5.1 which shows the level of switching that apply to the North American telephone system. All telephone networks have a hierarchical organisation, but the number of switching stages differs from one system to another. The telephone network in Britain, for example, has four levels of switches and Figure 5.1 could apply to a multi-national, European, network.

The apparatus which is in the user's location and which connects to the PSTN is frequently known as the customer premises equipment (CPE) and is described briefly in the next section.

If the connection from the CPE to the serving exchange is an analogue line then a nominal bandwidth of 4 kHz is allocated to each voice circuit, as shown in Figure 5.2. On the other hand which speech is transmitted in digital form, using the techniques which are described in the previous chapter, then bit rates of 64, 32, 16 or 8 kbit/s are allocated to each one way voice channel. Any method that is used to transmit data through the PSTN has to take account of the fact that digitisation of analogue signals may mean that the original waveform is not exactly reproduced at the destination. This may be a problem if the analogue output from a data modem is digitised at one of the lower bit rates. Many of the higher-speed modems will not operate satisfactorily with any form of pulse code modulation that uses a bit rate of less than 64 kbit/s to digitise a 4 kbit/s analogue signal.

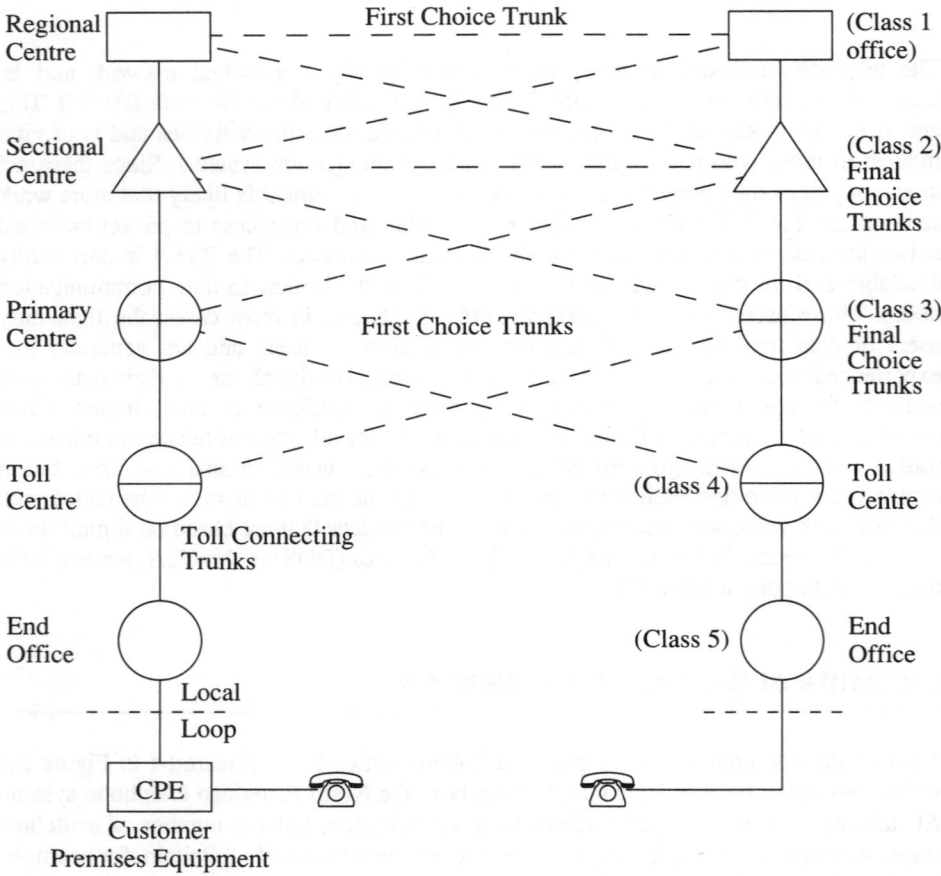

Figure 5.1 Hierarchy of Public Switched Telephone Network

The line between the serving exchange and the customer's premises is usually known as the *local loop*. In most cases this is a single pair of copper wires for each circuit. For the multi-channel digital systems, that carry 24 or more voice channels and are described in Chapter 5, a four-wire circuit (ie two pairs of wire) is generally needed. The equivalent of a four-wire circuit may be created over a physical two-wire circuit by frequency division techniques. Since 1986 fibre optic cable has been installed between the telephone company's equipment buildings and a rapidly increasing number of office buildings in most major cities.

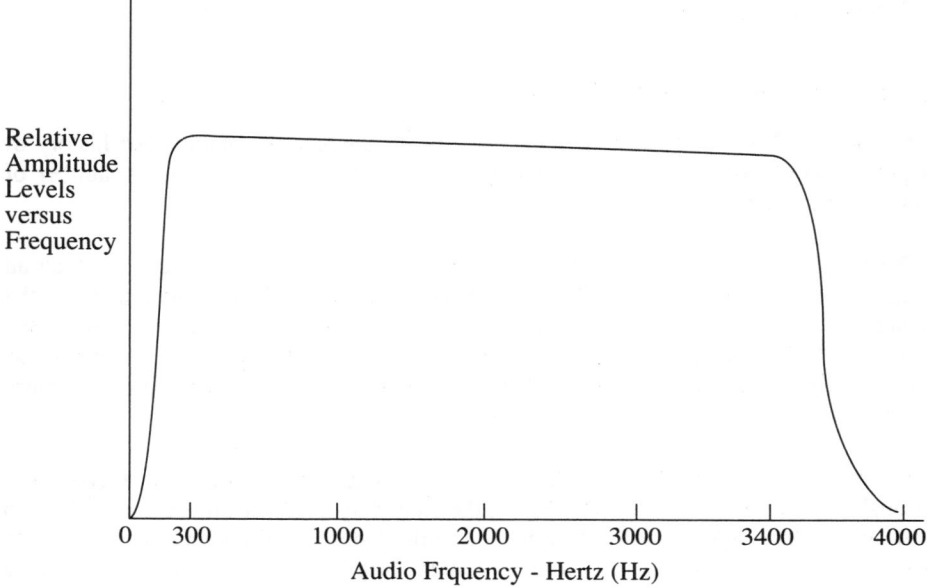

Figure 5.2 Bandwidth of standard telephone voice channel

Circuit-switching Techniques

We can see from Figure 5.1 that there are multiple levels of switching in most networks that use the PSTN. All switched voice networks are based on *circuit-switching* techniques, in which a connection is set up end to end before the conversation may start and that connection is retained, for the sole use of one conversation, until it is no longer needed. The switched circuit may consist of a dedicated physical circuit, across which the voice signals are carried in an analogue format, or the connection may include some switches (and transmission links) in which digitised voice signals are carried in specific time slots. In this case a number of time slots may share one physical link, but a given conversation will retain its allocated time slots from call set-up through to call completion.

At the customer's premises the in-house switch is generally a Private Automatic Branch Exchange (PABX), which is used for interconnecting internal telephones and for providing connections between those telephones and the PSTN. In North America and some other countries many small offices, especially those with fewer than 50 telephones, use an electronic key telephone system (EKTS) rather than a PABX.

The circuit-switching systems that are used as CPE or as public exchanges (central offices) may be broadly classified as analogue or digital systems. This classification really depends on whether the voice signal is carried through the switch in an analogue format or is digitised into a PCM form at, or before, the switch (eg in a digital telephone set). Some telephone switches are really hybrid systems, in which a digital computer is used to control analogue transmission paths.

Analogue Circuit-Switching

Analogue systems were used almost exclusively for the first 90 years in the history of telephone networks. During these years analogue switches went through a number of major stages of development.

The last of the old style manual public switchboards in North America, at which an operator inserted plug ended cords into a vertical array of jacks to complete the connection between two telephones, was taken out of service in the western USA in the spring of 1990. There are probably a few such manual switchboards still being used as in-house private branch exchanges in some of the industrialised countries and even more as local exchanges in less developed parts of the world.

A number of electromechanical telephony switching technologies were introduced into national and international networks during the first sixty years of this century. The two most widely accepted technologies were the step-by-step, or Strowger, system and the Crossbar system. Strowger technology, which originated in the USA, was very extensively used in Britain, and in countries heavily influenced by British manufacturing companies, until the middle of this century.

Crossbar switching systems were introduced simultaneously into North America and Scandinavia in the 1940s and it is still not really clear whether the design was originated by Bell Labs or by the L.M. Ericsson company.

Both of these technologies have enjoyed an exceptionally long life, although it is generally accepted that, for a given size of switch, a crossbar exchange requires half the floor space and less than half of the maintenance time than the corresponding step-by-step system. By mid 1991 about 30 percent of the telephone lines in Britain and over 20 percent of the lines in Canada were still served by electromechanical circuit-switching exchanges, but most of these systems will have been replaced throughout the world by the year 2000.

Less than 10 percent of the in-house PABXs in North America are now electro-mechanical switches and perhaps 20 percent are of the non-digital electronic type. It is most likely that the majority of medium sized offices in many parts of the world, such as South America, India and most countries in Africa, still rely on electromechanical PABXs.

Hybrid Circuit-Switches

In the evolution from electromechanical to fully digital circuit-switches a number of computer controlled systems come onto the market place over the ten year period from 1965 onwards. These are generally called stored program controlled (SPC) systems. A few major telephone equipment vendors introduced electronically controlled crossbar or crossbar-like switches. Northern Telecom sold its SP-1 central office for a few years in Canada, while Philips sold a large number of EBX systems for PABX applications, which used sealed read relays as the cross-points. Siemens used a somewhat similar technology for public exchanges for about a decade.

AT&T sold large numbers of central office and PBX systems that used pulse amplitude modulation (PAM) for the switched voice signals and were controlled by AT&T-made central processors. These switches, known as the IESS and Dimension, were in production for nearly 15 years and at mid 1991 over 30 percent of the telephone lines in service with Bell operating companies were still on IESS central office systems. Alcatel of France also produced large numbers of PAM-based switches and sold these systems extensively in Africa and Asia. The characteristic of this PAM technology is that a bit rate of only 8 kbit/s is needed to support one voice channel, since this corresponds to the sampling rate, but a number of voice channels can be carried on one physical connection. Interleaving and buffering techniques are used to make the connection between two stations in these software controlled analogue switches as in the later PCM-based switches. The main components of a computer-controlled analogue voice switch are shown in the block diagram of Figure 5.3

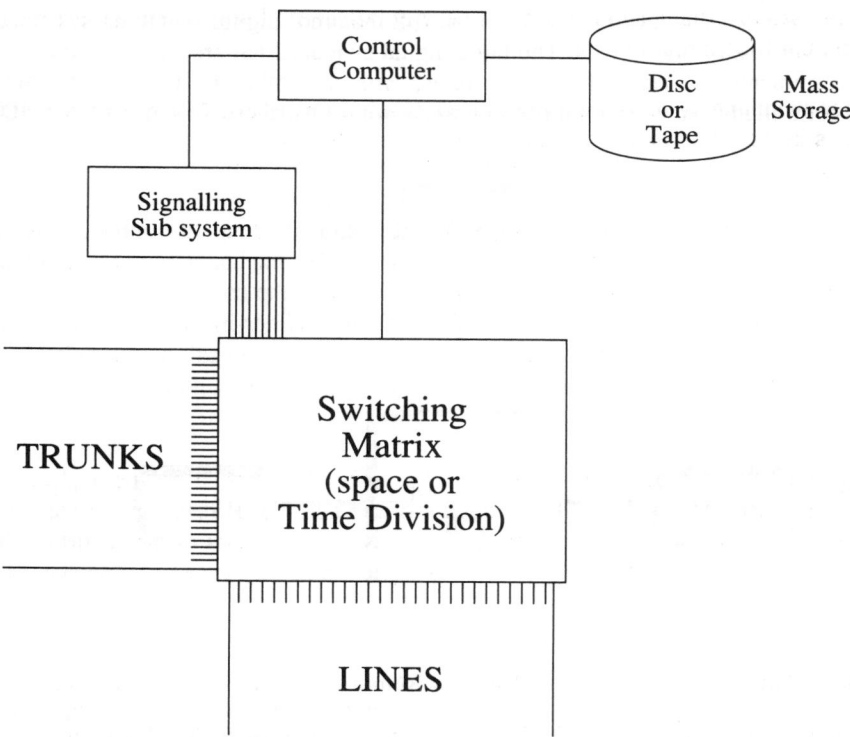

Figure 5.3   Block diagram of computer-controlled switch

Other manufacturers produced completely solid state telephone switching systems, where the cross-points were transistors in an integrated circuit and the voice signals carried through the switch in their original analogue form, without any use of time-slice sampling. Because the voice waveform is transmitted through the system on a physical path these are truly analogue systems from the data communications viewpoint. The best examples of these software controlled analogue switches are the small to medium PABXs that were produced by Mitel, which sold over 100,000 units in a ten year period. These extremely reliable switches, some of which were used as public exchanges in remote areas, are no longer being made, since it is less costly to produce fully digital PABXs. One reason for this cost differential is that at least 32 voice channels can be carried over one physical cross-point in a digital PCM system.

Digital Circuit-Switching

Fully digital software controlled telephone switches are now the only type being manufactured for all sizes of PABXs and public exchanges. By the year 2000 almost all of the telephone switching systems in Europe and North America will be digital.

The number of companies manufacturing telephone switching systems has been falling significantly over the past five years and it is likely that only five out of six will survive through the decade. The cost of developing a 'full featured' digital switching system is now well over one billion dollars. The large digital switches that are used in the PSTN are often implemented on a distributed basis, with a number of remote switching modules controlled from a central processor, as shown for Northern Telecom's DMS-100 in Figure 5.4.

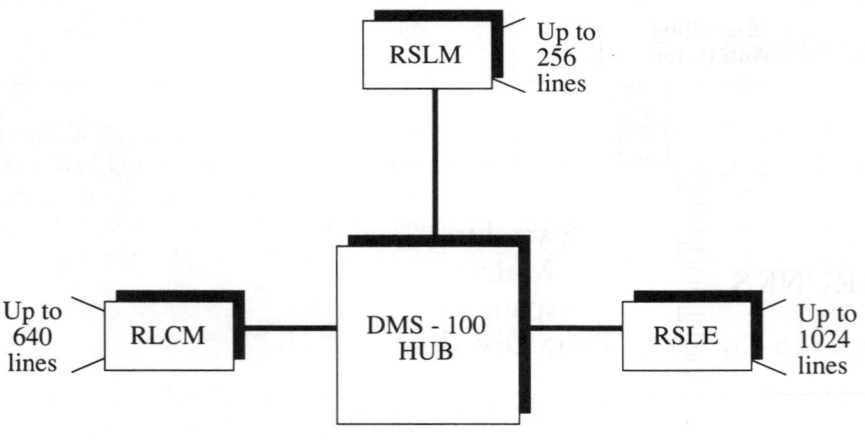

Busy Hour Call
Attempt capability
= 1 million (approx)

RSLE = Remote Subscriber Line Equipment
RLCM = Remote Line Concentrating Equipment
RSLM = Remote Subscriber Line Module

Figure 5.4 Architecture of the DMS-100

Circuit-Switches and Data Communications

The various types of telephone switching systems that are in the public network or on customers' premises have different effects on the quality (ie the error rates) of data that are communicated through the switch. The most serious problems of electrical noise are likely to be caused by step-by-step (Strowger) switches, because of the brushing motion of the electrical contacts and the extensive making and breaking of contacts that occur in these switches. This electromagnetic interference means that any data communication link should be routed away from a step-by-step system, if at all possible.

Since there is less mechanical movement in a crossbar system, with a more direct make-and-break contact, this technology is not as electrically noisy as a Strowger switch. If low-cost modems are used for speeds higher than 600 bit/s there may be an unacceptable error rate on digital signals if more than one crossbar system is encountered in a data link.

The effect of an electronic cross-point or PAM system on a data signal may not be noticeable, if the modems on the link use band that is well within the nominal 4 kHz bandwidth of these systems. However these are still analogue signals and thus the level of electrical noise can accumulate, since the signal is not being regenerated at each switching stage as it is in fully digital systems.

We should also note that in most countries only two or three PABX or central office systems dominate the marketplace. For example, in Britain over 80 percent of digital lines are on System X and the remainder are on AEX systems, while about one third of the telephone lines were still on Strowger switches in 1991.

Transmission Circuits

The sequence of circuits that link two telephone users on an intercity call is shown in Figure 5.5. In this diagram the British terminology is given in the top half and the American terms are below the line. In most cases the customers' premises (housing the telephones, data terminals and business communications system) are linked to the public telephone network by a two-wire circuit, within a cable of multiple copper conductor pairs.

Junction Circuits

The tandem trunks within an urban area, connecting together the various switches in the local hierarchy, are largely carried on multi-channel digital links, which provide virtual four-wire circuits. A physical four-wire circuit has one pair of wires for each direction of transmission and provides a much better quality of transmission than a two-wire circuit, with a lower level of electrical noise and of echo signals because of the physical separation between the two paths. A four-wire circuit can support full duplex transmission. Two-wire circuits can support full duplex given modems that provide suitable signalling.

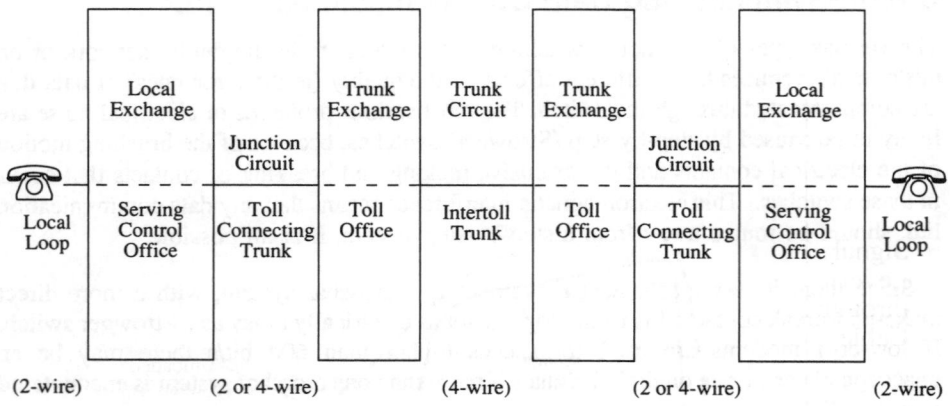

Figure 5.5 Intercity trunks and circuits

However, some of these intra-city trunks, or junction circuits, may still be physical two-wire circuits that are carried over copper wire.

For this reason the North American tariffs charged for leased, analogue data circuits for full duplex communication within an urban area are still set at twice the rate for half duplex circuits. For inter-city service the tariffs that apply to leased analogue voice band data lines are the same for full or half duplex service, since these trunk circuits are always effectively four-wire.

Many data communication networks in local urban areas rely on the use of *line drivers*, which transmit (and receive) a baseband digital signal. That is, a line driver does not modulate the data signal onto an analogue carrier wave and, therefore, requires a bandwidth that extends from direct current (ie 0 Hz) to much higher frequencies than the standard telephone voice bandwidth provides. Many line drivers are now used for speeds of 19.2 kbit/s over distances of at least 10 km, or at 9.6 kbit/s up to a 30 km range. Some line drivers can transmit data at 128 kbit/s for at least one kilometre over leased lines. For successful operation these devices need an end-to-end metallic four-wire circuit (ie with no intervening carrier systems or fibre optic cable) and an effective bandwidth that is at least twice the transmitted bit rate. These stringent requirements are becoming much more difficult to meet in larger cities than was the case five years ago.

Many of the original multi-pair copper cables incorporated *loading coils*, that were connected to each pair in the cable at regular intervals, typically of about one mile, to add inductance to each circuit. The effect of these loading coils was to improve the amplitude/frequency response of each circuit over the limited range that is needed for the 3 kHz voice bandwidth. This is illustrated in Figure 5.6, which compares the attenuation relative to audio frequency of the same length of loaded and unloaded cable pairs. It is clear that loading coils must be removed from any pairs in a cable that is to be used to transmit viable signals at frequencies higher than 4 kHz.

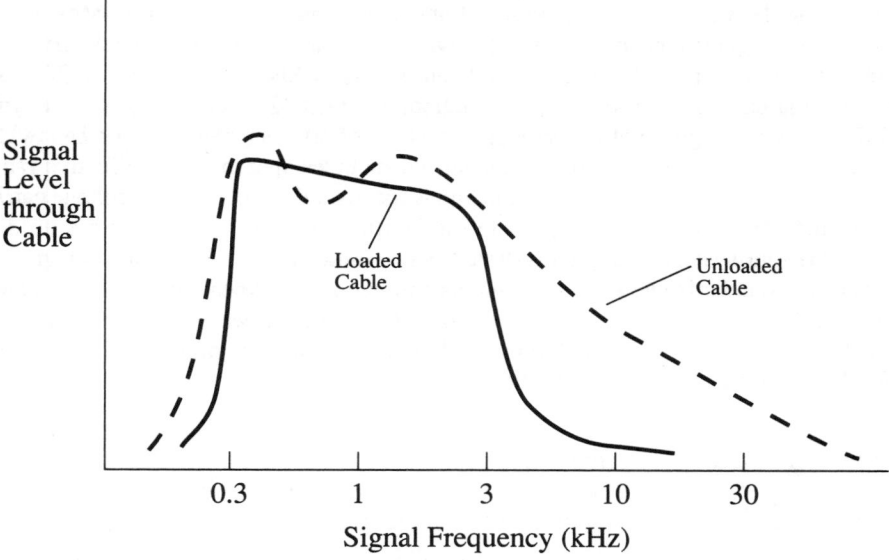

Figure 5.6 Attenuation of loaded and unloaded cables

Trunk Circuits

The telephone companies in some areas, such as North America, have installed *echo suppressors* on the longer voice circuits, in order to reduce the bothersome echo that may be noticeable to a speaker, if it appears more than 50 mS after the echoed syllable was spoken. This effect can certainly appear on copper wire circuits that are over 1500 miles long and in some cases, in the USA, echo suppressors are used on circuits that are no longer than 300 miles. Echo suppression is very desirable for telephone conversations that are carried by satellites, where two way delays of one half of a second are common. An echo suppressor switches a significant attenuation into the return path of the two way circuit.

The use of an echo suppressor means that a trunk circuit cannot be used for full duplex transmission and so totally negates the advantage of a four-wire circuit for data communications. Echo suppressors may be disabled by a continuous tone of approximately 2150 Hz that is transmitted for about one half second. All duplex modems that are to be used on switched trunk circuits in North America have to include this disabling feature. Echo suppression is not used in Britain, but may be encountered on international switched connections.

Analogue Multiplexing

The technology of transmitting multiple conversations over one pair of wires was first introduced on long distance, overhead-wire circuits in the late 1920s. With analogue multiplexing the voice signal is modulated onto a continuous carrier wave. Most of the systems use single-sideband transmission, with suppressed or reduced level carrier. In those standard systems the voice channels are spaced 4 kHz apart and the CCITT set recommendations for a hierarchy of modulation, in which 12 voice channels, occupying 48 kHz, constitute a group of five groups, for a total of 60 voice channels in a bandwidth of 240 kHz, from a Supergroup. Each sideband includes speech in the effective range from 200 to 3300 Hz. The frequency allocations for several widely used telephone carrier systems are shown in Figure 5.7, where the North American letter codes are used to identify the four systems that are illustrated. Since these carrier systems are designed to international standards, the same bandwidths and frequency allocations are used in most countries. For example, in Britain some heavy traffic routes were equipped with 600-channel systems over coaxial cable in the 1940s and extensive mileage of 12-channel two-cable systems was installed in the early 1950s.

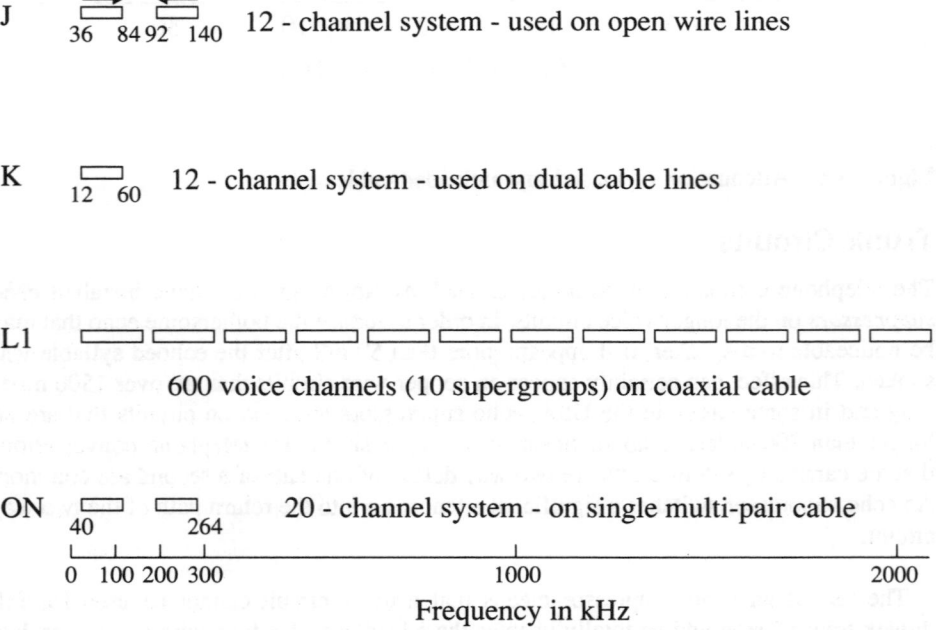

Figure 5.7 Analogue telephone carrier systems

The mastergroup of 600 channels which is produced at the third level of modulation, in the L1 carrier system, is used as the building block in some higher capacity systems. For example the analogue microwave systems in the 2 to 4 GHz range typically carry

600 voice channels per radio channel and higher frequency microwave links, in the 4 to 6 GHz band, often transmit 1800 voice channels per radio channel.

On the older, lower capacity systems it is feasible for every voice channel to be used for transmitting modulation data traffic, since the signals are transmitted at fairly high power levels. However in the higher capacity systems, particularly those that are used over microwave links, the proportion of audio channels that are used to transmit continuous data signals must be strictly limited, in order to avoid swamping the other channels with crosstalk.

Most data modems may be attached to a nominally 4 kHz voice channel at the channel bank level, as shown in the hierarchy diagram of Figure 5.8. Some applications have used higher speed group modems, which utilise a bandwidth of 48 kHz, and these must be connected into the group bank to occupy one fifth of the bandwidth of a supergroup.

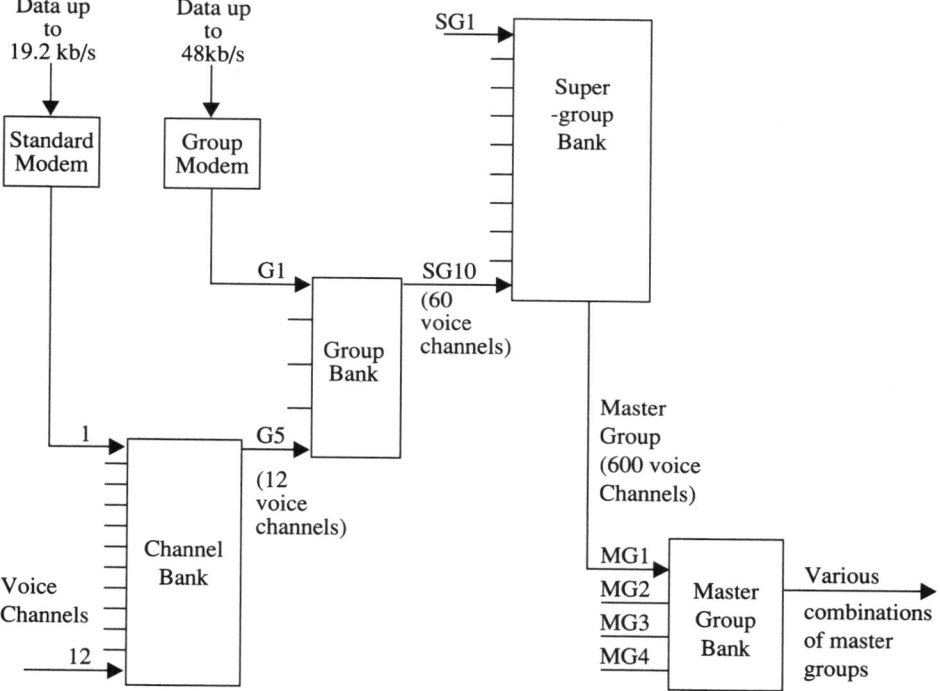

Figure 5.8   Frequency division multiplexing hierarchy

Digital Transmission

Pulse code modulation (PCM), as described in Chapter 4, is now used in digital transmission systems on the majority of trunk and junction circuits in industrialised countries, whatever medium is used to carry the signals. Digital transmission is being increasingly applied to the local loop into business premises, for connection to PABXs or digital multiplexers. Some telephone companies have also made extensive use of subscriber line carrier (SLC) systems which concentrate a number of residential telephone lines onto a multi-channel digital link.

Three different digital multiplexing hierarchies are used in North America, Japan and the rest of the world. The CCITT has published a large number of recommendations in the G.700, G.800 and G.900 series that apply to these pulse code modulation digital transmission systems. The European and North American digital hierarchies are shown in Figure 5.9, as these are defined in CCITT Recommendation G.702.

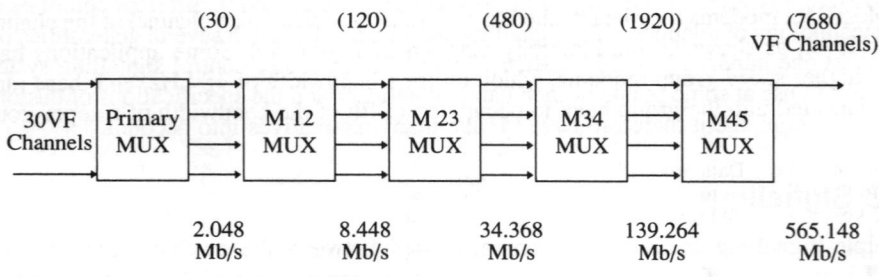

E (CEPT) Digital Hierachy

(VF = Voice Frequency)

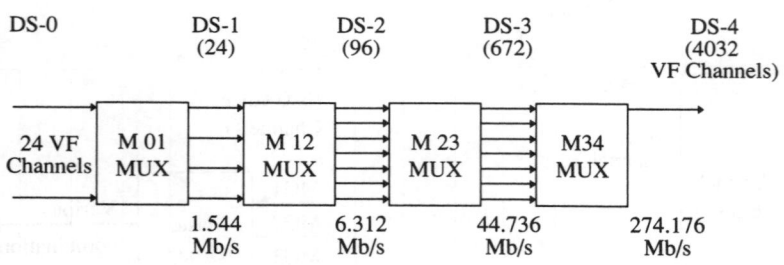

DS/T Digital Hierachy

Figure 5.9 Digital modulation hierarchies

For data communications applications some of the digital modulation levels, such as links operating as DS-1, DS-2 or DS-3, can be leased by the telephone company to a customer in an unstructured (ie non-channelised) format. The customer then provides the appropriate dynamically controlled multiplexing equipment to allocate the bandwidth, or approximately 1.5, 6.3 or 45 Mbit/s, in what ever way is best to carry the data, video or voice traffic of the moment.

Control Signalling

A variety of control signals, for call set-up, supervision and disconnection must be sent through the switched telephone network. When we plan to use the PSTN for data transmission these control signals have to be taken into account, to ensure that the possibilities of interference in both directions is minimised. The considerations are extremely important for the designers of modems, multiplexers and PABXs, but should be of no immediate concern to the users of data communications systems.

Different types of control signalling systems have been used in different countries and the technology of telephone line signalling has advanced through several generations over the past one hundred years. Unfortunately, most of the older control signalling systems are still in use at some points in each national network and the specifications of customer premises equipment therefore have to take many alternatives into account.

DC Signalling

Simple signalling methods are generally used between the customer and the local exchange (the serving central office), in order to keep the cost of the telephone set to a minimum. Some of these signals may be considered as direct current (DC) signals, since the signalling depends on the opening and closing of a circuit which carries direct current applied to the line termination in the public exchange from the central batteries, at about 50V.

When the user of a rotary dial telephone lifts the handset and dials a number the current flowing on the local loop varies from zero to a value of several hundred mA, as shown in the diagram in Figure 5.10.

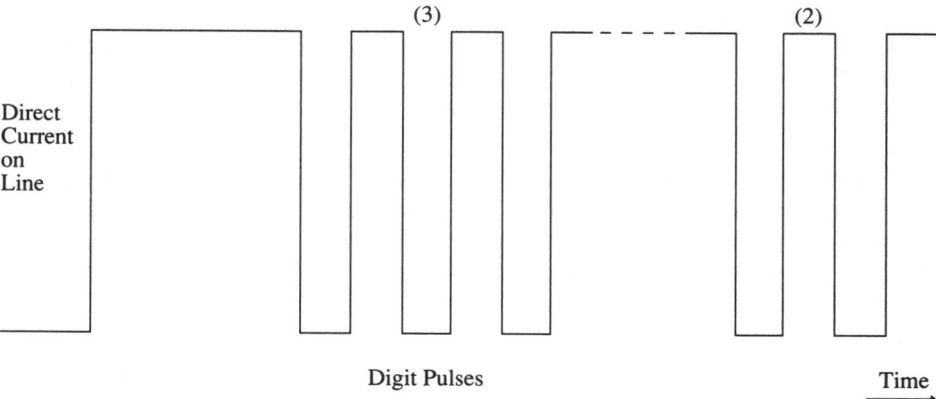

Figure 5.10   Rotary dial pulse signals

The make-and-break pulse rate from a well adjusted electromechanical dial should be ten pulses per second, but a deviation of +/– 20 percent can be handled by the exchange equipment.

Loop pulsing (which is often called loop start signalling) is normally used on local loops or trunks that have a maximum conductor loop resistance of 1200 ohms. Under some circumstances, when the receiving exchange has a long range pulse receiving relay, the maximum resistance may go to 2000 ohms. For longer circuits, eg with conductor loop resistances in the range from 1200 to 4200 ohms, battery and ground pulsing is used. This one exchange applies a battery voltage to the common ground which is of an opposite voltage to that applied by the exchange at the other end of the circuit. This is frequently known as *ground start signalling*.

Another type of DC signalling is used to provide simultaneous two way signalling over long switched circuits and over leased (private) lines. This method is known as *E and M signalling*, which separates the DC signalling from the voice path by having a separate M lead from the switching equipment to the signalling equipment on a trunk and an E lead to carry signals to the switching equipment. A popular interpretation of this terminology is that M stands for mouth, for transmit, and E for ear, for receiving. E and M signalling is used to interface with analogue and digital carrier systems and, therefore, has no distance limitations.

AC Signalling

Most of the signals that pass between the exchanges in the PSTN are carried by alternating currents (AC). The CCITT has published a series of signalling system recommendations over the years and these have been followed, with some local modifications, in Britain and the USA.

The AC signalling systems that are used between exchanges at the various levels of the public network are quite different from those that are applied between the customer's telephone and the local exchange. The 20 Hz ringing current generated at about 80V may be considered an AC signal, which is not audible because of its low frequency. The audible dial, busy and ringing tones which are returned to the customer are also AC signals which are carried through the network.

In most long-distance (ie toll or trunk) networks a continuous single frequency (SF) tone is sent over each idle trunk and this tone is interrupted when the trunk is seized by the sending exchange. The SF tone is resumed when the line is disconnected, at either end. If an active trunk receives a short burst of the SF tone then the trunk carrying that tone is automatically disconnected.

The trunk control tone is set at 2280 Hz in Britain, a tone of 2400 Hz is used in most European countries and the frequency of 2600 Hz is standard in North American networks. Any modem that is used in the PSTN must be set up not to transmit tones at these critical frequencies.

Multi-Frequency Signalling

Most control signalling systems, on both sides of the serving exchange, now employ multi-frequency (MF) tones. Two tones are sent to the serving public exchanges for each digit on the keypad of a dual-tone multi-frequency (DTMF) phone, which is known as the MF2 system in Britain. The DTMF tone combinations are shown in Table 5.1.

Table 5.1 Multi-frequency combinations from CCITT Recommendations

Hz	1209	1336	1477
697	1	2	3
770	4	5	6
852	7	8	9
941	*	0	#

Two tone MF signalling is also frequently used between the trunk exchanges (toll offices), with the digits and the start and stop signals using six different frequencies. In Britain two different sets of frequencies are used in opposite directions over a switched connection in the ranges from 540 to 1140 Hz and from 1380 to 1980 Hz. The North American toll network uses six signalling frequencies, spaced at 200 Hz intervals, between 1100 and 1700 Hz.

Signalling on Digital Links

For data communications applications the most significant difference between the PCM systems in North America and most other countries is that channel associated signalling is used in the North American DS/T carrier system. One bit, out of the eight bits in each time slot, is used to carry supervisory and signalling information. This means that seven bits per time slot, which is repeated at the sampling rate of 8000 times per second, are available to carry voice or data signals. This gives an effective data rate of 56 kbit/s for each of the 24 channels in the North American system.

By contrast in the E-digital carrier (sometimes known as the CEPT system) often used in other countries time slot 16, out of the 32 time slots available, is used for common channel signalling. The 64 kbit/s capacity of this time slot is used to carry four signalling channels, each with an effective rate of 500 bit/s, for each of the 30 information carrying channels. Since there is no bit-robbing in this PCM system for signalling purposes, the full 64 kbit/s throughput of each of the 30 channels in the E1 system is available for data.

The primary rate interface (PRI) of ISDN in North America is now providing 23 channels, each with a clear throughput of 64 kbit/s, since in this modified DS-1 scheme the 24th time slot is used to carry common channel signalling (known as the D-channel).

In fact, ISDN-PRI is being implemented in North America in such a manner that one D-channel on one DS-1 link can carry the common signalling for several, parallel, DS-1 links. In this way, for example, one DS-2 link used in the ISDN mode can provide 95 digital channels, each capable of carrying voice or data at a full 64 kbit/s throughput.

Out-of-Band Signalling

The SF and MF signalling techniques that have been discussed all employ frequencies that are well within the voice frequency bandwidth (between 300 and 2400 Hz) and are subject to interference from speech and other signals in that band. Most analogue telephone systems rely on these in-band signalling systems for circuit supervision and sending the called telephone numbers.

A number of commonly used signalling methods employ frequencies that are outside the standard voice bandwidth. Two familiar examples are the on off pulses from a rotary dial and the ringing voltage at around 20 Hz. Both of these are extremely low, inaudible, frequencies. Some modems employ a secondary channel, which is sometimes known as a *reverse or telemetry channel*, that provides a bit rate of 75 bit/s and is transmitted in a narrow sub-channel that is centred on 420 Hz, as illustrated in Figure 5.11. The low bit rate secondary channel is used for diagnostic and network control purposes.

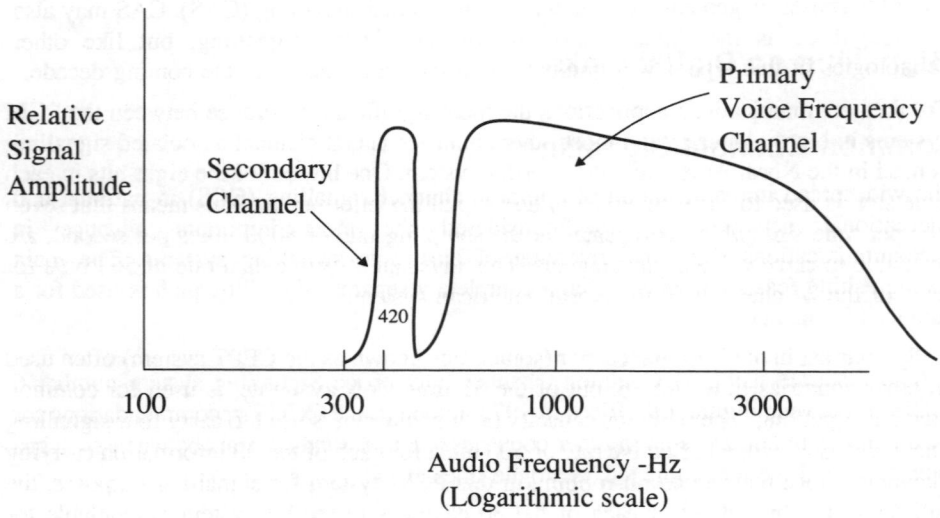

Figure 5.11 Secondary channel in 4 kHz band

The utilisation of previously unused space in the telephony bandwidth is generally referred to as data under voice (DUV) technology. Another example is that some widely-used microwave radio systems carry data traffic in part of the radio frequency band that is below that used for voice channels and, again, this is known as DUV transmission.

A second possibility is to make use of frequencies above the conventional voice band, but that are transmitted through the network, and this approach is called data over voice (DOV). A signalling sub-channel that is centred on 3700 Hz was incorporated into some carrier telephone systems. DOV technology was an essential feature of a number of low cost multiplexer products that enabled data and voice signals to share one pair of wires in local networks. This technology was commonly used with data switches that were popular, particularly in mini-computer installations, in the decade from 1975 and provided in-building and campus-wide data speeds of up to 19.2 kbit/s.

Several of the Bell operating companies have experimented with central office based DOV products to provide simultaneous voice and data services to their customers over the two-wire local loop. These trial DOV services simulated the two B-channels that will be provided by basic rate ISDN and gave the telephone companies an opportunity to test the viability of a number of applications that could be provided by ISDN-BRI. As it was an interim technology, none of these DOV products has been extensively used in the public networks.

The signalling technique of the current North American DS-1/T1 digital carrier systems, in which one bit out of each eight-bit word is reserved for signalling and control purposes, is generally called channel associated signalling (CAS). CAS may also be thought of as the digital equivalent of out-of-band signalling, but like other technologies of this type it will gradually become redundant over the coming decade.

Common Channel Signalling

The widespread implementation of common channel signalling (CCS) in national and international networks can be considered the most important advance in telecommunications since the invention of electronic switching systems. The main distinguishing feature of CCS is that a completely separate signalling path is used for a number of voice channels.

AT&T started using its Common Channel Inter-office Signalling system number 6 (CCIS 6) between digital toll offices in 1976, although the CCITT recommendations for CCS No.6 were not taken up in other countries. CCIS 6 sends signals between switches over dedicated modem equipped, analogue lines at 2400 bit/s.

The technique of CCS No.7, which are now being implemented in about 20 countries, will bring many improvements to the telephone network, when compared with the limited capabilities of in-band, or channel associated, signalling. The advantages of digital CCS include very rapid call set-up, greater security for sensitive data, user selection of the service required on a dynamic basis, complete internal network management and efficient use of bandwidth. CCS No.7, which is becoming known as signalling system 7 (SS7) in North America, works as a packet-switching network, across which signalling and control messages are sent as data. This capability is the essential base upon which the ISDN may be built. At the customers' premises these signalling messages are carried on the D-channel of the BRI or PRI of ISDN, although in most cases the SS7 network officially ends at the serving public exchanges. In some cases the telephone company may choose to extend SS7 directly to the PABX at the customers' site, over the PRI.

CCS No.7 uses a layered protocol which corresponds closely to the bottom three layers of the open systems interconnection (OSI) model. At the layer 2 level the signalling protocol is Link Access Procedure-D (LAP-D), which is an extension of the LAP-Balanced protocol that is used for public packet-switched networks that conform to the X.25 recommendations. In the LAP-D protocol many logical links may be established across one interface. The functional signalling protocol for the D-channels of ISDN, which link into the CCS No.7 inner network, is defined in Recommendation Q.931, from the CCITT. The corresponding link layer protocol is Q.921. These recommendations support the intelligent peer-to-peer communication that is required to provide feature networking.

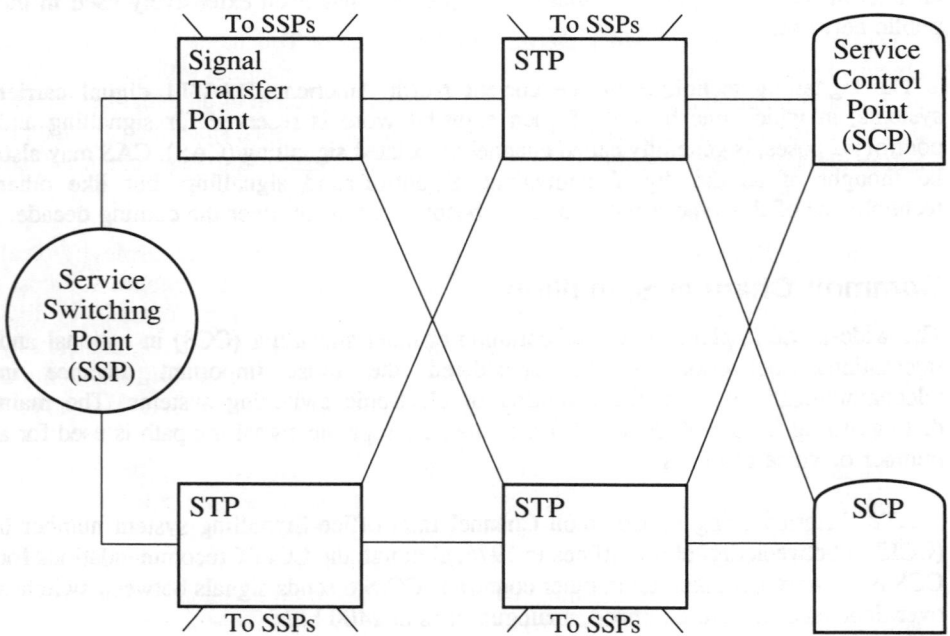

Figure 5.12 CCS No.7 signalling links

CCS No.7 is being implemented as a separate data network which is overlaid on the national telecommunications network. The intelligent database systems that are at the centre of CCS No.7 are resident in a few service control points (SCP). The packet-switching systems that are used to transport signalling messages between CCS No.7

nodes are called signal transfer points (STP), so that an STP serves as a hub in the signalling network.

A typical STP may be connected to several hundred fully loaded CCS No.7 links, each operating at 64 kbit/s, and future developments will lead to an STP with a fan-out of one thousand signalling links. Each STP supports many service switching points (SSP), which are tandem exchanges that carry all of the services switched through the public network. These functions are summarised in Figure 5.12, where the interconnection between STPs and SCPs is emphasised.

Terminal Equipment (Customer Premises Equipment)

Telephone Sets

The telephone is really the universal electrical terminal, and is widely available in almost all parts of the world, since over 700 million sets are in use. Most of these telephones are still electromagnetical devices, with rotary dials that are of limited use for data communications purposes.

DTMF telephones, for which the trademarked name of Touchtone is frequently used, are widely available in the USA and Canada, where they constitute well over 50 percent of the set population and are probably used to make 90 percent of all calls. Less than 10 percent of residential telephones in Britain are actually MF sets, since many of the push button phones work in the dial pulse mode. In some Western European countries, such as Spain, over 99 percent of the phones are dial pulse sets.

Electronic telephone sets, which incorporate a small alphanumeric display (typically for two 20 character lines), are just starting to be available for the public, at a cost of around $150. These phones are supported by features in a growing proportion of digital exchanges. Fully digital telephones have been widely installed in offices, attached by a proprietary transmission technique to an in-house digital PABX, and will come into use on the public network as the basic rate interface (BRI) to ISDN is made available.

Within the last few years many services have been developed using the telephone as a simple terminal in a data communications network. Tele-banking, tele-shopping and information retrieval are some of the more obvious applications for this technology. Some 25 tele-banking systems have been implemented with major banks and building societies in Britain and several hundred similar DTMF-driven systems are active in the USA and Canada. These services provide for account status inquiries, funds transfer and bill paying services. J.C. Penney handles well over 100 million tele-shopping, catalogue, sales each year, for a total of more than five billion dollars. These transactions are processed through several regional centres, each of which is equipped with a ROLM CBX and an IBM host computer. In those countries, such as Britain, that have few true tone signalling phones these integrated voice data applications must rely on voice recognition capabilities to process the input from the telephone user. With current voice recognition equipment a vocabulary of up to 30 recognised words, including the ten digits, is now generally available.

PABX Systems

For most businesses and other organisations the terminal to the public telephone network is a PABX or, in some countries for many small businesses (with up to 50 to 100 telephones) this may be an electronic key telephone system (EKTS). As described previously, only fully digital PABX systems are now being manufactured and the majority of offices in Western Europe and North America already have digital PABXs installed. The most successful EKTS are now fully digital, even though this trend has been much more recent than with PABXs. In those countries where EKTS are widely used, such as USA, Canada and Japan, these are generally replaced more frequently than PABX systems. So we can assume that in nearly all offices, in those countries with any significant amount of data communications, the customer premises equipment (CPE), linking the users to the PSTN will be digital by the middle of this decade. This does not mean that the ultimate user's terminal will necessarily be a digital device, as in most cases this is still an analogue, DTMF, telephone. However, if the switch is a digital system this makes it very feasible to connect data workstations and computers through the switch.

Each PABX or EKTS is linked to its serving public exchange (central office or CO) in the public network by a number of lines. In North America the tariff that is applied to a CO trunk to a PABX is usually twice the monthly rental for a line to a key system. This discrimination does not apply in Britain or in most of Europe. Currently most of these PBX-to-network links are analogue lines that are carried on copper wire pairs. There is a rapidly increasing trend towards the use of multi-channel digital links between medium to large PABXs and the serving public exchange. Over 1,000 or the 30 channel E1 links have been installed in Britain and several thousand DS-1/T1 links are used in Canada and the USA between PABXs and the public network. Some large PABXs in North America are now linked only by T1 circuits to their serving COs and so do not have any analogue trunks at all.

So far only a relatively few PABXs (eg less than 100 in either Britain or the USA) are linked by Primary Rate ISDN links to the public network, but this situation will change dramatically over the next five years, as many more digital public exchanges are equipped with ISDN capabilities for the primary rate interface (PRI).

The great advantage of using ISDN/PRI to carry the trunks between a busy PABX and the PSTN is that each channel may be dynamically assigned, under software control, to suit the instantaneous traffic pattern. This is particularly valuable in those few countries where the telephone companies provide a range of volume discounted services, such as WATS and 800 lines. Another great advantage of the common channel signalling that is associated with ISDN is that call set-up times are reduced by a factor of 10 compared with the older systems. With many of the multi-channel digital links between the CPE and the public network the 64 kbit/s channels in those links may be shared between voice and data traffic on a call-by-call basis, with the switching being done in the PABX on the customers' premises.

Most of the digital PABX systems on the world market can now be equipped with the necessary hardware and software to support ISDN/PRI. Unfortunately, there are still some incompatibilities between a PABX from one manufacturer and a public exchange (central office) from another.

As existing E1/DS-1 links are converted to ISDN/PRI and as more PRI links are cut over we can confidently guess that by 1995 over half of the traffic between PABXs and the PSTN will be carried over ISDN channels, in those ten countries where the telecommunications infrastructure is most developed. This trend has valuable implications regarding the ease and reduced cost of implementing data communications services to these many customers.

Very little progress has yet been made in the application of the basic rate interface (BRI) of ISDN to the lines, or trunks, that link the CPE to the PSTN. Few of the commonly used PABXs, and hardly any EKTS, support BRI, even though it would seem to be a highly desirable mode of linking the smaller business telephone systems to their serving public exchange.

Interfaces to link a local area network (LAN) to the PABX have been developed for a few systems, but have been used in very few installations. The 802.9 working party of the Institute of Electrical and Electronic Engineers (IEEE) is discussing a set of standards for PABX to LAN interfaces, but has not yet issued any firm recommendations.

Modem Pools

One of the most cost effective applications for a PABX in data communications is to use the switch as a gateway to wide area networks, in order to gain access to external computers or databases. Many PABXs are connected to a variety of lines and networks and in these situations it is often less costly to transmit data through a PABX than it is to acquire a data multiplexer. For cost and convenience reasons a modem pool has become the most popular data related application of a PABX.

A modem pool becomes especially attractive when medium to high speed modems are to be used with the PSTN. This applies to modem speeds of 4800 bit/s and up, since the prices for modems, or modem cards, for 1200 and 2400 bit/s are now so low that it may be less costly to allocate one such modem to each user.

Better utilisation of the more expensive dial-up modems is achieved by placing these modems in groups, or pools, on the trunk side of the PABX, as shown in Figure 5.13. The number of modems in a pool is decided on the basis of estimated traffic calculations for the expected usage of external trunks for data applications in the busy hour. In many organisations a ratio of one or two 4800 or 9600 bit/s modems for every 100 personal computers is sufficient.

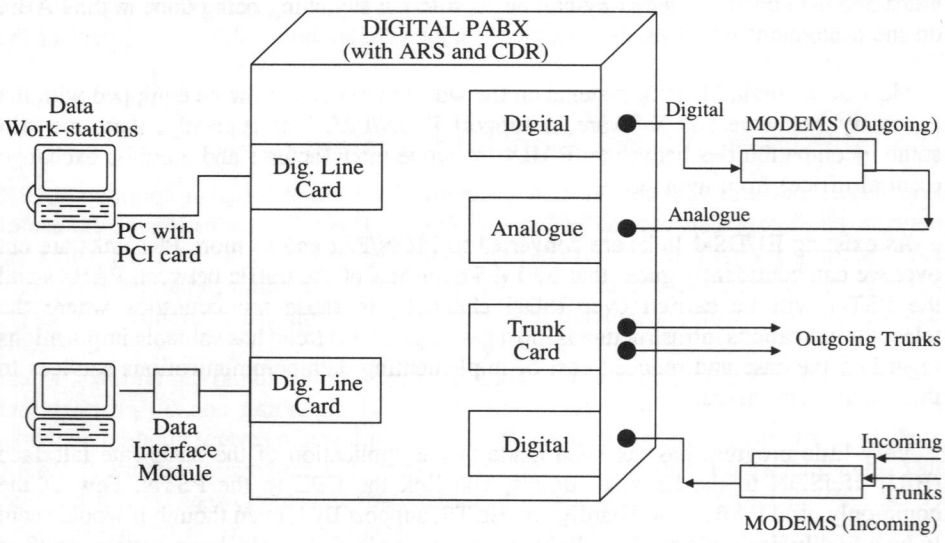

Figure 5.13 Outgoing and incoming modem pools

It is common to arrange modems into separate outgoing and incoming pools. The analogue sides of the outgoing modems are looped back through the switch to take advantage of the automatic route selection (ARS) software of the PABX. This configuration means that outgoing data calls share groups of trunks with voice calls and that the call detail reporting (CDR) package can record detailed traffic and usage statistics. The modems in the incoming pool may be connected directly to incoming trunks and then linked into the PABX on the digital side, as in Figure 5.13. An alternative for these incoming modems is to place that pool on the line side of the PABX so that they are in a hunt group to serve workstation users who are calling in from the outside over dial-up lines.

Network Software

Most PABX manufacturers provide a software package to facilitate the interworking of their own systems across wide area networks. Two examples of these packages are electronic tandem network (ETN) from AT&T and CorNet that was written and is sold by Siemens AG. The network-software packages for PABXs provide feature transparency across the network and also enable the users to set up a system wide uniform dialling plan. In other words, a network of remote PABXs appears to be one logical system to all of these who have access to the network, for both voice and data traffic.

Since the majority of these networking software products are proprietary in design they do not generally provide for the interlinking of PABXs from different manufacturers. One major exception has been that AT&T's ETN and Northern Telecom's electronic switched network (ESN) can communicate with each other, which is an essential development since each of these major vendors holds about 25 percent of the PABX market in the USA.

Digital Private Network Signalling System (DPNSS)

From 1985 onwards British Telecom encouraged development of a common PABX network package which became known as DPNSS. This software facilitates the use of PABXs from different manufacturers in a shared network. DPNSS was written in the absence of international standards and before the ISDN recommendations were fully defined, but has become a standard in several countries almost by default.

The DPNSS specification is written for 32 channel (2048 Mbit/s) digital links, where one 64 kbit/s channel is used for common channel signalling and 30 channels are available for voice or data applications. DPNSS is employed in several hundred networks in Britain and we estimate that at least one quarter of these have a mixture of PABX types. DPNSS allows about 20 commonly used PABX features, such as call transfer, central operator service, route optimisation and conferencing, to be carried across a network.

All the PABX manufacturers who hold any significant share of the British market now supply a DPNSS-complaint network package. Several of these vendors, such as Ericsson and Philips in Europe and Siemens in South Africa, now offer their DPNSS packages for sale in other countries.

Using a somewhat different approach Mitel, which did not have a proprietary PABX networking package before it developed DPNSS for the British market, has named its package Mitel Superswitch Digital Networking (MSDN) and has adapted this software for use with DS-1/T1 digital links, with 23 B channels and one D channel. Mitel has installed several major networks, carrying a significant proportion of data channels, in the USA and Canada with MSDN based on multiple SX-2000 PABXs.

Analogue Private Network Signalling System (APNSS)

Some PABX manufacturers that operate in the British market have also developed a networking package that is known as APNSS, as it became obvious that the concepts of DPNSS were becoming successful. These APNSS packages support the same range of networked features (such as corporate wide call transfer) as the earlier DPNSS.

APNSS is intended to be used on these PABX systems that are not large enough to justify the installation of 30 channel digital (E1) links across a private voice and data telecommunications network. APNSS can accommodate a variable number of analogue private lines in the network and uses one dedicated analogue line, as a data link, to carry the control signals between any two nodes in the network. In this way APNSS is analogous to CCIS No.6, as the shared inter-PABX signalling datalink usually operates with 4800 bit/s modems.

As it did with earlier developed DPNSS, Mitel is now offering its version of APNSS for use on integrated networks, based on the SX-2000, in North America.

Digital Access Signalling System (DASS)

A third network software development that has been encouraged by British Telecom is known as Digital Access Signalling System, version 2 (DASS 2). This package is the multi-access, common channel, signalling system to link a digital PABX to the public ISDN through a PRI. Because ISDN is not yet generally available in Britain, DASS 2 has not been used to any great extent and its availability is not so widespread on different PABX systems as DPNSS.

DASS 2 was specified in document BTNR 190, from British Telecom, before the latest standards were published and the software will need to be modified to bring it into line with the CCITT's latest recommendations for ISDN. However, DASS 2 does conform to the international HDLC standard, incorporating generally accepted link access protocol and call handling procedures. By the middle of this decade, when a large number of public exchanges will be able to support ISDN-PRI, the designer of data communication networks in Britain will then be presented with three alternatives for implementing flexible and powerful systems through digital PABXs.

Digital Centrex

Centrex (Central Exchange) is the generic name for the provision of switched, business, telecommunications services from the telephone company's central office, rather than from equipment on the customers' premises. In other words, *Centrex (CTX)* is a shared intra-company telephone service provided by the common carrier, instead of PABXs being installed on customers' premises. With Centrex the facilities of a large central switching system are logically partitioned by software and rented out to a number of business customers as *Virtual PABXs*.

Centrex has not yet become commercially successful in Britain, but is being seriously considered by several telecommunication administrations in Europe, since its delivery would be greatly facilitated by the availability of ISDN on digital exchanges. Over 10 percent of all business telephone lines in North America are based on Centrex services.

Since modern Centrex service is based on fully digital switching systems, providing a standard, two way, digital channel for voice communications, so Centrex systems are able to switch and transmit asynchronous and synchronous data streams, regardless of code or protocol, up to bit rates of 64 kbit/s. This use of Centrex is sometimes known as a central office local area network (CO-LAN). Very few organisations are using CO-LANs within one building, since the bit rate does not compare well with widely used LANs, that support from 4 to 100 Mbit/s. However the possibility of transmitting data through a city-wide Centrex network makes excellent economic and operational sense, when compared with the much higher costs of analogue leased lines or dedicated digital data networks.

An example of an organisation that uses a digital, Centrex-based, network from six different buildings in an urban area as an extremely cost effective means of linking 3174 cluster controllers to dual IBM 4381 is illustrated in Figure 5.14. In this case up to four terminal cluster controllers, each supporting up to 32 3270-type terminals, are multiplexed onto one 64 kbit/s *hot line* through the Centrex network.

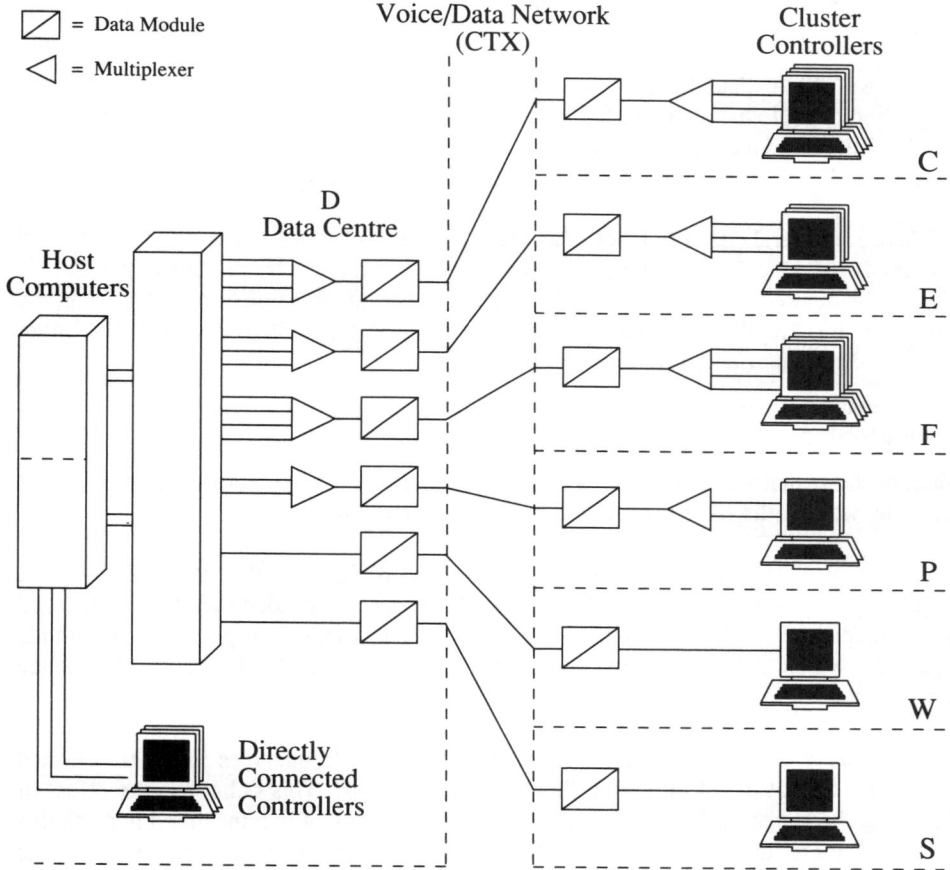

Figure 5.14 Data network through Centrex

Private Line Networks

Many telecommunication networks in Britain and North America have been implemented by using lines (or, more correctly, links) that are leased from the telephone company and are dedicated to the use of the leasing customer. These leased links are sometimes known as private circuits and are available for use at any time, without the call set-up procedures of the PSTN.

Analogue Leased Lines

Leased analogue lines may be used for either data or voice traffic, although certain parameters of a line may need to be optimised to achieve maximum utilisation for data communication purposes. Business machines, such as terminal controllers and computers, are much less forgiving towards signal impairment than the human ear, largely because we now attempt to send and receive digital data at a much higher rate than the spoken word.

Because digital data links provide a lower bit error rate and are less costly, the proportion of analogue circuits used in private data and voice networks is now decreasing rapidly, but leased analogue lines are still important to networks in the USA, where they were cost effective, as compared with digital links, until very recently.

There are several characteristics of an analogue voice channel which may be measured and modified when that channel is carried over a private circuit. Since the link is physically marked and dedicated (at least for a period of some months) to a given customer it is feasible to *condition*, or tune, a specific circuit to give the maximum possible throughput and a minimal error rate.

Conditioning

Most of the systematic impairments that adversely affect the operation of a voice circuit and thus worsen the bit error rate are known as *distortion*.

For a typical leased line the attenuation does not differ from that at 1 kHz by more than 10dB between 300 and 3400 Hz. If these limits were exceeded then the result could be attenuation distortion. Most of the higher-speed modems incorporate an amplitude equaliser, which reduces the attenuation deviation over the standard audio frequency range of telephony.

When delay distortion occurs, the signals at different frequencies in the transmitted bandwidth do not travel through a line at the same speed. This delay has the effect of distorting the shape of the original rectangular waveform that is transmitted and relative delays of more than two milliseconds can seriously upset the effectiveness of the modem, more complex, modulation schemes. Phase equalisers are sometimes used to delay the signals in the middle of the audio frequency band, so that the maximum relative delay between 300 and 3400 Hz is reduced to about 1 mS.

When attenuation and delay equalisation are applied to a private line by the telephone company in North America, this is known as *type C conditioning*. Most high speed modems that are now being used include automatic equalisers, so that it is not necessary to pay a premium over the regular line tariff for type C conditioning.

If signals of varying amplitude are attenuated differently through a line, this is known as *harmonic distortion*, since it has the same effect as adding voltages at second and higher harmonic frequencies at the signal. This type of distortion can be minimised by *type D conditioning*, which is also used to specify limits for the acceptable signal-to-noise ratio at the receiving end of a line. A typical specification, after the application of type D conditioning, would be for noise, second harmonics and third harmonics to be at least 40 dB below the received signal level over the standard audio frequency range. Type D conditioning is implemented by selective routing of the leased line, to avoid noisy and inferior circuits, and can, therefore, only be the responsibility of the telephone company.

Leased Digital Links

The world's first dedicated digital data network, known as Dataroute, was launched by Telecom Canada in 1972 and immediately offered higher quality transmission parameters than the equivalent analogue lines. At that time the tariffs were set at levels roughly corresponding to the bit rate used by the customer, so digital leased links were much more cost effective than analogue lines, at least up to 9.6 kbit/s. Following on the success of the Canadian network, British Telecom introduced its Kilostream service of point-to-point and multipoint leased digital links, for data rates up to 64 kbit/s, and AT&T began to offer its equivalent Accunet digital links, in the early 1980s.

By the late 1980s many of the major, high volume, data communication networks in Britain and Canada, used by such large customers as the airlines, banks, governments and manufacturers, were based on leased digital links. Within a decade the use of digital data circuits, at speeds up to 56 or 64 kbit/s, had become the accepted way to implement backbone wide area networks (WAN). The move to all digital WANs was not nearly so rapid in the USA, primarily because the tariffs for analogue leased lines were still attractive compared to the digital alternative and, in most cases, because analogue circuit quality was being maintained at a good level.

It should be noted that these all digital services, such as Kilostream and Dataroute, are carried out the same physical transmission facilities, such as microwave links or cables, as the other telecommunication services, whether for voice or video traffic. However, these digital services may be restricted to data traffic only and then customers are not allowed to interconnect these data services into their private integrated networks.

Integrated Digital Links

Following the court controlled divestiture of the Bell System in the USA on 1 January 1984, the telephone companies started to lease digital T1 links (each supporting an effective bit rate of 1536 Mbit/s) directly to users. These leased digital links were, in most cases, attractively priced and hundreds of organisations in the USA were using private T1 networks for voice and data traffic, with some applications for facsimile and video, by 1988. A large market for T1 digital multiplexers had also grown up by that time to facilitate the creation of cost effective, flexible and reliable networks. Generally speaking, over most routes in the USA, it was financially advantageous to lease a T1 link if about one half of its capacity could be filled with voice traffic during the working day.

The network diagram in Figure 5.15 shows a multipurpose digital network, based on intelligent, switching, multiplexers (eg the D/VNX) supplied by Timeplex, where four PABXs are connected by 1544 or 2048 Mbit/s links and some of the voice channels are compressed to 32 kbit/s ADPCM, as described in Chapter 4. This frees capacity on those links to handle digitised video signals and/or data streams between computer systems.

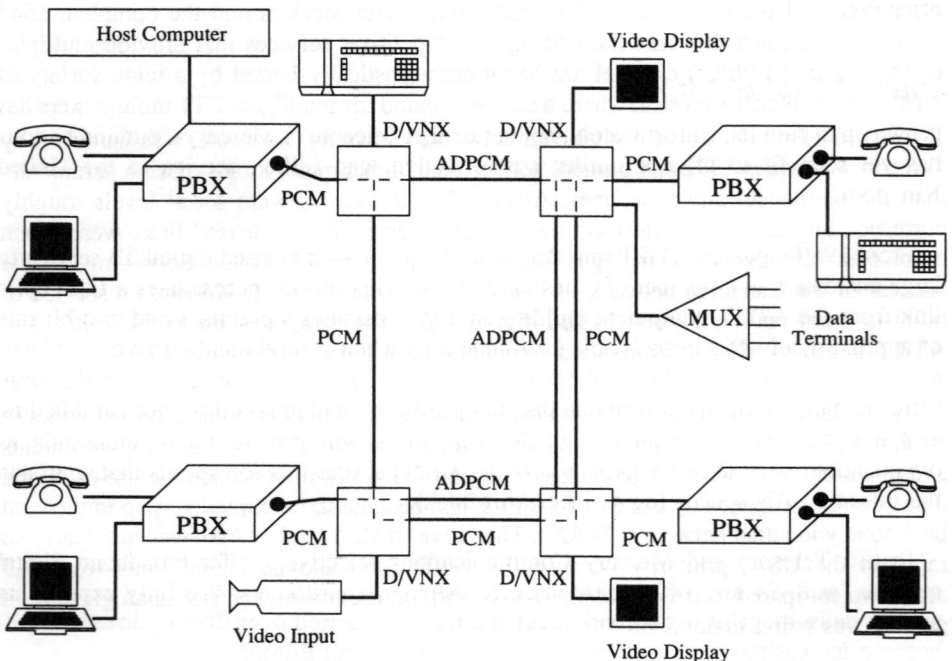

Figure 5.15 Digital multipurpose network

The initial tariffs that were set in Canada for similar DS-1 multi-channel digital services (eg Megaroute from Telecom Canada) were not nearly so competitive with the alternative leased analogue lines, for voice, or with leased digital links, for data, as were the T1 rates in the USA. This meant that only a few large corporations that could utilise three or four 56 kbit/s channels for data communications and the remaining 20 channels for telephony were able to justify the use of DS-1 services. Until the tariffs for DS-1 services were reduced significantly to competitive levels in mid 1989 the use of these integrated digital leased links languished in the Canadian marketplace.

British Telecom had considerable success with its Megastream service, providing leased digital links at 2048 Mbit/s and cut over hundreds of such circuits for its customers each year from 1985 onwards. One significant difference in these markets for high speed, integrated, digital links is that in Britain a high proportion of the links are for intra-city networks, (ie between a PABX and its local public exchange or between high speed multiplexers in buildings that are only a few miles apart), while in North America the majority of such links are used for inter-city traffic, on a regional basis. The local equivalent to Megastream service, based on CEPT's E1 standard, is available for leasing in several Western European countries.

In 1988, as the demand for T1 links began to peak in the USA, the competing carriers there began to offer leased fractional T1 (FT1) services over those routes and to those customers where full 24 channel T1 service could not be cost justified. FT1 services effectively fill the gap between 56 kbit/s digital data services and the complete 1544 Kbit/s digital links. Because the pricing is competitive, services that provide multiples of the DS-0 (56 kbit/s) channel are being enthusiastically leased by a wide variety of customers in North America, where a strong demand for intelligent FT1 multiplexers has grown up in parallel. Unfortunately, neither British Telecom or Mercury Communications has yet seen fit to offer a similar cost effective, gap filling, service in the United Kingdom.

From 1987 onwards AT&T introduced very high capacity leased digital T3 service to large customers in a number of cities in the USA. This service necessitates a fibre optic link from the AT&T equipment building into the customer's premises and at a bit rate of approximately 45 Mbit/s is cost justifiable only when several hundred voice trunks or aggregate data streams of several Mbit/s are required. There are probably over 500 large office complexes in North America that can justify these high speed services at this time and this number of potential T3/DS-3 customers will rise to many thousands as organisations need to link together corporate LANs at transmission speeds that approach the internal LAN speeds (eg 4, 10 or 16 Mbit/s).

British Telecom and Mercury Communications selectively offer broadband digital services, at 8, 34 and 140 Mbit/s rates, to their large customers. We fully expect that similar multi-megabit per second leased digital links carried over fibre optic cables will become increasingly available in major cities throughout Europe.

Private Networks

Some experts in the telecommunications field believe that the use of private, leased, networks for data and voice transmission has now reached a peak and the increasing availability of high-capacity switched networks will make private networks redundant. The concept of virtual (or software defined) networks, in which one broadband link is shared by the digital traffic from a large number of users, is quickly gaining ground. The technologies of fast packet transfer and frame relay will become available in public data networks, for both metropolitan and wide area networks, during the next five years in all of those countries where private data networks are now widely deployed.

By the year 2000 it may not be economically worthwhile for most organisations, with the possible exception of very large telecommunication users such as national governments and defence departments, to implement and manage their own dedicated networks. The competing switched multi-megabit data services and broadband ISDN may well provide the networks for all types of voice, data and video communications.

6

Data networks

The book so far has been almost entirely concerned with the principles and underlying transmission technology which support data communication. A data communications system, in which terminals communicate with computers and computers with computers in order to achieve specific results, will comprise of more than the transmission media. It may involve components such as modems, multiplexers, packet-switches and local area networks (LANs) connected together in various network configurations. The user, when she/he accesses files and application programs, should have a transparent view of the communications system. This result can be achieved by configuring networks of many physical appearances. The chapter covers these physical structures.

Network Topologies

Networks consist of two or more locations (nodes) which are connected together using communication links. A node may contain any number of communication and computing devices. The basic structures are described below.

Point-to-Point

Point-to-point is the simplest and is extensively used (see Figure 6.1). It may be transitory and exist only for the duration of a call on the switched network, or exist permanently as a leased circuit. Point-to-point configurations are commonly used to connect:

    —   asynchronous (start-stop) terminals to host computers (or FEPs);

    —   multiplexers to each other, so that a number of terminals can share a single, physical link to a host.

A ———————————————————————————— B

Figure 6.1 Point-to-point data link

Multidrop

Where a large number of locations have to be connected, and these can be broken down into geographical clusters, the multidrop form of configuration is generally more cost effective (see Figure 6.2).

All transmissions from Node A can be received by Nodes B, C and D. Similarly, only Node A can receive data from B, C and D, only one of which may transmit at a time. Multidrop circuits provide a way of reducing line costs by using a single branched circuit to connect Node A to Nodes B, C and D respectively rather than the three point-to-point circuits that would otherwise be required. It must be stressed that Nodes B, C and D cannot communicate directly with each other, only with Node A. Node A will use a protocol mechanism, known as polling, to address each of B, C and D in turn, accepting input data from each before moving on to the next. The mechanism is explained later in the chapter.

Multidrop networks are mainly used to connect host computers (at Node A in our example) to terminals or terminal clusters at several remote locations. Multidrop configurations are only available on leased circuits. Point-to-point and multidrop circuits are the basic network components from which other types of network can be built up.

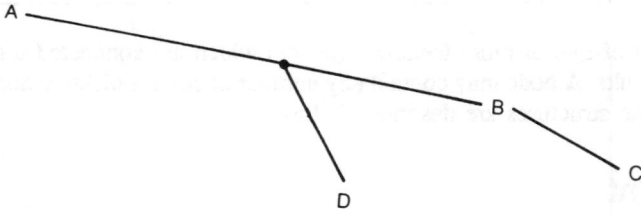

Figure 6.2 Multipoint or Multidrop data links

Star Configurations

Figures 6.3 and 6.4 depict two star configurations, the former employing point-to-point links and the latter multidropped links. In the figures, A represents a central computer site, and the other letters represent remotely situated terminals. At the present time configurations of this type are still in common use. The star configuration has two major limitations. First of all, the remote devices are unable to communicate directly and must do so via the central computer which functions like a switching exchange in addition to carrying out its primary processing tasks. In the second place such a network is very vulnerable to failure, either of the central computer, or of the transmission links and may require back-up (redundant) components configured in the network.

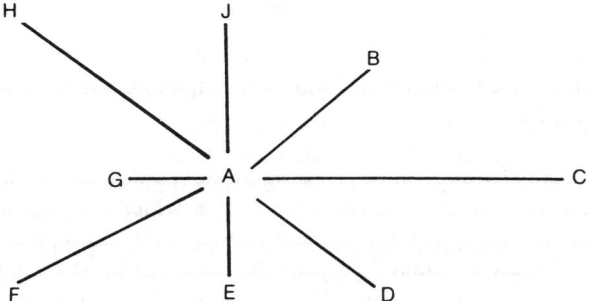

Figure 6.3 Start network using point-to-point circuits

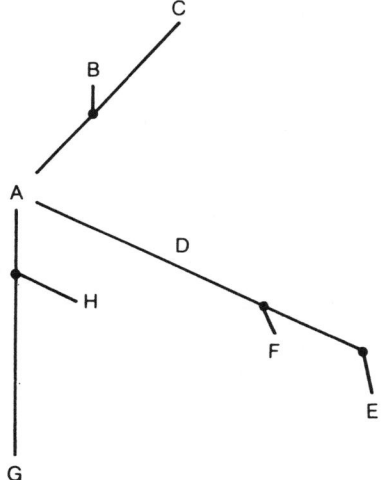

Figure 6.4 Multidropped star network

Loop (or ring) Networks

There are several variations of loop networks:

— a closed loop with the data links being provided by two-wire leased circuits. Messages are passed between the nodes in the network in one direction only; the host computer (A) controls communication using a mechanism known as list polling. The failure of a single data link will halt all transmission on the loop;

— a closed loop capable of supporting transmission in both directions. In the event that a single data link is broken, the host computer (A) will be able to maintain contact with the two sectors of the network. Two data links will need to be broken before one or more nodes are isolated from the host;

—   token ring is a local area network (LAN) closed loop technique based on a user station eventually receiving its own transmission. A single token circulates from station to station. When a station wishes to transmit, it waits for the empty token, puts in addressing information and data, marks the token busy and sends it to the next station. A station receiving a token addressed to itself, copies the data and sends on the token which is eventually received by the sending station. It discards the information and sends on a free token (Figure 6.5).

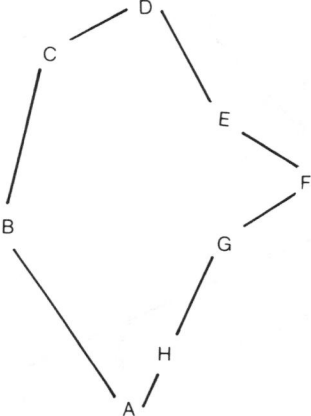

Figure 6.5   A closed loop network

Mesh Network

While star networks are suited to linking host computers to slave terminals or computers on a one-to-many basis, mesh networks are primarily used where multiple hosts need connection to multiple slaves (see Figure 6.6). In many cases the idea of host and slave is inappropriate as the nodes connected are of equal status.

Mesh networks are very resilient to failure, with alternative data routes being available when data link failure occurs. In many cases this resilience means that users are unaware that a network failure has occurred. Mesh (eg packet-switching) networks have been expensive to implement, but effective costs are now reducing as organisations share this type of network between many applications.

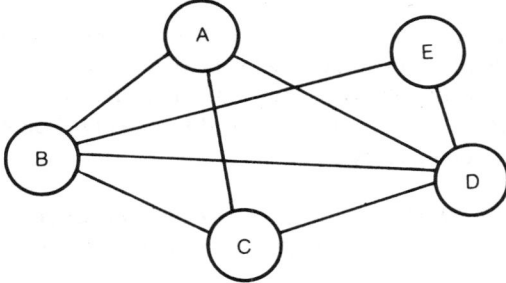

Figure 6.6   A mesh network

Bus and Tree

A bus is a single strand of cable to which all the user devices are connected. In this it is similar to a multidrop, but bus networks are characterised by each user being able to transmit messages to every other device. Ethernet is the best known example of this type of network, where transmission of messages is controlled by a technique called carrier sense multiple access with collision detect (CSMA/CD). A transmission from any node will be heard by all the others; however, only one node should transmit at a time, a single simultaneous transmission from more than one node could result in collision and data corruption.

Any node wishing to transmit, first listens to check whether another node is already transmitting (ie whether a carrier signal is present on the network). Once the network is free, the node will transmit its addressed message, but it will continue listening to ensure that it is the only node transmitting. In the event that a collision is detected, the node will wait for a random period of time before trying again. The random nature of this delay avoids the possibility that the same two nodes will transmit repeatedly at the same intervals, with each attempt resulting in a collision and neither node ever being able to transmit successfully.

Buses may be connected to each other to form a structure called a Tree.

Classification of Networks

Wide area networks (WANs) and local area networks (LANs) are terms that are in common usage. They are not precise classifications, however, they will tend to have the following characteristics:

Wide Area Networks

—   A network using the public switched telephone network (PSTN), integrated systems digital network (ISDN), a packet-switched public data network (PSPDN), any circuit switched public data network (CSPDN) or private circuits rented from a common carrier (eg British Telecom or Mercury).

—   Although 2Mbit/s is available, speeds tend to be relatively slow (ie 64kbit/s and less), as high speed usually means high cost.

—   It may be necessary to use analogue media, because of cost or availability, which introduces a relatively high error rate. Error correction techniques have to be employed which add to network delays and increase costs.

Local Area Networks

—   They cover a limited geographical area, for example to enable information and resources to be shared by a group of users in a single location. PC users, in particular, can share files and printers. Where sharing with other groups is required, LANs may be interconnected. Repeaters can be used to extend the network within a building and routers and bridges to interconnect over wider distances which can effectively make them into wide area networks.

—   The media costs are relatively cheap.

—   They have high availability and are relatively error free. There is still a requirement for error protection to be provided by higher level protocols to ensure data integrity.

The main LAN types are:

—   *Ethernet*, which was mentioned above. It uses a bus structure and the CSMA/CD technique for controlling media access. The following are the three common forms of implementation:

•   10 Base 5. This runs on thick coaxial cable at a speed of 10Mbit/s. It uses Baseband transmission with a maximum segment length of 500m.

•   10 Base 2. This runs on thin coaxial cable with a maximum segment length of 185m.

- • 10 Base T. This runs on twisted pair cable with a maximum segment length of 100m.

— *Token rings*. This was also mentioned above. The most common implementation is the IBM system running at 4Mbit/s on screened twisted pair cable.

— *Token bus*. This uses Broadband transmission on a coaxial cable bus. The attached stations are organised into a logical ring for token passing. A variety of speeds are used, up to 10Mbit/s.

— *Fibre distributed data interface (FDDI)*. This uses optical fibre cable, a ring topology and token passing media access control. It can operate at 100Mbit/s and up to 100 km, making it a little more than local.

Baseband and Broadband

These are two transmission techniques used on local area networks:

— Baseband where a device sends an unmodified digital signal on to the media; and,

— Broadband which refers to a modulated analogue signal. Here the network may be divided, by frequency division, to allow more than one signal on the cable. Modems are required to connect devices to the cable.

Alternative Data Switching Approaches

There are three well established switching approaches for providing connections between end users:

— circuit-switching;

— message-switching;

— packet-switching.

The approach is not something that should be of concern to an end user, so long as the facilities he requires are provided and he can compare prices.

Circuit-switching

The chief characteristics of a circuit-switched network are illustrated in Figure 6.7 to 6.9 and summarised below:

— a temporary sequence of fixed point-to-point circuits joined by switching exchanges is created;

- all the links and switches must be available for the full duration of the call or interaction;

- following an initial call set-up delay, interaction is almost immediate (see Figure 6.7);

- the called and calling parties must be simultaneously available;

- the interaction between the two parties occurs at a fixed transmission speed (see Figure 6.8);

- the mechanics of the network are transparent to the users, who only need to be acquainted with the procedures for making the set-up call;

- as the load of the network increases, acceptance of calls is increasingly delayed (see Figure 6.9).

We can cast further light on the discussion by referring to how people conduct a telephone conversation over the existing PSTN, although they could equally well be sitting at terminals. The way two people communicate and interact depends almost entirely upon the delay inherent in the communication channel between them.

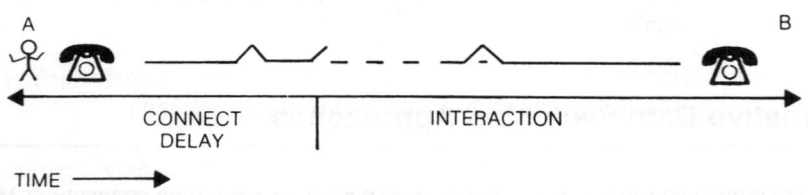

Figure 6.7 The sequence of events in making a switched call

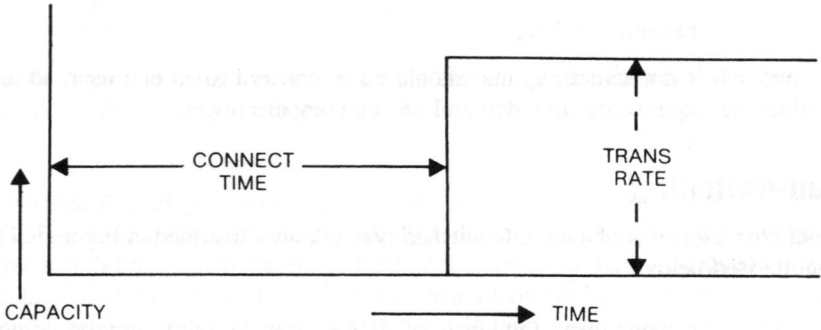

Figure 6.8 Capacity utilisation

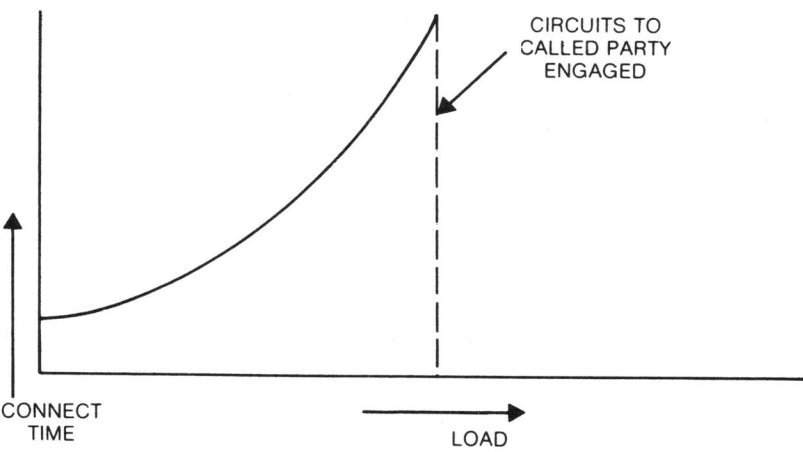

Figure 6.9 PSTN behaviour under increasing load

This type of interaction involves:

—   close conscious attention by both parties;

—   minimum delays in exchanging information. Excessive delays would result in poor feedback and wandering of attention.

For a telephone call, the PSTN characteristics do not unduly restrict verbal communication. The call set-up procedure and connect time are acceptable. The direct connection permits almost immediate interaction (see Figure 6.7).

But suppose one or other of the following circumstances applies:

—   the called party is engaged (engaged tone);

—   the network circuits and plant are heavily loaded or congested, so that one subscriber is unable to call another even though he is able to accept it.

In those cases the user will have to try again, thus increasing the call connect time. The diagrams further illustrate a number of the features described above. Following successful connection there is an immediate start to the interaction and this occurs at a fixed transmission rate (full channel capacity see Figure 6.8). As the total load on the network increases, the connect time also increases, until saturation is reached and the network will hold off further calls by transmitting busy tones (see Figure 6.9).

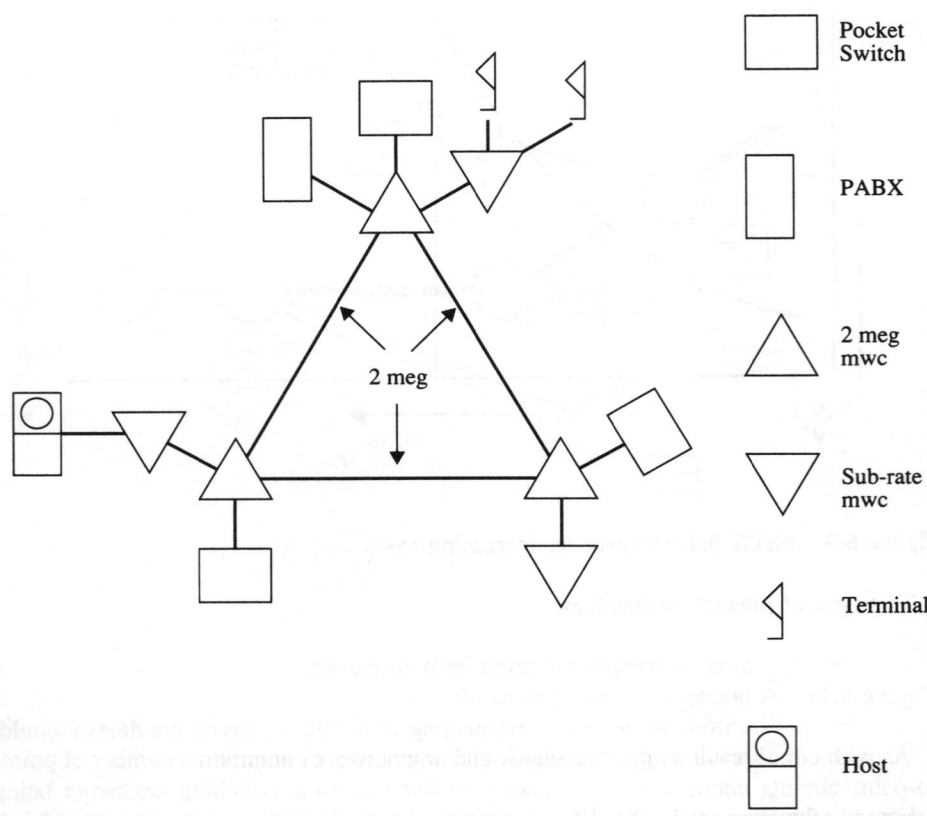

Pocket
Switch

PABX

2 meg
mwc

Sub-rate
mwc

Terminal

Host

2 meg

Figure 6.10   An example of a switching network

An example of circuit switching, in many organisations, is the use made of multiplexers which operate on 2Mbit/s links, rented from the PTT. Each multiplexer is connected to one or more other multiplexers (see Figure 6.10). The multiplexers will typically split each link into 30-64 kbit/s channels. Dedicated channels are then provided to other equipment such as PABXs, packet-switches and sub-rate multiplexers (which further divide the channel into lower speed channels). In this example, the allocation of channels will remain relatively static.

Message-switching

Message-switching is an entirely different form of communication from circuit-switching. By exchanging messages the participants need not be simultaneously involved with each other, and the message transfer can take place when it is more convenient.

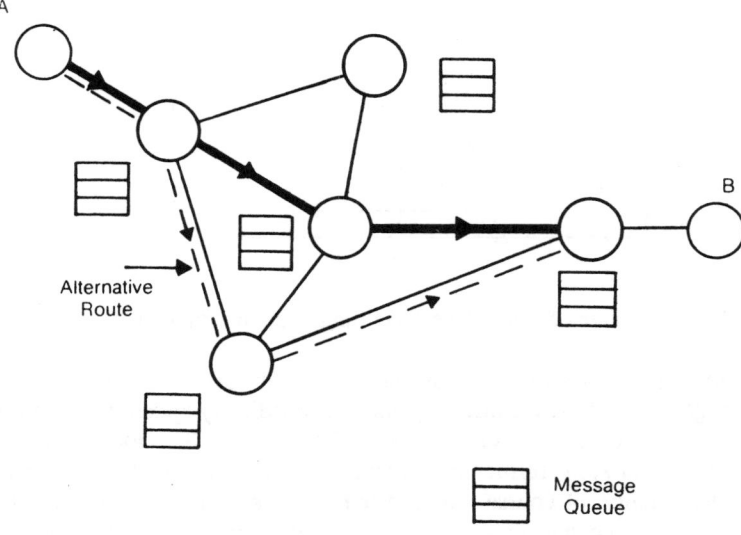

Figure 6.11   A message-switching network

As with circuit-switching, a message-switching network comprises a number of point-to-point circuits connected by switching exchanges, each switching exchange being equipped with computer intelligence and memory within which messages are stored (see Figure 6.11). Each message contains within it the address of the destination and the address of its origin apart from the information to be transmitted.

Each exchange inspects the address of the message's destination, and, providing that an outgoing circuit is free, forwards it on to the next exchange. If there is no outgoing circuit available, or the destination is unable to accept it, the message is stored in a queue of messages, and is subsequently transmitted when circumstances permit. For this reason the method is sometimes known as *store-and-forward* message-switching.

A message-switching system using store-and-forward switching obviates the need for repeated attempts at establishing a call, by accepting the message and undertaking to deliver it when this becomes possible. There is a minimum time necessary to transfer a message through the network. This comprises the time to travel between switches at the transmission speeds of the intervening links and a minimum message handling time at each switching point, including the time taken to read a message into and out of the stores at the switches. This is illustrated in Figure 6.12.

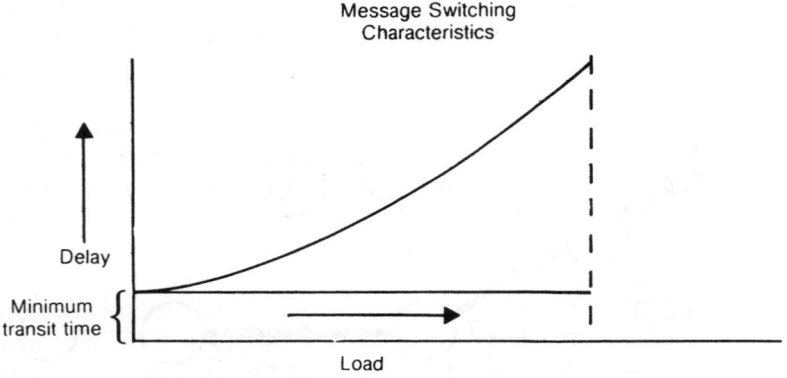

Figure 6.12 Behaviour of a message-switching system under load

With a store-and-forward system, an increase in the overall network load naturally results in congestion. However this may not be immediately apparent to the user, as the system continues to accept messages, at least for a while. The local switching centre usually can take and store further information, but may not be able to forward it for some time. Eventually, of course, no further messages will be accepted, until there is some relaxation of the congestion (or the destination is free to accept further messages). The upper limit to capacity is clearly set by the total storage available and the number and capacity of the transmission circuits.

Another way of looking at it is to regard the network as a large distributed set of stores which enable peak loads to be effectively spread out through increasing the transmission delay. By contrast, in the telephone network, the smoothing occurs because subscribers obtain a busy tone when they dial, and so try again later. In the store-and-forward network, messages will be accepted (if the store at the first switch is large enough) but delivery will be delayed. In the latter case, the originator of the message is spared the task of repeatedly trying to begin a conversation with another subscriber, but the conversation will proceed more slowly as the network load increases. With the circuit-switch, the start of the conversation will be delayed as the load increases, but once begun it may proceed at a speed unaffected by loading on the network.

In summary, the fundamental properties of message-switching are:

- no connect time delay;
- accept, store, deliver more of operation;
- equipment speed matching;
- retransmission on error;
- alternative routing;
- different grades of service are possible;
- opportunities for improved network control.

In addition to the features described above, eg accept, store-and-forward, and loading behaviour, there are a number of other characteristic properties and other features which can be incorporated. The following can be carried over into packet-switching networks:

- *Speed matching*: The store-and-forward capability enables the behaviour of the equipment involved in an exchange to be more closely matched. The network accepts messages at the rate at which they can be prepared and delivers them at the rate at which the destination can accept them;

- *Error retransmission*: Each originating point, destination and switching centre can hold a copy of the message so that it can retransmit in the event of a faulty transmission being signalled. This applies over each component link of the transmission path, whereas on PSTN it can only be applied end-to-end;

- *Alternative routing*: Since each message contains its own destination address, all messages can be treated as independent entities so far as the network is concerned. All that matters is that it guarantees to deliver a message to its correct destination irrespective of the actual physical route taken. Where more than one distinct physical route exists between two points on the network, we can turn this to advantage by arranging for the switching centres to select the best alternative in the event of congestion or failure of a circuit;

- *Different grades of service*: The previous discussion assumed, in effect, that the messages in a queue are released for onward transmission on a FIFO basis. In fact it is possible to specify different grades of service, so that for example messages given a high priority would receive preference over other lower priority messages. This is a common feature in a number of existing message-switching networks;

- *Improved network control opportunities*: The presence of computer processing power in the switching exchanges clearly provides opportunities to exercise a level of overall control of the network and its performance.

- *End-to-end multiplexing*: Because messages are individually addressed and a *call* or interaction between two subscribers does not require a dedicated physical transmission path, it is perfectly legitimate to interleave messages originating and terminating at different addresses (see Figure 6.13). This has three main consequences. First of all it means that the network itself provides multiplexing as an inherent property instead of an add-on facility supplied outside the network through a variety of devices called multiplexers/concentrators. Two other effects are described below;

— *Flexible connection/interconnection*: When we use the expression end-to-end we imply that the network multiplexing, in principle at least, extends to the computer and terminal line interfaces. Since the network itself helps to match the speed and response characteristics of the communicating parties and their equipment, there is no longer the same requirement for multiple line access ports distinguished by speeds and other transmission characteristics. Traffic originating at different speeds and having different transmission modes (eg synchronous or asynchronous) can share lines and ports in the network and at the destination. Coupled with the ability to engage in simultaneous multiple conversation or interactions with either the same or different subscribers, the arrangements also provide a high level of interconnection flexibility;

— *Improved circuit utilisation*: The within-network multiplexing clearly results in far higher circuit utilisation than is possible using traditional circuit switching.

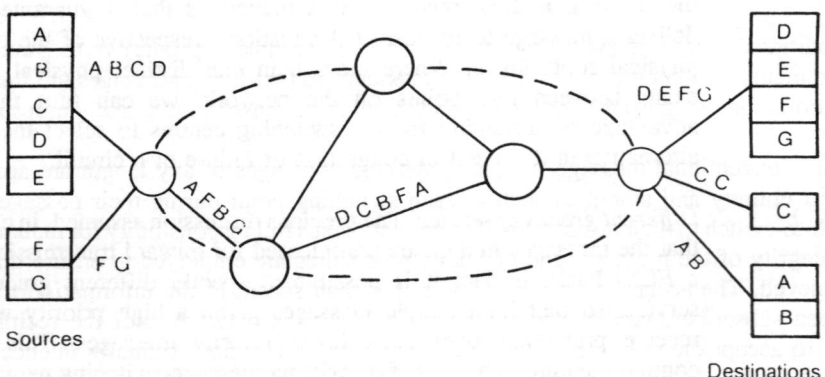

Figure 6.13 Message interleaving and multiplexing

Message-switching principles are well established in both the public and private domains. Telegraphy and telex are important examples of public networking, and there are numerous private systems operated by individual companies and other organisations. The SWIFT international banking network is an example.

We have so far described message-switching in relation to the provision of a person-to-person message service, which provides some message storage capability. For this type of application the storage facility is generally limited in extent, and there is no intention of providing long term storage. If however, the transitory storage function is replaced or augmented by longer term storage in the form of individually addressable mailboxes, the capability for a message system is considerably enhanced. For instance, individual users can consult their mailboxes, not only on demand, but after a considerable elapse of time. It is also practicable to use such a facility for purposes other than simple message transmission, including applications like holding conferences, either immediate or extended over time, exchange of information on co-operative ventures, and even the editing of material by geographically dispersed authors for publication.

Messaging and electronic mail services are widely available. An efficient message and mailbox service is a vital component in the electronic office, and there are many products which can provide such services.

A number of widely used proprietary electronic mail systems exist, but systems which conform to CCITT Recommendation X.400 are widespread. This is described in more detail in Chapter 13. In the UK, BT's Telecom Gold electronic mail service has been in existence for a number of years and GOLD 400 is a new service based on X.400.

The Transition to Packet-switching

The traditional message-switching approach contains the germs of packet-switching, subject to some important modifications. These are designed to overcome certain features which limit its application in computer communications.

Traditional message-switching systems were designed primarily to meet the requirements of human users exchanging messages in text form. Messages are the units of information recognised by the users, and their length must be moderately unrestricted.

With conventional message-switching systems, messages of any length are accepted in their entirety and stored as such at each switching point during their passage from switch-to-switch to their destination. The network takes full responsibility for maintaining the integrity of the message, and elaborate procedures are employed to ensure that this is achieved. The accent is on reliability, rather than speed, in the information transfer between subscribers, and the messages are held by the network until the recipient is ready to accept them, however long this may take. It was not originally intended that subscribers should interact rapidly with each other through a message-switching network, so the type of interaction common between telephone users to overcome errors introduced by the network, or indeed their own mistakes, is not possible.

We find therefore, that in traditional message-switching networks the transit delay (transmission time plus queueing time) for a message can be of the order of minutes. This is clearly unacceptable for many terminal-to-computer or computer-to-computer applications.

Call Identification and Message Sequencing

We have noted earlier that the message interleaving and multiplexing property not only simplifies the local connection arrangements but permits multiple streams of messages to be despatched to different locations. Clearly, a computer could indulge in more than one conversation at the same time, providing the message of one can be distinguished from those of another.

One solution is to carry both the source and destination addresses in each message, and this would distinguish the different sources/destination paths. However, a computer could be engaged in several simultaneous conversations each relating to a different job. Therefore, to identify which conversation a message belonged to, it would still be necessary to include additional information in each message. Also, since different messages belonging to the same application may experience different delays, they may arrive at their destination in a sequence which differs from the original.

Implementation

It should be evident that the operations of a network of this type are far more complex than in the case of the PSTN. Substantial intelligence and appropriate software are required at the exchanges, and, although the local connection arrangements are visibly simpler, and more flexible, they still have to be implemented. Universality of access to switched networks has to be done in such a way that end-to-end compatibility between widely different types of equipment is possible.

This is achieved through the adoption of the X.25 and associated protocols which are described in Chapter 10.

Packet-switching Principles

A major difference between packet- and message-switched networks is that the data (or message) is broken down into standard fixed maximum length packets, where the length is negotiated at call set-up time. Large messages can be readily handled by breaking them into packets before transmission and reassembling them at their destination. The term packet is appropriate because the data is carried within an envelope of control and error checking information.

The control field contains a number of items such as:

-     the addresses of destination and origin;
-     a process number indicating the facilities required at the destination by the user, this will be a specific application or a sub-address such as another computer;
-     other information to allow the user to reconstruct the data stream.

Every packet is terminated by an error checksum which is used to ensure that packets have been correctly transmitted across the network interface. The checksums are verified on reception and only if these are found to be error free is a packet accepted, and an acknowledgement transmitted.

Unlike a message-switching network, a packet-switching network is designed primarily for computer-to-computer communications. It has a much more rapid response which matches the internal speeds of computers, and handles information in much the same way as does a computer. At the same time it can readily match the speed of attached computers to that of the terminal users, by virtue of its internal storage.

The improved response time is achieved by the following means:

— The fixed packet structure permits efficient handling, and the absence of indefinitely long messages prevents the blocking of transmission links and keeps the queues at switching points small. Storage at switches is made sufficient only for a few packets and the total amount of information stored in the network is low. The result is that the delay through a packet-switching network is much smaller than through a normal message-switching network and the rate of throughput of information can be much higher. Where message storage is required as a service to subscribers, it must be provided externally to the network, rather than as an integral part of the switches in the manner usual in a conventional message-switching network.

Packet-switching technology can provide an efficient vehicle for message and mailbox services, as well as applications involving human interaction.

In order to accommodate the lengthy messages and queues at an acceptable cost, traditional message-switching systems have been compelled to use mass storage such as disks and drums. Packet-switching in contrast uses fast-access memory. A time-shared computer system provides a useful analogy.

A time-shared computer system is able to serve many users apparently simultaneously because it is inherently much faster in operation than any one user and, by switching rapidly between them, is able to share its resources among them, serving each one at a rate convenient to him. This is achieved by using storage to smooth out the traffic flow: the users are allowed to fill and empty buffer stores at rates suitable for them, while the computer communicates with the same storage at very much higher speeds. The use of storage in this way provides a match between the users and the much faster computer.

The storage in this case is located within the computer and each user has a separate connection with it, but a very great advantage is gained by distributing some of the storage within the communications network that connects users with a computer, because it can be used to share high speed communications links between users in much the same way as the computer itself is shared. This can make better use of the links; but far more important to the user is the rapid response obtained by sharing a high speed channel rather than having sole use of a lower speed one.

Multiplexing and Packet Interleaving

In the discussion of message-switching we referred to the message interleaving property and the multiplexing capability that this provides. Packet-switching networks also employ packet interleaving.

Strictly speaking the form of multiplexing used is called *statistical time division multiplexing*. What happens is that a user is only allocated real transmission capacity when he is sending packets. The trunk network will then allocate the maximum bandwidth it can along each link. The effect is that on the trunk network, the bandwidth used by an individual user can vary from zero (no packets to transmit) to the maximum speed of the trunk circuits (normally up to 64 kbit/s). The same applies in principle to the user's local access circuits up to the maximum speed of the circuit.

In application terms, if there are no packets to transmit for that application, then the capacity released could be used for packets belonging to another application or associated with another user located elsewhere.

Datagrams and Virtual Circuits

Two methods of operation are used in packet-switched networks. This refers to the generic case rather than X.25 packet-switching. The first is referred to as the *datagram mode*. This means that each packet is a self-contained entity and there is no explicit relationship between successive packets, and for many categories of application this is perfectly adequate.

However, there are many situations where the following requirements must be met:

— the interaction involves the transmission of multiple packets and the sequence must be preserved;

— there is some requirement for initial negotiations of, eg packet size, number of packets that may be sent before an acknowledgement should be returned.

The X.25 method of dealing with this is to establish in some way a temporary liaison between the two communicating applications within the computer or terminals.

The way this is done is to arrange for a preliminary exchange of packets between the two parties, equivalent to the procedure for establishing a call on PSTN. These packets contain reference numbers (logical channel numbers) and, assuming the call is accepted and information transfer takes place, all subsequent packets contain those reference numbers identifying the call to which they belong. For reasons which will become apparent this is called a *virtual call*. It should be noted that the virtual call facility can be implemented either within the network or external to it within the user's equipment.

Now that we can distinguish between packet streams in this way, it is also possible to arrange for packets belonging to a call to be delivered in the correct sequence. In order to create, maintain and control virtual calls, packet-switched networks which provide the facility employ a concept called the *virtual circuit*. It should be noted that the circuit is purely conceptual and there is no corresponding physical circuit. To the user, however, it has an appearance very similar to a conventional call.

A liaison can either be permanent or temporary. If the association is permanent, it is rather like a leased line running directly from one subscriber to another. Packets bearing the appropriate logical channel number are transmitted by the sender and routed by the network directly to the receiver, where they are delivered bearing the logical channel number appropriate at the end. This facility is called a *permanent virtual circuit.*

More detailed information on the mechanics of packet-switching, such as the call establishment procedure and the arrangements for connecting subscriber equipment to a packet-switching network, is given in Chapter 10 in the context of X.25.

Terminal Handling

The move in networking is towards balanced interfaces between end user systems, as opposed to hosts with subordinate terminals. This latter type of networking has been used for many years and is still very widespread and cost effective. The two main methods use either asynchronous or block-mode terminal handing:

Asynchronous Data Transmission

Asynthronous data transmission is in widespread use, characters being sent to line on each terminal key depression. The pre-eminent manufacturer employing this technique has been DEC, typically between VAX processors and VT100 type terminals. The other large manufacturers have supported asynchronous transmission but have persuaded their users towards the block-mode type of terminal (eg IBM3270).

Block-mode Data Transmission

This type is characterised by messages being blocked into a buffer before transmission. Multipoint lines are used and terminals are polled for their messages. The first protocols used for block-mode were character oriented which use special (normally ASCII) characters for synchronising, indicating start and end of message as well as using vertical and longitudinal character checking methods to check for errors in transmission (see Table 6.1).

Table 6.1   Character oriented message using ASCII

Supplier		Protocol
IBM	—	BSC
Unisys (formerly Burroughs)	—	Poll/Select
ICL	—	CO3
:SYN:SYN:SOH:ADDR:STX: DATA :ETC:BCC:		

The protocols are all of the polling type, in which the controlling end (ie host) invites each secondary (eg terminal), in turn, to send a message. A message is sent by the polled end, or an indication that there is no message. The mechanism of the protocol is used to retransmit messages in error. These protocols have in-built inefficiencies because of the number of control characters and number of acknowledgements (see Figure 6.14). Transparency may also be a problem with certain types of data. Transparent protocols have to take special precautions to overcome the problem of bit patterns in the data which have the same format as control characters. The problem is overcome by preceding the data characters concerned with an ASCII control characters, usually *DLE*.

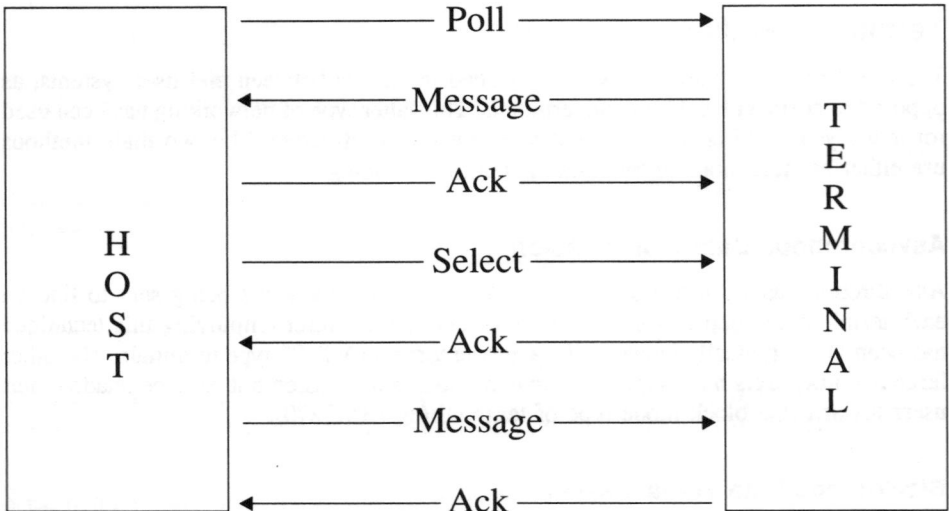

Figure 6.14 Transmission sequence using a character oriented protocol

Bit oriented protocols such as HDLC, SDLC, UDLC were a later development for block-mode transmission, using the polling mode of each one. These are in fact the same bit oriented protocol except for a few, little used, functions. In these protocols only one control character is used 01111110, which marks the start and end of each message. Chapter 10 contains a fuller description of HDLC and related protocols.

The characteristics of the two transmission types are:

    SENDING: *Asynchronous:* Each character sent to line as a key is depressed.

                            Block-mode: Characters stored in a buffer until RETURN is pressed. The message is sent when invited by a POLL.

    RECEIVING: *Asynchronous:* Characters are displayed as they are received.

Block-mode: Characters held in a buffer until ETX or ETB (end of text or end of block character) received before being displayed.

ERROR CHECKING:

Asynchronous: Parity on each character by using the eighth bit to make the number of 1 bits per character odd or even. Echoing of input back to the terminal. No automatic retransmission. Very limited but see 'Statistical Multiplexers'. To provide any quality of service, especially if dial-up is used, requires error protection.

Block-mode: Each block is checked to be free from errors before being displayed. The block can be retransmitted on receipt of a NAK or on timeout.

FLOW CONTROL:

Asynchronous: X-on/X-off (ie special characters to start and stop transmission) or CTS raised and lowered.

Block-mode: 1. Stopping polling.
2. Sending 'RNR' or 'reject' frames.

CONNECTION CHARACTERISTICS:

Asynchronous: Each asynchronous port at the host end can only support one terminal at a time (see Figure 6.16).

Block-mode: One host port can support a number of terminals/clusters (see figure 6.15). This is achieved using multipoint lines, terminal concatenation, etc.

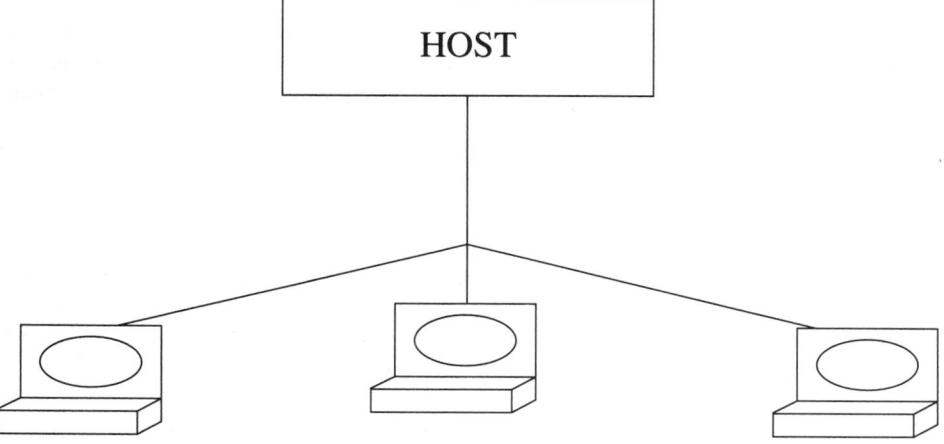

Figure 6.15 Polled block-mode terminals on a multipoint line

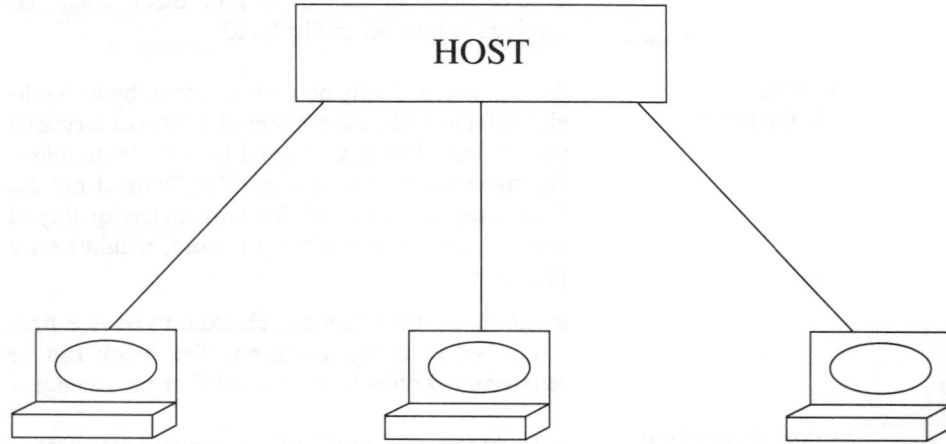

Figure 6.16   Asynchronous terminals each on its own line

Despite the appearance of a poorer set of functions, the asynchronous terminal has more than held its own over a number of years, for instance, when PCs and Apple MACS appeared, so did software to provide them with asynchronous (eg VT100 type) terminal emulation. Examples of emulation software are Smarterm, Procomm and Crosstalk. The shortcomings of asynchronous terminals can be overcome by the use of technology (see Figure 6.17) which still allows the final configuration to be cost effective:

— No error correction. This is overcome by error correcting modems (see Chapter 8) or statistical multiplexers;

— inefficient use of the communications line. A statistical multiplexer can be used to share one physical line between several terminals where the line speed may be less than the aggregate speed of all the terminals;

— one host port per terminal. This may be overcome by the use of contention devices so that the host ports are less than the number of terminals.

Bit oriented protocols have helped overcome one of the weaknesses of block-mode working but cutting down the number of message acknowledgements required compared with character oriented protocols. The other inefficiency comes from having to poll all of the terminals on the line. Methods used to overcome this problem include:

— using cluster controllers. IBM use this method, connecting terminals in the same location to the controller by coaxial cable. Only the controller then needs to be polled;

— using concatenation (ie linking together). Burroughs have used a method in which the linked terminals are group polled. The first terminal in the link, if it has a message it sends it and prevents the others from seeing it, otherwise it allows the poll through to the next terminal.

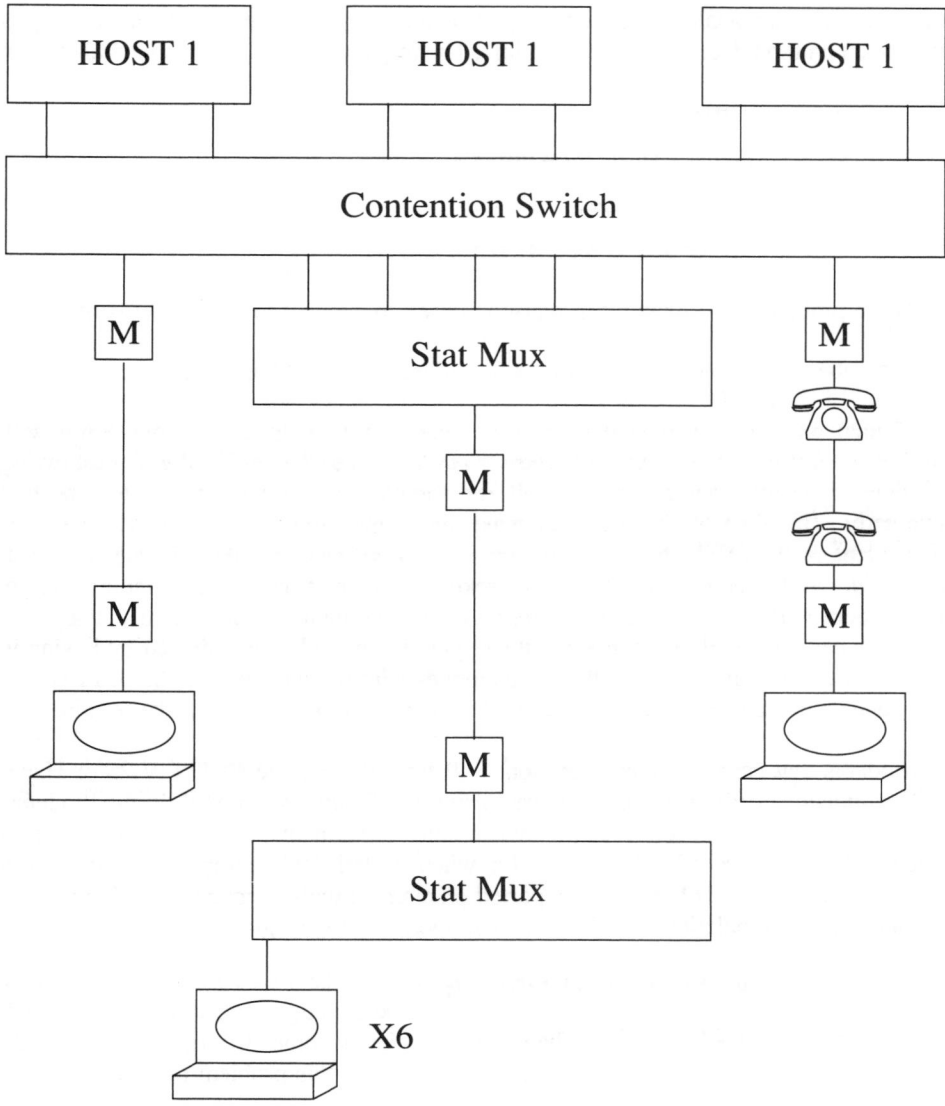

Figure 6.17 Examples of contention switching and statistical multiplexing

Lease and Dial-up (PSTN) Links

Even though leased lines are more reliable and do not require a call set-up, dial-up can be successfully used for interactive communications, eg credit card verification. If usage is low, dial-up is much cheaper.

Leased lines are the inherently more reliable medium, that is they are less prone to dropping out (loss of carrier) and to transmission errors (bit errors). Digital lines are more reliable than analogue, but countryside availability is limited. Analogue dial-up is the most common data communication link. The major problems with using dial-up are:

- carrier loss;

- higher transmission errors;

- call set-up time;

- higher access security risks.

These problems still exist when dial-up is used to connect to packet networks.

A number of techniques are available to reduce these problems in call set-up.

Manual answering and switching from telephone to data can be overcome with *automatic dial-up modems and auto-answer modems* (see Chapter 8). The biggest saving in time is through not having to wait for someone to answer and switch the call manually. The time to dial out and make the connection is not reduced as this is dependent on the PSTN, however *multifrequency dialling* can make this relatively fast. *Auto-dial* can be reduced to a few key strokes on the terminal. At the simple extreme, the modem will prompt the user to input a telephone number. Alternatively, telephone numbers can be stored in the modem, the required number being activated by keying in a code at the terminal. With a PC, the dialogues with the modem and the other end of the connection can be automated and the telephone numbers stored in the terminal.

Modems that carry out these functions will typically use the CCITT V.25 or Hayes AT command set (see Chapter 8). Customising the functions required, in the modems, can be performed from a terminal. Asynchronous communications normally demand a duplex link, because of echoing and for when a statistical multiplexer is used. For practical purposes, this has to be supported on a single dial-up connection. The CCITT recommended modulation techniques that support duplex, single dial-up are:

- V.32 at 9600 and 4800 bit/s;

- V.22bis at 2400 bit/s.

7
Modems

In Chapter 3 we discussed the techniques of modulation and demodulation necessary to transfer data messages over a channel which otherwise would constitute an unsuitable transmission medium. These revolved around various modulation schemes and the ways in which they can be adapted for the special case of digital signals. From Chapter 3 it should be apparent that for communication to occur using such techniques, equipment is required at either end of the channel to perform modulation at the source and demodulation at the destination. If the communication is to be anything other than simplex, this equipment will be required to support both these functions and hence the term *modem*, combining the initial three letters from the words *modulator* and *demodulator*. Modems are available for all kinds of transmission scenarios using a variety of media, but because of the massive expanse of installed telephony plant and its inherent accessibility, by far the most popular modem technology is that which permits transmission over normal voice quality telephone line.

Until recently, the respective national telecommunications operators have traditionally exercised tight control over the attachment of modems to their country's PSTN, and have, with a few notable exceptions, insisted that the modem must always be supplied by themselves.

In the UK, before liberalisation, this was true of British Telecommunications plc (BT) and the Post Office Telecommunications before it, except in circumstances where an alternative product provided a superior performance or facility than the BT product. For private circuits, the regulations were less strict. In all cases though, BT approval was required for attachment whether to the public network or to leased circuits. However, under the telecommunications liberalisation policies stemming from two Acts of Parliament (1981 and 1984), the regulations governing equipment attachment have been greatly relaxed and this has resulted in greater freedom of choice for users.

Approval and certification of equipment is carried out by the British Approvals Board for Telecommunications which vets equipment and certifies that it conforms to standards specified by the British Standards Institution in association with BT.

In the rest of Europe this is not the situation and equipment is still very much in the control of the Postal Telegraph and Telephone Authority (PTT), as is, in many cases, the equipment supply itself. However, more recently there have been moves to relax the equipment market as part of a wider objective to remove the trading barriers between all the member countries of the European Economic Community (EEC).

A wide range of modems are available to meet varying requirements such as synchronous/asynchronous, full duplex/half duplex transmission in addition to matching the circuit characteristics. There are also a number of data rates which are available. Hence, the strict controls on modem supply and approval just described have been beneficial to the user in standardising modem technology. The Comité Consultatif International Télégraphique et Téléphonique (CCITT), which is largely comprised of the global PTTs, produce the V-series of recommendations as part of an ongoing process. These have been almost single-handedly responsible for the nucleation of commercially available modems around an uncomplicated selection of appropriate technologies.

This chapter begins by examining some of the ways in which modem technology can be, or has been, used. After presenting the range of equipment that is available, the importance of a standard data interface is discussed. Included in this is a description of the most common of these interfaces. Finally, the chapter looks at the standards which operate beyond this interface and have a bearing upon the transmission rates achievable using modems.

Modem Technology

Modems are one of the most enduring types of communications technology. Year after year predictions are made to the effect that with the rapid rise in digital transmission the modem's days are numbered. Despite these prophecies, more modems, running at faster speeds and advertising enhanced features continue to be produced. The evidence indicates that new market opportunities for modem technology are still being created and filled.

From its humble beginnings, advances in microelectronics have helped to fuel a tremendous increase in modem power and these developments are dealt with at the end of this chapter. To begin with, we shall look at some of the variations in modem technology, which should illustrate how it has developed from being simply a method of data transfer and moved towards becoming an essential ingredient of the entire communications system.

Acoustic Couplers

One of the earliest forms of the technology, an acoustic coupler is a modem equipped with audio transducers (a loudspeaker and a microphone) so that it can interface to the telephone handset instead of the telephone line. It can be used at low data rates where the standard modulation method is frequency shift keying.

An acoustic coupler accepts data from the terminal and converts it into audible high and low frequency tones, which are then fed to the microphone in the handset. In the reverse direction, the coupler converts the audible tones from the telephone earpiece into binary data signals for transfer to the terminal. Some terminals had in-built acoustic couplers.

The great advantage of an acoustic coupler used to be that it permitted a terminal to use any convenient telephone. No wiring or other modifications were necessary. However, two advances in the design of the standard telephone have negated these positive attributes. Firstly, since the advent of the wall socket and plug approach the attachment of normal modems has become trivial, and secondly, many of the new handset designs are unsuitable for use with the acoustic chambers of a coupler. In any case, the performance of an acoustic coupler is inferior to a normal modem in that the error rate is higher although this is still not a serious problem for many users.

Acoustic couplers are (reluctantly) permitted by the PTTs, and have to be approved to ensure they do not interfere with other users or with telephony plant. They are not favoured for permanent terminal installations.

Auto-Dial

Smarten, X-talk, Procomin all provide auto-dial functions from PCs, as well as asynchronous terminal emulation. The auto-dial functions are built into the modems, eg Hayes type (see description later in this chapter).

The auto-dial facility allows the computer or terminal to 'dial' a telephone number. Compatible interfaces are commonplace on computer and terminal systems.

Auto-Answer

An auto-answer option will allow a dial-up connection to be made without human intervention at the called site. It is important that the computer or terminal at the called site can handle the auto-answer modem interface.

The auto-answer has particular use in a time-sharing computer bureau, eg credit reference operation. Here a series of modems can be installed in a hunt group associated with a contiguous series of exchange lines. The bureau customer will dial the number of the hunt group and will be automatically connected to the first available modem.

Further use for the auto-answer feature is found in BT and Mercury's Packet Switched dial-up service which has benefits to customers in that they would no longer require banks of modems and hunt for (or even hunting) groups.

The auto-dial and auto-answer facilities can be used in tandem to provide a security function. when a remote terminal dials into a computer and a link is established, before the terminal may proceed it must identify itself and leave its own telephone number. Having identified itself, the dial back is to a number associated with the user ID. These functions are usually in the modem management system. Then the management system terminates the communication whilst it goes away and checks the access rights of the terminal. The management system will only ring back those terminals it has confirmed are authorised users. This method of working is known as *dial-back*.

Double Dial-Up and Split-Stream Modems

Double dial-up modems are designed for use on the public switched telephone network. Two telephone connections are established to the required destination, and the two paths are then used as a four-wire connection with full duplex connection (see Figure 7.1).

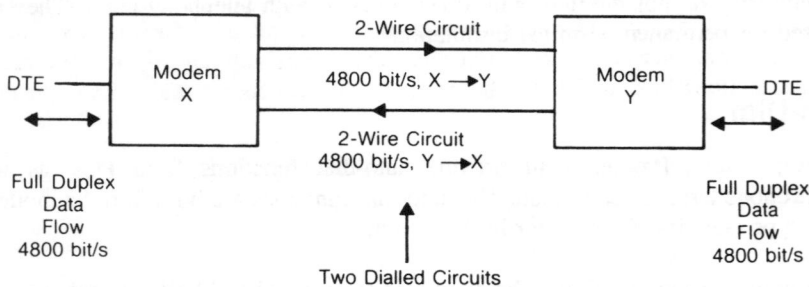

Figure 7.1 Double dial-up (4800 bit/s)

Split-stream modems also use two separate circuits, but in this case both circuits carry data in the same direction. By transmitting at 9.6 kbit/s over two high grade leased lines operating in parallel, a split-stream modem can provide a data throughput rate of 19.2 kbit/s (see Figure 7.2). These have become obsolete since higher speed digit (eg Kilostream) circuits became available.

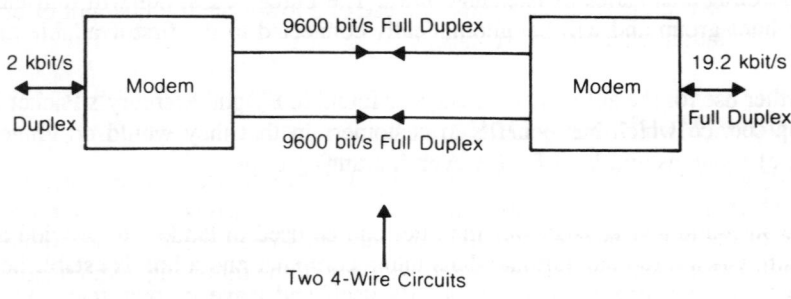

Figure 7.2 Split-stream modem (19.2 kbit/s)

The electronic circuitry which splits the data-stream and handles all the synchronisation problems can take the form of a separate unit, when it is known as a *biplexer*.

Modem Sharing Unit

A modem sharing unit, also called a fan-out unit, allows several terminals to share one modem. All the terminals receive data from the modem simultaneously, but only one terminal can transmit at any one time.

Modem-sharing units are economical where several co-located terminals require access to the same information and where the data traffic generated by each is low. They may be used at a computer centre to permit more than one computer port to have access to a communications link, which can be useful in circumventing faults. Examples of use are:

- line sharing for synchronous polled terminals;

- contention unit for asynchronous terminals;

- cluster controller, eg as in IBM configurations.

Modems for Leased Lines

The modem supplier will specify the quality of the leased circuit required for the satisfactory operation of their equipment. All the telephone circuits have the same nominal electrical characteristics. However, tolerances are applied to those nominal values in defining the guaranteed characteristics of a line. The higher the stated quality of the line, the more closely it will conform to the stated nominal values, ie the tighter will be the tolerances allowed. The term *conditioning* is used in describing the quality of a telephone circuit.

It may be that, because of a modem's design, it requires a lower line quality than others operating at the same data rate. The use of a lower quality line will offer savings in line installation, charges and rentals.

Multiport (Multinode) Option

Many higher speed synchronous (4800 to 14400 bit/s) modems incorporate a multiport option which allows the capacity of the modem to be divided between several ports with the sum of the port data rates equalling the modem's rated speed. For example, a 9600 bit/s multiport modem can be configured with any of the following combinations of port speeds:

- 1 x 9600 bit/s ports;

- 2 x 4800 bit/s ports;

- 1 x 4800 bit/s plus 2 x 2400 bit/s ports;

-        4 x 2400 bit/s ports;

-        1 x 7200 bit/s plus 1 x 2400 bit/s ports.

This in effect provides the facilities of both a modem and a time-division multiplexer.

The facility can offer significant cost savings by avoiding the need for multiple point-to-point circuits. The ports can serve co-located devices, or alternatively the links may be extended to further locations, as shown in Figure 7.3, where the tail circuits can also be multipoint. In either case the overall circuit lengths can be greatly reduced. Compared to point-to-point, it is important that clocking between higher speed tail circuit modems are synchronised.

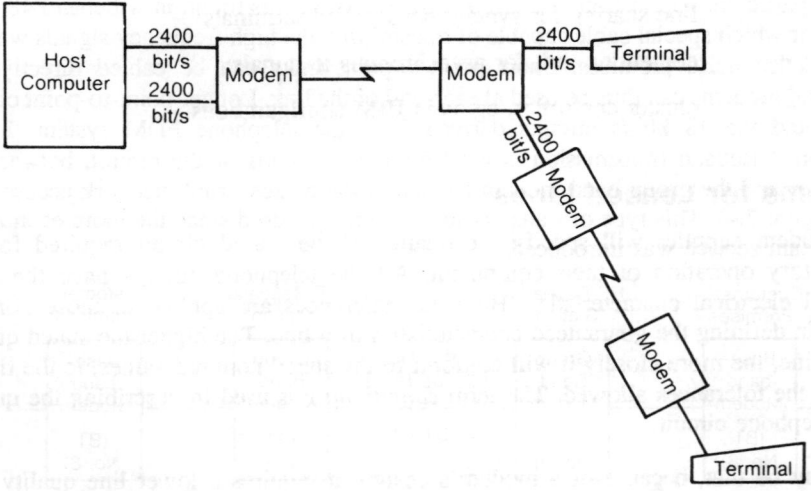

Figure 7.3   Multiport (multinode) modems

Dial Back-Up

Where a data link is crucial to an organisation's day to day work, it is important that an alternative service should be available. One approach would be the provision of a dial back-up capability making use of the PSTN network. Full duplex on two wires can be provided by using a V.22 bis (2400 bps) or V.32 (9600 bps) modem.

Some dial-up circuits may not be of sufficient quality to support the original data rate. Many modems have controls which allow them to operate at lower speeds to cover this eventuality. For example, it may only be possible to operate a 9600 bit/s modem at its *fall-back* data rate of 4800 bit/s.

Baseband modems

A baseband modem does not modulate or demodulate a carrier signal in the way that a true modem does. It merely takes the data binary signal and transmits it to line in a modified form, as a baseband signal. (A baseband signal, it will be recalled, is a signal containing frequencies down to DC.) The modification takes the form of pulse shaping — rounding off the square pulses to reduce the high frequency components and scrambling to give the baseband signal a satisfactory frequency spectrum independent of the data binary signal.

Baseband modems require a physical pair of wires over which to transmit, since the baseband signal contains frequencies outside the normal 300-3400 Hz speech band. They are therefore only suitable for short-haul applications where an unloaded physical pair is available.

One particular application of baseband transmission was found in wideband 48 kbit/s links, for which special cable capable of transmitting the high frequency signals was laid to the customer's premises. Short point-to-point links can be cabled directly, and baseband modems can thus be used at each end of the link. Longer point-to-point circuits are routed via 48 kbit/s circuits derived from the telephone FDM system. In this situation, baseband transmission is used for the two ends of the circuit, between the customer and the group band modem located at the nearest trunk network access point (see Figure 7.4). This type of transmission is no longer used since the more economical Kilostream service was introduced.

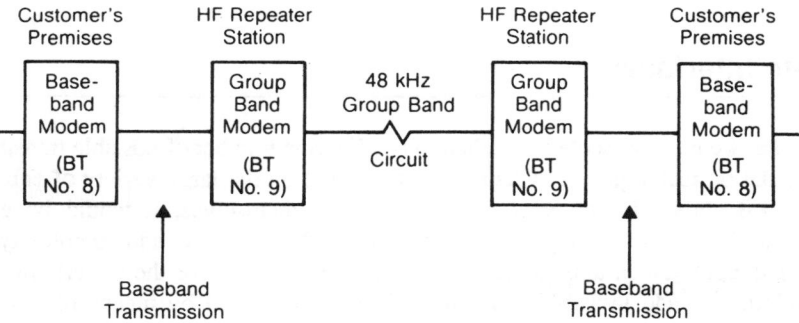

Figure 7.4   Modems in a 48 kbit/s link

Limited Distance Modems

The modems standardised by the CCITT are designated for international and intercontinental use. Over shorter distances it is possible to achieve equivalent performance using less sophisticated modulation/demodulation and equalisation techniques, and there is now a number of limited distance cheap modems offering very good price/performance ratios. Typically these modems have a range of about 50 miles, and operate up to 4800 bit/s. These can be run on a cheaper grade of BT line and are very economical.

However, not all 'short-haul' or 'limited-distance' modems are true modems. Many are baseband modems and can only be used when unloaded physical pairs are available, which limits their range considerably.

Modem Eliminators and Cross-Over Cables

Modem eliminators are employed for in-house transmission up to a mile or so, where the use of full modems is not justified. A modem eliminator is a line driver/receiver packaged to look like a modem and providing the same data interface circuits as a modem, including the clock information that a synchronous modem would normally supply, should this be the method of operation. Thus, eliminators require external power, although they consume so little that it is possible to supply this over one of the other interface circuits.

When two devices that would normally communicate via modems are brought into closer proximity they can be connected by a simple cable. However, this does have to be specially adapted to ensure that the *data transmit* line from one device is attached to the *data receive* of the other (and vice versa), as well as crossing over some of the other control circuits, eg RTS to Carrier Detect. (A full description of the function of interface circuits is given later in this chapter.) Hence, this type of cable is called a *cross-over* cable, or is sometimes referred to as a *null modem*. If the devices (DTEs) are synchronous a clock is required, which would normally come from the modem. A simple 'clock' can be introduced into the circuit or one device may be capable of clocking the other.

Modem Interfaces

So far, what we have presented is perhaps a bewildering number of possible transmission and modulation techniques (in Chapter 3), supported on the large variety of equipment as described above. In terms of the data devices themselves, it would be entirely undesirable if all the technological permutations available were to add complexity to the task of data communications. Ideally, these devices should have the underlying details of modulations made invisible to them, and this is achieved by means of a standard communications interface with the modem. Furthermore, in order that a myriad of different interfaces do not substitute one complexity for another, a set of standard interfaces require definition.

The concept of an interface is dealt with more completely in the next chapter, but standard modem interfaces are discussed here. In the context of an interaction between a data device and a modem, the interface is the point past which the data device need no longer worry about the intricacies of the actual transmission. The data device is only required to comply with the rules governing the interface and this being the case, can expect the modem to perform the necessary function.

The primary standardisation body for telecommunications (particularly in the area of modems) is the CCITT, which does not publish standards as such, but issues them in the form of recommendations. Adoption of these is not mandatory, but the PTTs generally do so, and so do the manufacturers. In addition, it makes good commercial sense.

CCITT have devised the following nomenclature for use in the context of data communications interfaces:

- DTE stands for data terminal equipment, which could be a computer, a communication processor terminal controller, PC or terminal, etc.

- DCE stands for data circuit terminating equipment (the modem) which provides the interface between the DTE and the physical transmission circuit, or network.

The standard modem interfaces all refer to the DTE and DCE functions. During the discussion of DTE/DCE interfaces that follows it is worth bearing in mind that this is for the case where the path between the two DCEs is a straightforward link. The situation is slightly different (particularly with regard to link control) when the path is actually a dynamically routed network such as is described for X.25 (in Chapter 10). Finally, it is also worth mentioning that the standard interfaces presented here are not exclusively restricted to modem equipment and are, in fact, to be found on any number of communications devices. Mainly, however, it is DTE to modem.

The DTE/DCE Interface

The boundary between the DTE and the DCE is at the connector by which they are linked. However, a complete description of this boundary goes beyond the physical attributes of the connector and includes protocols governing the exchange of electrical signals as described later in this section. The electrical signals fall into two main categories: those which perform control functions such as causing some event to happen or seeking or communicating status information to the other party; and signals which are transporting data. So far as the DCE and the communications link are concerned, the meaning and content of the data are irrelevant. The interfaces and relationships for two communicating DTEs are shown in Figure 7.5. This figure also illustrates that at the signalling level the total path comprises three segments, along each of which there has to be exchange of signals according to prescribed rules. These are:

- two DTE/DCE paths; and

- DCE/DCE path spanning the transmission circuit.

Also shown is the end-to-end link, the control of which, in this scenario, is the responsibility of a link level protocol (see Chapter 8).

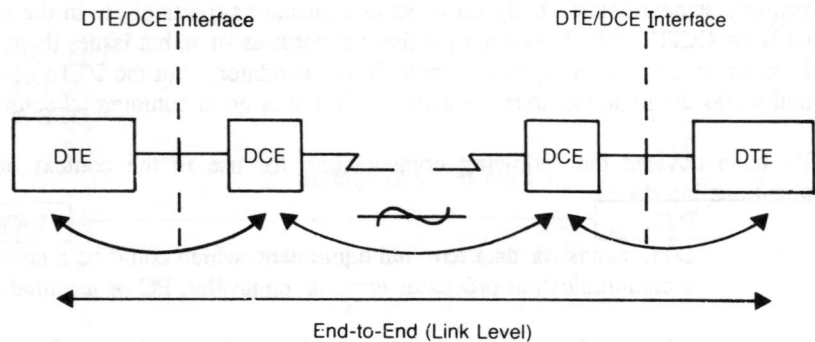

Figure 7.5 The physical DTE/DCE path

A Simple Example

Before discussing the CCITT recommendations we present a simple example in the form of the conversation of the signalling interchange or handshaking procedures involved in transmitting data over the PSTN.

Step 1

The modems at each end must be advised that the data terminal equipment (DTE) is ready to operate. The modems must also be switched to the line. These two requirements could be achieved in a number of ways, but the two most common methods are shown below as steps 1(a) and 1(b).

(a)   A single signal can be sent from each data terminal equipment to its associated modem.

        A and B

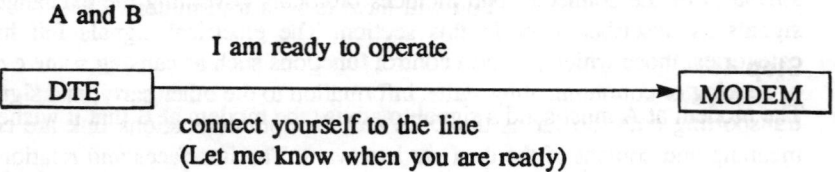

(b)   The DTE can merely advise its modem that it is ready to operate. The switching of the modem to the line will not take place until it receives a signal from another source (eg from the DTE or, in earlier days, a switch on an associated telephone).

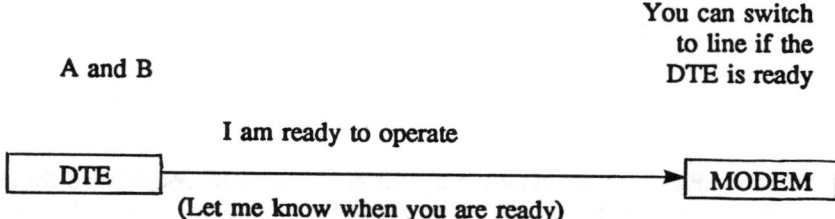

Step 2

The modems at each end advise their associated DTEs that they have switched to line and are ready to accept further instructions.

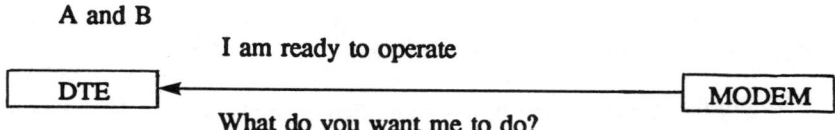

Step 3

No indication has yet been given to the direction of the transmission which is to take place. The modem at A must be conditioned to transmit and the modem at B conditioned to receive. The first action must be from the DTE at A.

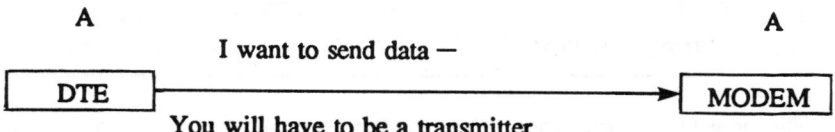

Step 4

The modem at A must send a signal to advise the modem at B that it wishes to transmit.

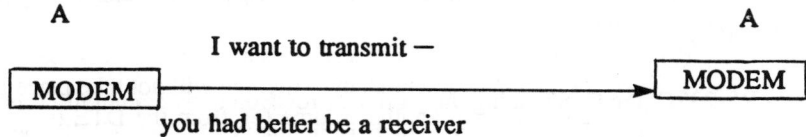

Step 5

When the modem B is satisfied that it is receiving a satisfactory line signal from A it will advise its DTE to organise itself as a receiver.

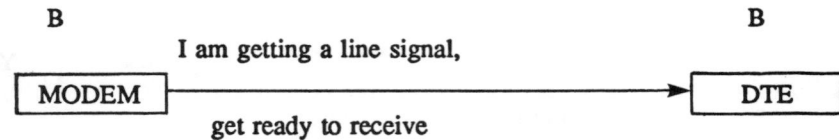

Step 6

After a preset time to allow the distant end time to be conditioned for receiving, the Modem at A can give an affirmative response to requests to send data from its DTE.

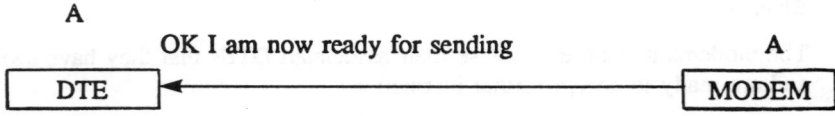

We will assume that a block of data is successfully passed from A to B. The DTE at B will assemble a receipt message and steps 3-6 will be followed again in the reverse direction beginning with a request to send.

It will be seen that, even in this simplified example, a strict discipline is necessary to bridge the demarcation successfully between modems and the terminals which work with them. It will also be clear that standardisation of the interface is essential if there is to be compatibility between a wide range of different equipments. (Note that this example relates to half duplex transmission. With full duplex, modems can transmit simultaneously.)

The Modem Interface and the CCITT V-Series Recommendations

The interfacing functions of a modem are additional to the basic modulation/demodulation function. The various electrical signals are carried on a number of *interchange circuits* and these, together with a number of other features summarised in Table 7.1, comprise the specification of the interface.

These features have all been standardised internationally, the standards being known generally as V.24 and RS-232.

RS-232C (the C indicating the particular revision most commonly found, although the current revision, D is the one which is in line with the CCITT specifications) is a recommended standard of the US Electronic Industries Association (EIA). The standard is in widespread use in America, and formed the basis of CCITT Recommendation V.24, which is common in Europe. RS-232C defines all the features listed in Table 7.1, and is therefore a complete interface specification. Recommendation V.24, however, lists only the DTE/DCE interchange circuits and their functions; the electrical characteristics are defined in another CCITT recommendation, V.28, and the connector pin allocations in an international standard (ISO 2110). Nevertheless, conformity to these other two standards is usually implied when referring to a V.24 interface. For most purposes, V.24 and RS-232C can be regarded as synonymous.

Table 7.1   The modem interface

Physical Attributes	*Logical Attributes*
Dimensions and construction of connector	Meaning of the electrical signals on each pin
Number of pins in connector	Interrelationship between signals
Electrical signals on the pins	Procedures for exchanging information between DTE and DCE

It is neither practicable nor necessary for our purposes to give a detailed and comprehensive account of these interface specifications, and our primary objective will be to highlight their main features, and convey a broad picture of how they work. Those readers who are interested in a more extended treatment, and many of the details not covered here, are referred to the bibliography.

Interchange Circuits

In a V.24 interface, signals between the terminal and the modem are carried on separate interchange circuits. One interchange circuit is provided for each function. There are more than 40 interchange circuits in all, which seems excessive at first sight, but then V.24 is a general purpose collection of interface circuits covering a wide range of modem applications, and no single modem would use all the interchange circuits. (Note that the usual V.24 configuration is in association with a standard ISO 2110 connector, which only has provision for 25 pins.)

There are in fact two sets of interchange circuits, the 100 series used for data, timing and control circuits, and the 200 series used for automatic calling, the procedure of which is detailed in the CCITT V.25 Recommendation. Of the 100 series circuits, there is a core of eight circuits which is common to many applications, and these are listed in Table 7.2.

The operation of these circuits can best be understood using the earlier example relating to data transmission over the PSTN. This is a simplified description aimed at giving an overall view of the procedure; a detailed description of the operation of each circuit is given later.

When a data connection is required over the PSTN, the first step is to dial the required number on the telephone in the usual way. When the call is answered and a communication path exists between the two parties, the telephone line needs to be switched at each end, from the telephone to the data terminal. This is done by the modem once the data terminal has turned on circuit 108. (There are two slightly different ways of using this circuit as described later. The circuit is designated Connect Data Set to Line (108/1) or Data Terminal Ready (108/2) to distinguish the two.)

Table 7.2 100 Series core interchange circuits

Circuit V.24	Designation	Direction	
		To Modem	*To Terminal*
102	Signal ground or common return		
103	Transmitted data	X	
104	Received data		X
105	Request to send	X	
106	Ready for sending (Clear to send)		X
107	Data set ready		X
108/1	Connect data set to line	X	
or			
108/2	Data terminal ready	X	
109	Data channel received line signal (ie carrier) detector		X

When connected to line, the modem informs the terminal by turning on the Data Set Ready circuit 107. If the distant modem is already connected to line and is transmitting, the local modem will turn on the Carrier Detect circuit 109 to indicate that it is detecting a carrier signal. This is often indicated by a light on the terminal. Any data that is received will be passed to the terminal over the Receive Data circuit 104, Figure 7.6 shows this sequence of events.

If a terminal wishes to transmit, it turns on the Request to Send circuit 105. When the modem is ready to accept data for transmission, it replies by turning on the Clear to Send circuit 106. Data can then be transmitted by the terminal on the Transmit Data circuit 103.

The one circuit not mentioned so far is the Signal Ground circuit 102. This provides the essential common return lead for all the other interchange circuits.

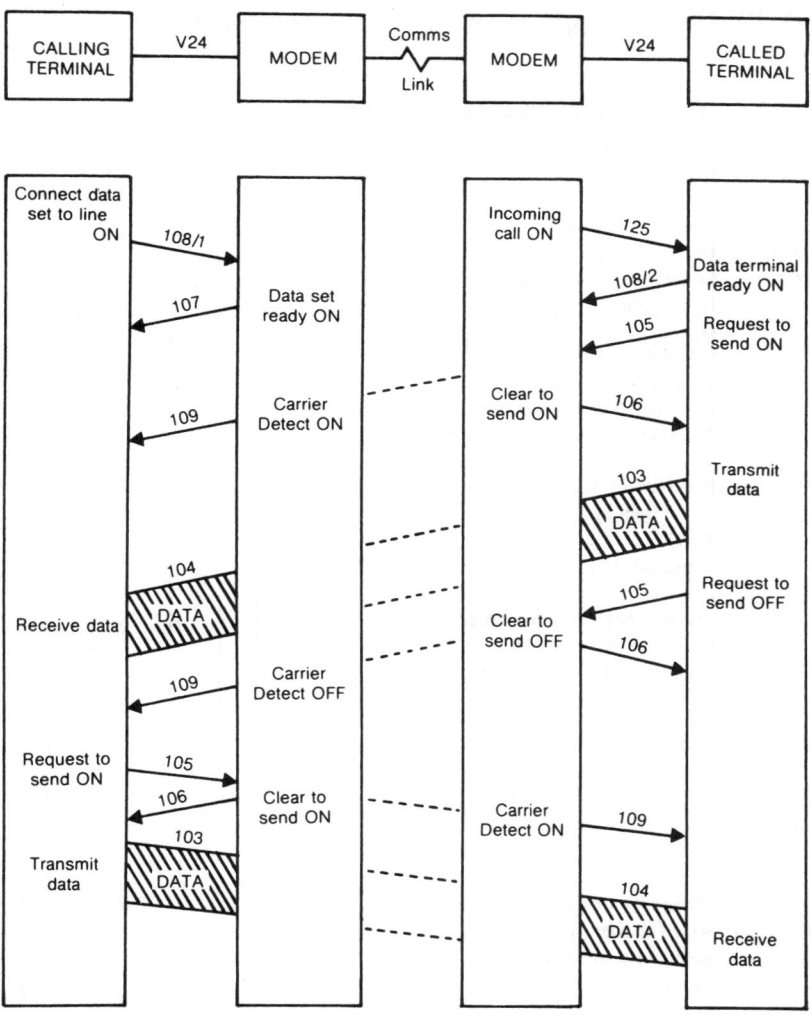

Figure 7.6 V.24 interface procedure

Table 7.3 100 Series interchange circuits by category

Interchange Circuit Number	Interchange Circuit Name	Ground	Data		Control		Timing	
			From DCE	To DCE	From DCE	To DCE	From DCE	To DCE
102	Signal ground or common return	X						
103	Transmitted Data			X				
104	Received Data		X					
105	Request to send					X		
106	Ready for sending				X			
107	Data set ready				X			
108/1	Connect data set to line					X		
108/2	Data terminal ready					X		
109	Data channel received line signal detector				X			
110	Signal quality detector				X			
111	Data signalling rate selector (DTE)					X		
112	Data signalling rate selector (DCE)				X			
113	Transmitter signal element timing (DTE)							X
114	Transmitter signal element timing (DCE)						X	
115	Receiver signal element timing (DCE)						X	
116	Select standby					X		
117	Standby indicator				X			
118	Transmitted backward channel data			X				
119	Recived bacward channel data		X					
120	Transmit backward channel line signal					X		
121	Backward channel ready				X			
122	Backward channel recieved line signal detector				X			
123	Backward channel signal quality detector				X			
124	Select frequency groups					X		
125	Calling indicator				X			
126	Select transmit frequency					X		
127	Select receive frequency					X		
128	Receiver signal element timing (DTE)							X
129	Request to receive					X		
130	Transmit backward tone					X		
131	Received character timing						X	
132	Return to non-data mode					X		
133	Ready for receiving					X		
134	Received data present				X			
136	New Signal					X		
140	Loopback/ Maintenace test					X		
141	Local Loopback					X		
142	Test indicator				X			
191	Transmitted voice answer					X		
192	Received voice answer				X			

Table 7.4 Interchange circuits essential for V.23 modems when using the public switched telephone network (PSTN)

(Note 2)

No.	Designation	Forward (Data) Channel One-Way System				Forward (Data) Chennel Either Way System	
		Without Backward Channel		With Backward Channel		Without Backward Channel	With Backward Channel
		Transmit End	Receive End	Transmit End	Receive End		
101a	Protective ground or earth	x	x	x	x	x	x
102	Signal ground or common return	x	x	x	x	x	x
103	Transmitted data	x	-	x	-	x	x
104	Received data	-	x	-	x	x	x
105	Request to send	-	-	-	-	x	x
106	Ready for sending	x	-	x	-	x	x
107	Data set ready	x	x	x	x	x	x
108/1 or	Connect data set to line	x	x	x	x	x	x
108/2 (Note1)	Data terminal ready	x	x	x	x	x	x
109	Data channel received line signal detector	-	x	-	x	x	x
11	Data signalling rate selector (DTE)	x	x	x	x	x	x
114 (Note 3)	Transmitter signal element timing (DCE)	x	-	x	-	x	x
115 (Note 3)	Receiver signal element timing (DCE)	-	x	-	x	x	x
118	Transmitted backward channel data	-	-	-	x	-	x
119	Received backward channel data	-	-	x	-	-	x
120	Transmit backward channel line signal	-	-	-	x	-	x
121	Backward channel ready	-	-	-	x	-	x
122	Backward channel received line signalling detector	-	-	x	-	-	x
125	Calling indicator	x	x	x	x	x	x

a May be excluded if so required by local safety regulations.

a) May be excluded if so required by local safety regulations.

NOTE 1   This circuit shall be capable of operation as circuit 108/1 (connect data set to line) or circuit 108/2 (data terminal ready) depending on its use. For automatic calling it shall be used as 108/2 only.

NOTE 2   Interchange circuits indicated by X must be properly terminated according to Recommendation V24 in the data terminal equipment and data circuit-terminating equipment.

NOTE 3   These circuits are required when the optional clock is implemented in the modem.

The 100 Series Interchange Circuits

Table 7.3 presents a list of the 100 series interchange circuits. This is to be regarded as a shopping list, and which circuits listed are actually required will depend upon the modem used and the facilities to be provided. The table classifies the signals into categories, and for each circuit indicates whether the DCE is either the source or the recipient of the signal. If the former, then the DTE will be the recipient and if the latter, the DTE will be the source.

Each recommendation also stipulates which of the 100 series interchange circuits are needed. For example, Tables 7.4 and 7.5 show the interchange circuits which are essential for the 600/1200 bit/s modems covered by CCITT Recommendation V.23 when these are used on the general switched telephone network and on non-switched leased telephone circuits.

It will be seen that between 8 and 18 of the interchange circuits are necessary depending on the facilities required.

Notes on the 100 Series Interchange Circuits

Some explanatory notes on the thirty seven 100 series interchange circuits defined in the recommendations are given below. Although it is hoped that these will be helpful to the reader in understanding the functions of the various circuits, they should not be taken as a substitute for the recommendation itself.

Because the data communications equipment (DCE) referred to in the recommendation is usually a modem, the term modem has been used in the explanatory notes on this series. The on and off conditions referred to are logical conditions; on is given the binary value 0 and is expressed by a positive voltage; off is given the binary value 1 and is expressed by a negative voltage.

Table 7.5 Interchange circuits essential for V.23 modems when used on non-switched leased telephone circuits

(Note 2)

No.	Designation	Forward (Data) Channel One-Way System				Forward (Data) Channel Either Way or Both Ways Simultaneously System	
		Without Backward Channel		With Backward Channel		Without Backward Channel	With Backward Channel
		Transmit End	Receive End	Transmit End	Receive End		
101a	Protective ground or earth	x	x	x	x	x	x
102	Signal ground or common return	x	x	x	x	x	x
103	Transmitted data	x	-	x	-	x	x
104	Received data	-	x	-	x	x	x
105	Request to send	-	-	-	-	x	x
106	Ready for sending	x	-	x	-	x	x
107	Data set ready	x	x	x	x	x	x
108/1 or	Connect data set to line	x	x	x	x	x	x
108/2 (Note1)	Data terminal ready	x	x	x	x	x	x
109	Data channel received line signal detector	-	x	-	x	x	x
111	Data signalling rate selector (DTE)	x	x	x	x	x	x
114 (Note 3)	Transmitter signal element timing (DCE)	x	-	x	-	x	x
115 (Note 3)	Receiver signal element timing (DCE)	-	x	-	x	x	x
118	Transmitted backward channel data	-	-	-	x	-	x
119	Received backward channel data	-	-	x	-	-	x
120	Transmit backward channel line signal	-	-	-	x	-	x
121	Backward channel ready	-	-	-	x	-	x
122	Backward channel received line signalling detector	-	-	x	-	-	x

a May be excluded if so required by local safety regulations.

NOTE 1 This circuit shall be capable of operation as curcuit 108/1 (connect data set to line) or circuit 108/2 (data terminal ready) depending on its use. For automatic calling it shall be used as 108/2 only

NOTE 2 Interchange circuits indicated by X must be properly terminated according to Recommendation V24 in the data terminal equipment and data circuit-terminating equipment.

NOTE 3 These circuits are required when the optional clock is implemented in the modem.

Circuit 101 — Protective ground or earth

This circuit was used to extend the protective earth condition from the DTE to the modem but it is no longer specified.

Circuit 102 — Signal ground or common return

This circuit is the common earth return for the signals on all interchange circuits (except 101). At the modem, this circuit should be connected to protective ground or earth by a wire strap.

Circuit 103 — Transmitted data (DTE ———→ modem)

Data signals transmitted from the DTE to the modem are passed over this circuit.

When no data is being transmitted, circuit 103 is held in the off condition (binary 1). The DTE cannot transmit data on this circuit unless an on condition (binary 0) is present on all the following circuits, where these are implemented:

- Circuit 105   — Request to send
- Circuit 106   — Ready for sending
- Circuit 107   — Data set ready
- Circuit 108/1/2   — Connect data set to line/data terminal ready.

This condition is the result of the hand-shaking procedure shown in Figure 7.7. Transmission of data continues until one of these circuits is turned off. Normally the terminal would switch off either 105 to signify that it had finished transmitting and was ready to receive or 108/1, 108/2 to signify the end of a call.

Circuit 104 — Receive data (modem ———→ DTE)

Analogue signals received from the line are converted into digital signals and transmitted from the modem to the DTE on this circuit.

NB: To prevent spurious signals being sent to the DTE due, for example, to excessive noise on the line, this circuit may be held in the off condition until an on condition on circuit 109 indicates that a signal within appropriate limits is being received from the line (see circuit 109). This procedure is termed *clamping*.

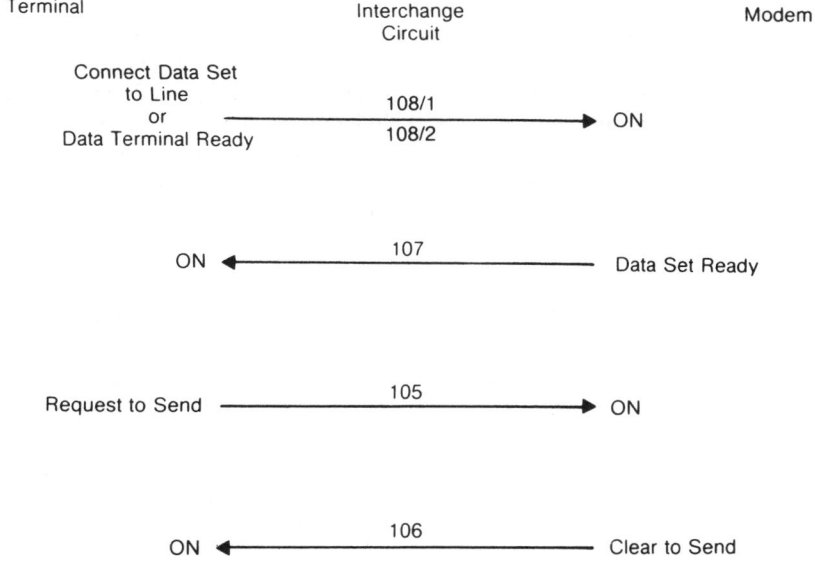

Figure 7.7 V.24 Initial handshake

Circuit 104 may also be clamped to binary 1 in half duplex operation wherever the terminal is transmitting, ie whenever 105 is on. This prevents the transmitted signal being fed back to the terminal. Removal of this clamp may be delayed for up to 175 ms after 105 is turned off, to allow for completion of transmission and to protect the terminal from false signals such as synchronisation sequences.

Circuit 105 — Request to send (DTE ⟶ DTE)

When the DTE wishes to send information, it applies an on condition on this circuit causing the modem to assume the transmit code. The off condition signifies that the DTE does not wish to transmit.

Circuit 106 — Ready for sending (modem ⟶ DTE)

Replies from the modem to the DTE, in response to a request to send, are passed on this circuit. An on condition indicates that the modem is ready to accept data from the DTE on circuit 103 (transmitted data).

The delay between an on condition being applied by the DTE to circuit 105 and the answering on condition being sent from the modem to the DTE is termed the *ready-for-sending delay*. This delay is built into the transmit modem to give time for a distant modem to condition itself to receive signals from the line.

There are number of factors which can influence the ready-for-sending delay required, and the number of times that this delay occurs:

      —   If automatic calling and answering are used, a time allowance is required to allow the calling condition to be detected at the receive modem; the detection of the calling signal (typically ringing current) allows the call to be automatically answered. For example, on the 600/1200 bit/s modem designed to CCITT Recommendation V.23, the initial ready-for-sending delay is between 750 and 1400 ms; after the call has been established, this time is reduced to between 20 and 40 ms.

      —   If uninterrupted, transmission is maintained in one direction (simplex) or both directions simultaneously (true or full duplex), there are no ready-for-sending delays after the first one. In these circumstances, whenever transmission restarts, there is further ready-for-sending delay.

      —   If the modems and/or the lines cannot give full duplex facilities and a half duplex method of transmission is adopted, it is necessary to 'turn round' the modems at each end whenever there is a change in the direction of the transmission. Ready-for-sending delays are therefore incurred for each change of direction. To avoid this delay, many applications use full duplex circuits even though the data flow is essentially half duplex. On multipoint circuits 'carrier' can be kept high from the primary end to avoid a 'turn round'. Of course, from the secondaries it must be switched. However, this procedure does not make optimum use of the full duplex channel capacity. As we shall see later, this limitation can be overcome using HDLC Link control protocols, and then only with the point-to-point version.

      —   Some modems require longer ready-for-sending delays than others. Generally speaking, the slower, asynchronous or unlocked modems merely require a carrier to be on the line for a brief period (typically 20-40 ms) to ensure that the receive modem is receiving it at the right power level.

The ready-for-sending delay is, therefore, the same whether the request to send is the first one in a transmission or one which occurs subsequently. The higher speed modems (over 2400 bit/s) are usually synchronous and may employ adaptive equalisers. An *initial training pattern* is sometimes sent which enables the receive modem to synchronise with the transmit modem and to allow time for the adaptive equalisers in the receive modem to adjust to the line conditions. Subsequent training patterns between blocks may be considerably shorter as the timing and equalisation elements in modern modems remain reasonably stable and require little adjustment once they have been set.

Circuit 107 — Data set ready (modem ⟶ DTE)

The signals on this circuit indicate to the DTE that the modem is ready to receive its next instruction. The transmission of line signals for equalisation, etc, will not take place unless this circuit is switched on.

Circuit 108 (DTE ⟶ modem)

There are two options for the use of this circuit to meet different user requirements:

— *108/1 — Connect DTE to line.* This circuit gives the terminal direct control over switching the modem to the telephone line, for connections set up over the PSTN. The call is dialled with circuit 108/1 off, and when the call is answered the circuit is turned on by the terminal. The on condition will immediately connect the modem to line, which results in circuit 107 (data set ready) being switched to on. An off condition of 108/1 disconnects the modem from the line when data on circuit 103 has been transmitted.

— *108/2 — Data terminal ready.* This circuit provides the terminal with indirect control over switching the modem to line. An on condition on this circuit from the DTE informs the modem that the data terminal is ready to operate. It is not in itself an instruction to connect the modem to line and before this can be done a subsidiary signal is necessary. This signal may be given by an operator pressing a data button on the telephone associated with the modem. This alternative would also be implemented when the modem has automatic answering facilities. Under this arrangement a terminal ready to receive incoming calls maintains 108/2 on, so that when a call is received and Calling Indicator circuit 125 is turned on, the modem is automatically switched to line. This happens at the end of the first cycle of ringing tone (see description of 125) and is indicated to the terminal by circuit 107 coming on. Circuit 108/2 cannot be permanently strapped on, because the DTE needs it in order to clear down PSTN calls.

Circuit 109 — Data channel received line signal detector (modem ⟶ DTE) (or Carrier Detect)

An on condition on this circuit indicates to the DTE that the received signal is within the appropriate limits. These limits are specified in the CCITT recommendations for the type of modem being used. For example, V.23 (600/1200 bits/s) will apply an on condition on circuit 109 when a signal greater than 43 dBm is received. This circuit may be *clamped* to circuit 104 to avoid false signals being passed to the DTE.

Circuit 110 — Data signal quality detector

Data signal quality detection is a method of error control whereby the line signal is checked for certain characteristics which are likely to cause errors. An on condition on this circuit indicates that there is no indication from the line signal that an error has occurred. An off condition indicates that there is a reasonable probability that the distortion detected on the line signal will cause an error.

Circuit 111 — Data signalling rate selector (DTE ⟶ modem)

When modems offer a choice of two fixed data signalling rates or a choice between two ranges of data signalling rate, the selection is usually made at the DTE. An on condition on circuit 111 from the DTE directs the modem to adopt the higher rate, or range of rates and an off condition indicates that the lower mode is selected. Either (but not both) circuit 111 or circuit 112 can be used for data signalling rate selection.

Circuit 112 — Data signalling rate selector (modem ⟶ DTE)

This is an alternative to circuit 111, choice of data signalling rate being made from the modem.

Circuits 113, 114, 115, 128

These circuits provide alternative ways of maintaining bit synchronisation in synchronous mode operation. For convenience we describe their operation under the separate section headed synchronisation.

Circuit 116 — Select standby (DTE ⟶ modem)

When standby communication facilities are provided, such as standby exchange line, this circuit may be used for selection between the normal and standby facilities. For example, some modems operating at 2400 bit/s over leased lines offer fall-back operation at 1200/600 bit/s over PSTN. Circuit 116 is used to select normal or fall-back operation. For the latter circuit 111 would be used to determine whether to operate at 600 or 1200 bit/s.

Circuit 117 — Standby indicator (modem ⟶ DTE)

The signals on this circuit indicate to the DTE whether the normal or standby facilities which have been selected are conditioned to operate. For both circuits the on condition is used for standby, and the off condition for normal operation.

Circuits 118 to 124 inclusive

These are used only when *backward channels* are provided by the modems. A backward or *supervisory* channel operates at a lower data signalling rate (typically 75 bit/s) than the data channel and is intended to be used for the return of short supervisory or error control messages; it is also used extensively to provide a slow speed (75 bps) channel from a terminal to a videotext host, whilst the 1200 bps channel is used for the host information flowing to the terminal. The interchange circuits used are equivalent to other circuits described above except that they are associated with the backward channel rather than the data channel. Another major use has been for modem management systems, etc, so that, for example, alarms from modems are sent over the same network to a central management system.

Circuit 118 — Transmitted backward channel data (DTE ———→ modem)

Equivalent to circuit 103 (transmitted data).

Circuit 119 — Received backward channel data (modem ———→ modem)

Equivalent to circuit 104 (received data). This circuit may be clamped to circuit 122 just as circuit 104 may be clamped to circuit 109.

Circuit 120 — Transmit backward channel line signal (DTE ———→ modem)

Equivalent to circuit 105 (request to send).

Circuit 121 — Backward channel ready (modem ———→ DTE)

Equivalent to circuit 106 (ready for sending).

Circuit 122 — Backward channel received line signal detector (modem ———→ DTE)

Equivalent to circuit 109 (data channel received line signal detector).

Circuit 123 — Backward channel signal quality detector (modem ———→ DTE)

Equivalent to circuit 110 (data signal quality detector).

Circuit 124 — Select frequency groups (DTE ———→ modem)

Signals on this circuit are used to select the desired frequency groups available in modems designed for parallel data transmission.

Circuit 125 — Calling indicator (modem ———→ DTE)

An on condition on this circuit notifies the DTE that a calling signal is being received from the line. This circuit is used for automatic answering and alerts the terminal to an incoming call. The circuit reacts in sympathy with the ringing current, turning on during rings. Figure 7.8 indicates the UK ringing tone. Its use in conjunction with 108/2 is described above.

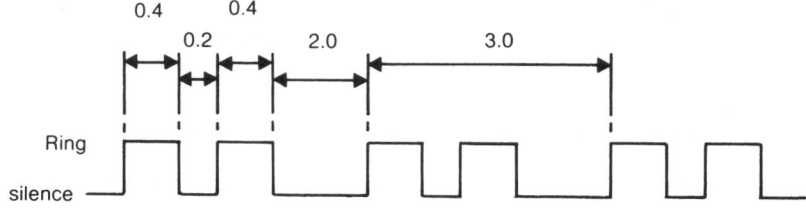

Figure 7.8 UK Ringing tone (times in seconds)

Circuit 125 is independent of the other interchange circuits, and remains operational during modem testing.

Circuit 126 — Select transmit frequency (DTE ———→ modem)

Circuit 127 — Select receive frequency (DTE ———→ modem)

These circuits were designed for the standard 20-300 baud V.21 modem, which uses different transmit frequencies for the two directions of transmission. Usually on PSTN connection, however, the choice of frequencies is made automatically by the modem, depending on whether it is the called or calling party, and these circuits are not required. On some multipoint applications control of the modem frequencies by the terminal may be required, but often circuit 126 will be used to control both transmit and receive frequencies and circuit 127 will not be implemented. For both circuits, on condition signifies the higher and the off condition the lower frequency.

Circuit 128 — Receiver signal element timing (DTE ———→ modem)

Use of this circuit is described under a separate heading — Synchronisation.

Circuit 129 — Request to receive (DTE ———→ modem)

Signals on this circuit are used to control the receive function of modems used for parallel data transmission. The on condition causes the modem to assume the receive mode.

Circuit 130 — Transmit backward tone (DTE ———→ modem)

An on condition on this circuit conditions the modem to transmit a single tone on the backward channel. A potential use of this circuit is in a system using push button MF telephones. The in-station modem could be conditioned to transmit a single tone as an audible acknowledgement to a person listening on the telephone that data had been received correctly.

Circuit 131 — Received character timing (modem ———→ DTE)

Any signals on this circuit provide the DTE with character timing information. This information cannot normally be provided. Most modems transmit and receive data serially bit by bit and do not know when characters begin and end. The circuit is only used in conjunction with the parallel modems which accept data from the line and pass it to the DTE a character at a time.

Circuit 132 — Return to non-data mode (DTE ———→ modem)

An on condition on this circuit instructs the modem to return to a non-data mode (eg a telephone) without losing the line connection to a remote station.

Circuit 133 — Ready for receiving (DTE ————→ Intermediate equipment)

This circuit is optional when there is intermediate equipment between the DTE and the modem (eg error control equipment to CCITT Recommendation V.41). An on condition on circuit 133 is an indication to the intermediate equipment that the DTE is ready to receive a block of data on circuit 104 (received data).

Circuit 134 — Received data present (Intermediate equipment ————→ DTE)

This circuit is only used when error control equipment is provided between the modem and the DTE. The intermediate equipment notifies the DTE on circuit 134 which of the bits in a block transferred on circuit 104 are information (on condition on circuit 134) or supervisory (off condition on circuit 134).

Circuits 140, 141, 142

These circuits enable loopback tests to be carried out, in which the transmitted data is looped back so that it appears on the received data path. This provides an effective means of fault isolation. On a simple point-to-point connection, there are four locations at which loopback can be conveniently arranged to supply helpful diagnostic information, and these are shown in Figure 7.9.

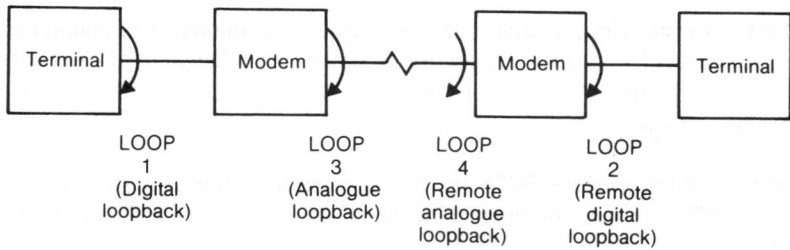

| LOOP 1 (Digital loopback) | LOOP 3 (Analogue loopback) | LOOP 4 (Remote analogue loopback) | LOOP 2 (Remote digital loopback) |

Figure 7.9 Loopback points

The method of operation is as follows:

Loop 1 checks the terminal, loop 3 the local modem, and loop 2 the remote modem. These loops may be activated by a switch on the modem, or via a terminal-modem interchange circuit. Loop 4 (which is only possible on four-wire circuits) is for PTT use only. The circuits provided in V.24 are:

— Terminal to Modem:

    — activates loop 2 in Figure 7.9

— Local loopback circuit 141

    — activates loop 3 in Figure 7.9

—   Modem to Terminal:

Test indicator circuit 142

The modem switches this circuit on in response to a loopback command from the terminal on circuit 140 or 141. Circuit 142 is also switched on when the modem is tested from a remote location. Data transmission is impossible when circuit 142 is on.

As data networks have grown in complexity, the need for adequate monitoring and diagnostic aids has grown in importance. Many modem manufacturers now provide centralised network control equipment which is able to communicate with all the modems on a network. With such equipment it becomes possible to monitor the V.24 interface circuits at each modem perform loopback tests, switch in standby modems, change over to spare circuits or alter data transmission rates, and all remotely. Communication can take place over the modem's low speed secondary or backward channel, or over a separate dialled connection, and so can occur while data is being transmitted on the main channel.

The 200 Series Interchange Circuits

The 200 series of interchange circuits are all related to the operation of automatic calling over the public switched telephone network. A similar interchange of circuits is provided for automatic calling over the Telex network; these are covered in CCITT Recommendation V.2.

Automatic calling over the PSTN involves a disciplined interchange of responsibility between the DTE (which is computer related equipment of some kind) and the automatic calling equipment.

Both automatic calling and automatic answering present particular problems in connection with the transmission of data on the public telephone network. This arises partly because the speech network was not designed for transmitting data, and also because of the need to ensure that, on the one hand signals are not generated which interfere with the operation of the network, and on the other, both nationally and internationally agreed signalling schemes are adhered to.

A good illustration is echo suppression and echo suppressor disablement on long distance (for example, transcontinental and transoceanic) circuits. Echo suppressors are introduced on long distance circuits to minimise the effect in which a person talking hears an echo of his own voice. For data transmission the effects are not as serious as the countermeasures. These take the form of echo suppressors, which, although preventing echoes on speech circuits, also prohibit full duplex data transmission. Modern echo suppressors are therefore fitted with a disabling mechanism which can be activated by a signal from a remote modem. By international agreement a signal frequency of 2100 Hz is used, both for answering and echo suppressor disablement.

There are twelve interchange circuits in the 200 series: four for data, seven for control and indication, and the common return circuit. These are listed in Table 7.6.

Table 7.6 200 Series interchange circuits in V.24 for automatic calling

		Direction	
Circuit	*Designation*	*To Terminal*	*From Terminal*
201	Common return		
202	Call request		X
203	Data line occupied	X	
204	Distant station connected	X	
205	Abandon call	X	
206	Digit signal 2^0		X
207	Digit signal 2^1		X
208	Digit signal 2^2		X
209	Digit signal 2^3		X
210	Present next digit	X	
211	Digit present		X
213	Power indication	X	

Method of Operation

The automatic calling unit allows calls to be set-up over the PSTN without manual intervention. The unit is usually part of the modem system.

The DTE passes the digits to be dialled to the automatic calling unit, and the unit converts these to dial pulses (or to multi-frequency tones for telephone networks so equipped). Having sent all the digits, the unit causes the modem to transmit a calling signal to line to announce the fact that the call is being originated automatically. The calling signal comprises short bursts of 1300 Hz or other binary 1 tone repeated every 1.5 to 2 seconds.

The automatic calling unit relies on detecting a 2100 Hz answering tone before it will connect the calling DTE to line. If no such tone is received within a specified period the unit advises the DTE to abandon the call.

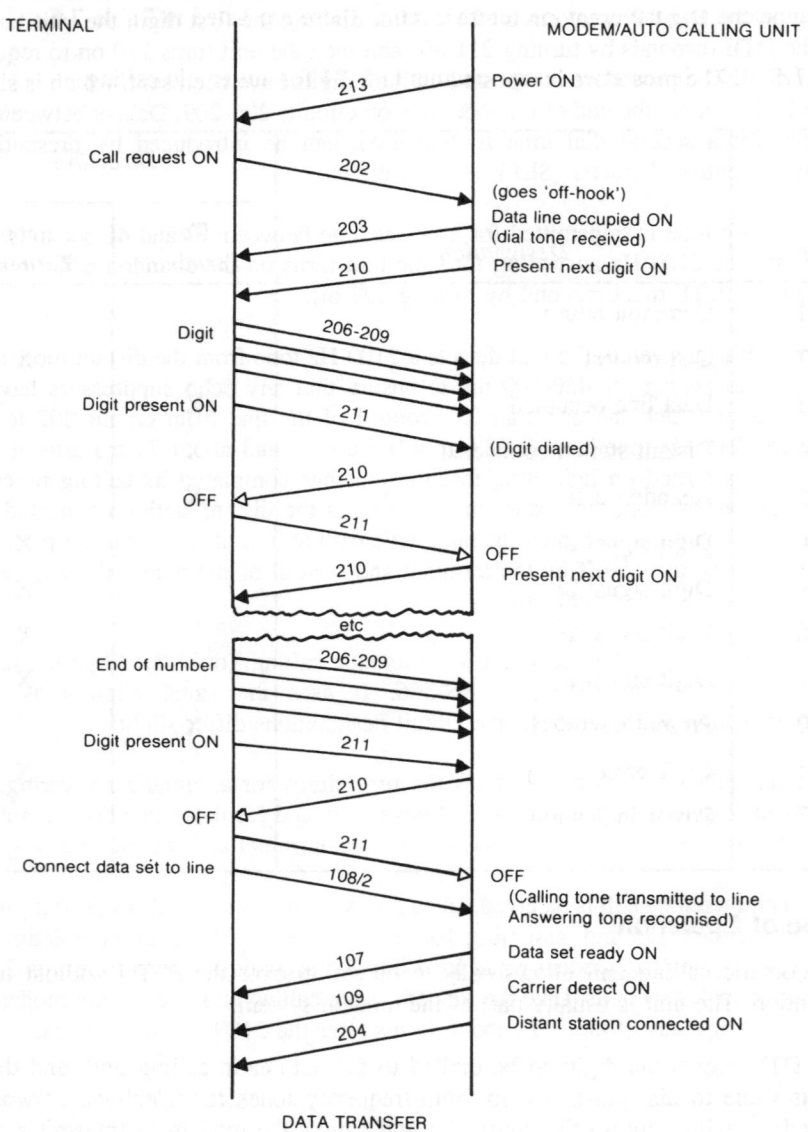

Figure 7.10 V.25 Call establishment procedure

The procedure for making a call is illustrated in Figure 7.10 which is based on CCITT Recommendation V.25. The DTE turns on the Call Request circuit 202 causing the automatic calling unit to go *off-hook*, which is indicated by the Data Line Occupied circuit 203 coming on. When dial tone is received the unit invites the DTE to present the next digit via circuit 120. The DTE presents the first digit to be dialled parallel form on the four circuits 206-209, coded as shown in Table 7.6 and informs the unit by turning on the Digit Present circuit 211. After dialling the first digit, the unit turns 210 off. The DTE responds by turning 211 off, and then the unit turns 210 on to request the second digit. The procedure is repeated until all digits have been sent, which is signified by the DTE placing the end of number code on circuits 206-209. Delays between digits, to allow for a second dial tone for example, can be introduced by presenting the separation control character (SEP) between digits.

The calling tone is transmitted for a preset time between 10 and 40 seconds, and in the absence of 2100 Hz answering tone the unit turns on the abandon call circuit 205, to which the DTE must respond by turning 202 off.

If the call is answered, the unit detects a 2100 Hz tone from the distant modem. This is allowed to persist for 450-600 ms to ensure that any echo suppressors have been disabled, after which the unit transfers control of the line from circuit 202 to circuit 108/2. The 210 Hz tone lasts for 2.6 to 4.0 seconds, and about 75 ms after it ceases, circuit 107 is turned on indicating that the unit has completed its calling procedures. Receipt of carrier from the distant modem causes the distant station connected circuit 204 to be turned on, after which the automatic calling unit plays no further part. Circuit 202 can then be turned off by the terminal and control of the connections is vested in circuit 108/2 as for a normal call.

In the US, an equivalent standard for automatic calling, formulated by the Electronic Industries Association (EIA), is RS-366. It uses the same circuits as CCITT Recommendation V.25, although the circuit designations differ slightly.

Recommendation V.25 also defines the procedures for automatic answering. Many modems have automatic answering facilities today, and for those that do not, automatic answering devices can be interposed between the terminal and the modem.

When an incoming call is received, an auto-answer modem signals to the terminal over the calling circuit 125 and, assuming 108/2 is on, goes off-hook on completion of the first ringing cycle. The modem waits for 1.8 to 2.5 seconds and then transmits 2100 Hz tone for 2.6 to 4.0 seconds. Like the modem at the calling end, the calling modem turns on the data set ready circuit 107 about 75 ms after the 2100 Hz tone ceases.

Synchronisation

For both asynchronous and synchronous transmission bit synchronisation is maintained either within the modem or the terminal. In asynchronous transmission, character synchronisation is also achieved within the terminal. All other synchronisation requirements, including character synchronisation in synchronous transmission, and block and message synchronisation, are the responsibility of *link level* protocols and *higher* protocols.

We consider the asynchronous and synchronous cases separately.

The Asynchronous Case

In asynchronous transmission, characters can be sent (and therefore also received) at any time. The human operator types characters at an indeterminate rate, and there will also be periods when no characters are being typed. Similarly, the terminal may receive a few characters from the remote computer, and there may be an interval, during which the user thinks about the message. In each case neither party knows when the next character is going to arrive and how many characters there will be.

Bit synchronisation in the modem is achieved, as we have already said, by a clocking mechanism common to both ends, either within the modem or the terminal. In fact, the traditional teletype machine maintained bit synchronisation by means of synchronous drive motors in the send and receive machines.

Character synchronisation is achieved by framing each transmitted character between a start bit and one or more stop bits, the format being depicted in Figure 7.11. Synchronisation is achieved in the following way.

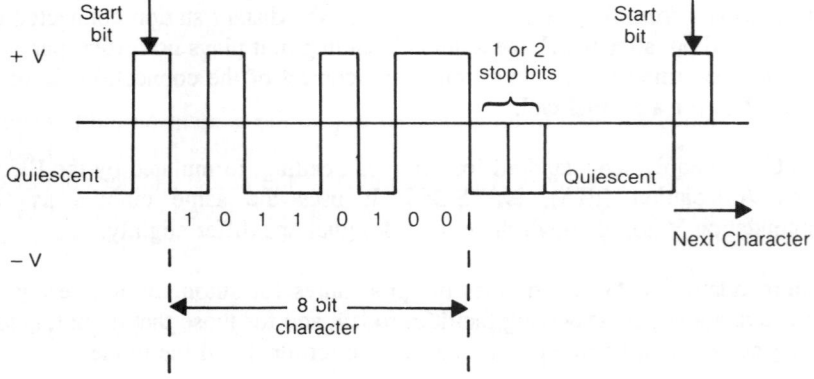

Figure 7.11 Asynchronous character framing

During a quiescent state, when there are no characters to transmit, the voltage is kept at −V (usually 12 volts) a logical 1. Depression of a key indicates that a character is ready for transmission, and this causes the voltage level to be changed to +V (logical 0) which it maintains for a one bit time duration, and this is called the start bit. Following the start bit, the bits which make up the character are then transmitted by using the appropriate voltage level, +V for a 0 and −V for a 1, the voltage level being maintained for one bit time duration on each occasion. When all the character bits have been sent, the transmitter then transmits a stop signal, by holding the voltage at −V for at least one bit duration.

The start bit serves to warn the receiving device that a character can be expected, and the receiver uses the −V to +V transition to start a local clock to read in the character bits. The stop bit is not directly concerned with marking the end of the character. By making it −V, it ensures that the −V to +V transition (the start bit of the following character) will be properly recognised regardless of what the last character bit was. Also, on mechanical terminals, two stop bits are used to allow the mechanical parts to prepare for the next character, whereas for teletype compatible VDUs one might be adequate. The bit duration is determined primarily by the transmission speed.

The Synchronous Case

In synchronous operation the data to be transmitted is clocked into the modem from the terminal at a steady rate. The clock which provides this timing may be located within the modem or may be external to the modem, in the terminal. The modulated waveform transmitted by the modem contains timing information which allows the destination modem to clock out the data to its terminal at the same steady rate.

There are four V.24 interchange circuits for conveying clock signals between modem and terminal.

Circuit 113   Transmitter signal element timing — (DTE source ———→ modem, clock in terminal)

Circuit 114   Transmitter signal element timing — (DCE source ———→ terminal, clock in modem)

These two circuits are used to time the data sent to the modem on the transmit data circuit 103. Either terminal timing (circuit 113) or modem timing (circuit 114) would be provided but not both. Usually data is transmitted to the modem under control of the modem clock, using the transmit clock timing circuit 114. External transmit timing is used in applications where modems are connected back-to-back, and timing is derived from one single source.

Circuit 115   *Receiver* signal element timing — (DCE source ———→ terminal, clock in modem)

Circuit 118   *Receiver* signal element timing — (DTE source ———→ modem, clock in terminal)

These two circuits are used to time the received data on circuit 104. The receive clock signal on circuit 115 tells the terminal when to sample the received data on circuit 104. This timing is derived from the incoming modulated signal and is therefore synchronised to the timing used at the transmitter.

Circuit 128 is rarely implemented. It enables the terminal to clock in the received data on circuit 104 in its own time. This circuit could be used in conjunction with a synchronous modem which had asynchronous standby facilities.

On the timing circuits, the on to off transition nominally coincides with the centre of the data bits on the transmit or receive circuits, and the off to on transition with transitions in the data signals (see Figure 7.12).

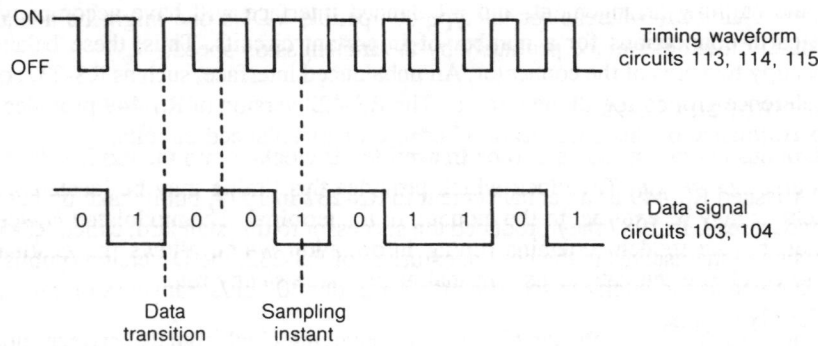

Figure 7.12 Relationship of timing and data circuits

Timing from the modem clock is normally provided to the terminal whenever the modem is powered up, although signals may be suspended for short intervals during modem testing. Receive data clock on circuit 115 may not be available when the carrier detect circuit 109 is off.

Other Interface Standards

The V.24 Recommendations specify a large number of interchange circuits, not all of which are used at the same time. Although colloquially referred to as a V.24 interface, it is only when a selection of these circuits are put together with electrical specifications from V.28 and a 25-pin connector detailed in ISO 2110 that the description is complete. This may be a source of confusion, since it is equally valid to use V.24 circuits with different electrical specifications and alternative connectors, and thus form a completely new interface specification. In fact, two other interface standards, RS-449 and V.35, are precisely this sort of mixture.

RS-449

As its name suggests, this interface was formulated by the EIA, and is meant as an alternative for their RS-232 specification. The latter is only capable of being used with fairly short cable lengths (15m or so) and data rates of up to 20 kbit/s. The maximum values achievable with RS-449 are 2 Mbit/s over a distance of 60m, although at slower data rates greater cable lengths are possible.

There are two connectors specified in RS-449, one having 37 pins and the other 9. The 9-pin connector is used when a secondary channel is required. All the interchange circuits used in the RS-449 standard are essentially V.24 compatible.

The standard is further complicated by the existence of two permitted electrical specifications. In RS-449 these are denoted as RS-422 and RS-423 and they also have CCITT equivalents which are X.27 (latterly V.11) and X.26 (latterly V.10) respectively. The major difference between the two electrical specifications is that RS-422

(X.27/V.11) is *balanced*, whereas RS-423 (X.26/V.10) is *unbalanced*. This terminology refers to the earthing arrangement, and a balanced interface will have accompanying reference earth connections for a number of important circuits. Thus, these balanced circuits occupy two pins of the connector. An unbalanced interface, such as RS-232, only has one reference ground for all the circuits. The RS-422 version of RS-449 provides the highest performance of this specification because of its balanced circuits.

The EIA issued RS-449 as an enhancement to RS-232 in 1977, but its take up has not been so widespread. In 1987 they produced the successor to this standard, called RS-530. Having resigned themselves to the continued usage of RS-232, the D-version (published in 1986) is given as the interface specification for under 20 kbit/s transmission rates and RS-530 for above this.

The V.35 Recommendation

The CCITT Recommendation V.35 was developed as a specification for a particular piece of modem technology which made use of wideband circuits (12 analogue FDM speech circuits) to transmit at rates up to 48 kbit/s. As part of this specification it also calls on the use of V.24 interchange circuits to describe the communication between the modem and the data device. However, the electrical characteristics of these interchange circuits are given as a mixture of V.28 and balanced circuits specific to V.35. Furthermore, the connector usually used with V.35 modems is a 34-pin one, which is based on ISO 2593. As well as the increased number of pins, it is square in form, further differentiating it from RS-232. The maximum interface cable length is not specifically defined, but 60m is an accepted safe upper limit.

Summary of V-Series Modem Specifications

So far we have dealt exclusively with interface standards, which is the way the data device views the modem. However, a large number of the CCITT V-series of recommendations actually deal with the specification of the underlying modem technology, ie how it performs transmission, a process which should remain as invisible as possible to the user.

There are, obviously, a number of different sets of circumstances which can be catered for, leading to several different permutations. For example, transmission can be full duplex or half duplex, it may take place over two-wire circuits or four-wire circuits and there are also a variety of transmission rates. Given this spread, it is not surprising that a selection of modem specifications exists. The numbers can at times cause confusion (although it is certain that without the standardisation effort by the CCITT the situation would have been many times worse). Tables 7.7 and 7.8 break down the applicable V-series recommendations into the configuration of those options which apply. In CCITT parlance, a *bis* appended to the end of a recommendation refers to a second version (either an enhancement or a permissible alternative) and *ter* to a third version.

Table 7.7 Combination of options for V-series modems

Recommendation	Data rate	Clocking	Flow	Circuit	Type of Modulation
V.21	0 — 300	A	F	LS 2	FSK
V.22	0 — 300	A	F	LS 2	PSK
	600	AS	F	LS 2	
	1200	AS	F	LS 2	
V.22 *bis*	2400	AS	F	LS 2	PSK/QAM
V.23	0 — 600	A	H	LS 2	FSK
		A	F	L 4	
	1200	A	H	LS 2	
		A	F	L 4	
	75/1200	A	F	LS 2	
V.26	2400	S	F	L 4	PSK
V.26 *bis*	1200	S	H	LS 2	PSK
	2400	S	H	LS 4	
V.26 *ter*	2400	AS	F	LS 2	PSK + echo cancelling
V.27	4800	S	F	L 4	PSK
V.27 *bis*	4800	S	F	L 4	PSK
V.27 *ter*	2400	S	H	LS 2	PSK
	4800	S	H	LS 4	
V.29	4800	S	H	LS 2	PSK/QAM
		S	F	LS 4	
	7200	S	H	LS 2	
		S	H	LS 4	
	9600	S	H	LS 2	
		S	F	LS 4	
V.32	4800	AS	F	LS 2	PSK/QAM + echo cancelling
	9600	AS	F	LS 2	
V.33	14400	S	F		

Table 7.8 Data rates for wideband V-series modems

Recommendation	Data rate (kbits/sec)
V.35	48
V.36	48 − 72
V.37	96 − 168

We will now give a brief description of each of the modem types mentioned in Tables 7.7 and 7.8. All the modulation techniques used by these modems have been dealt with already in Chapter 3. As well as these recommendations, other relevant documents from the V-series which apply to the data transmission aspect of modem technology are also discussed. Included in this category is the V.42 Recommendation, which is used in conjunction with some of the others mentioned to enhance the power of the modem.

V.21 Modems

The V.21 modem was the first to be introduced which was capable of full duplex operation over standard two-wire telephone circuits. It works by dividing the frequency band available into two channels, nominally number 1 and number 2 channels. The use of these by the two ends for sending and receiving signals is distinguished purely on the basis of who initiated the call. The originator uses channel 1 to send and receives on channel 2. The modem in answer mode will use these channels in the opposite manner.

Within each of these channels, two frequencies are allocated, one to represent a binary 0 and the other a binary 1. Thus, the full duplex operation relies upon two simultaneous frequency shift keying (FSK) processes. The two sets of frequencies specified in V.21 are:

—   Channel 1 — 980 (binary1)/1180 (0) Hz
—   Channel 2 — 1650 (1)/1850 (0) Hz.

Note that the equivalent Bell standard modem in the USA (Bell 103) employs two different sets of frequencies and is thus incompatible with V.21.

Originally it was assured for data rates of up to 200 bit/s, but capable of being driven from terminal equipment at 300 bit/s. It is only capable of asynchronous transfer.

V.22 and V.22 *bis* Modems

These modem recommendations can be used for both synchronous and asynchronous transmission, providing in all cases a full duplex channel. Using the phase shift keying

(PSK) technique, V.22 caters for data rates up to 1200 bit/s and V.22 *bis* uses quadrature amplitude modulation (QAM) to achieve 2400 bit/s. In both these specifications differential encoding is expected. The full duplex channels provided by splitting the bandwidth in two (as in V.21) and employing an originate/answer mode scheme. In order that the channel operating in the upper band should not suffer more degradation than the lower (due to the uneven frequency characteristics of the voice grade circuit) adaptive automatic equalisation is also called for in the recommendations.

The V.23 Recommendation

This Recommendation stems from the earlier days of modem technology, and is for use with asynchronous transmission. It can be operated at two different data rates, 600 and 1200 bit/s, and included in the 1200 bit/s scheme is the option of including a low speed (75 bit/s) reverse channel. In the UK, the 1200/75 configuration is used for the public videotex service known as Prestel. This reverse channel method is the only recommended way of achieving duplex transmission on a standard two-wire circuit (although obviously this is not a problem with 4-wire circuits).

The V.23 modem employs FSK to transmit the binary data, and the tones used differ for the two data rates. In some cases, where the channel imposes extra filtering above 2 kHz, the 1200 bit/s cannot be used, since it uses 2.1 kHz as one of the FSK tones. The selection of V.24 interchange circuits used by V.23 has already been used as an example in Tables 7.4 and 7.5.

V.26 and the *bis* and *ter* Variants

The V.26 Recommendation was initially developed as a high speed modem specification for use on four-wire leased circuits (although by current standards the 2400 bit/s specified falls more into the medium speed category) using synchronous transmission. In order to achieve the 2400 bit/s given in the recommendation, quadrature PSK is used, and on top of this, differential encoding is employed. In the first document (straightforward V.26), two alternative approaches to differential encoding are given (known as alternatives A and B), although the American equivalent (Bell 201) uses only the second of these.

Since V.26 assumes the existence of four wires to achieve full duplex operation, using this specification over the PSTN will limit transmission to half duplex. This activity is covered by the *bis* extension to the recommendation, which also permits a slower speed mode of 1200 bit/s. Simple bi-phase PSK is used for this alternative. At the higher rate, only the second (B) of the two differential encoding schemes is specified for V.26 *bis*.

Full duplex operation at the higher data rate and over the PSTN is achievable by following the V.26 *ter* option. In order to get full duplex working over just two wires echo cancellation techniques are employed. Another feature unique to the *ter* version of

V.26 is the mandated use of scrambling in the bit-stream. Only the first (A) of the two differential encoding schemes is allowed for V.26 *ter*.

V.27 modems and the *bis* and *ter* Options

Both V.27 and V.27 *bis* modems provide full duplex operation over four-wire leased circuits at a rate of 4800 bit/s. There are, however, important differences between the two. Amongst these differences, V.27 relies upon manual equalisation where V.27 *bis* employs an automatic adaptive one. The *bis* version offers an additional facility of fall-back to the lower data rate of 2400 bit/s.

The fall-back facility is one also provided on V.27 *ter*, except that this variation in the standard is for half duplex use on the PSTN. The modulation method for both these fall-back modes is QPSK, and is differentially encoded using alternative A (as in V.26). At the full data rate, eight-phase PSK is specified, and a differential encoding scheme given for the eight combinations of tri-bits, which is the symbol length used for 4800 bit/s.

The V.29 Recommendation

The V.29 modem provides data rates of up to 9600 bit/s to be achieved over four-wire leased circuits. It offers fall-back modes at 7200 and 4800 bit/s and for each of these three rate options, an appropriate form of differential encoding is given. To obtain such high data rates four bits/symbol is used and translated into a form of QAM which consists of eight possible phases with two amplitudes for each phase. The signal space diagram for 9600 bit/s V.29 is given in Figure 7.13.

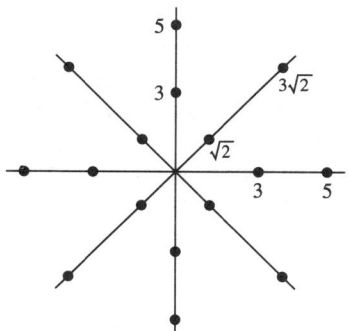

Figure 7.13   Signal space diagram Recommendation V.29, 9600 bit/s

A unique feature of V.29 is the availability of a multiplexing option, which permits the 9600 bit/s capacity to be shared amongst a maximum of four ports. There are several permissible combinations of data rates for each of these ports ranging from one port at 9600 bit/s (ie, no multiplexing at all), through to four ports at 2400 bit/s.

V.32 Modems

One of the latest inclusions in the V-series is the V.32 Recommendation, which provides a technique for achieving as much as 9600 bit/s full duplex over a standard two-wire PSTN connection. The full duplex working is attained by using echo cancellation techniques as in V.26 *ter* and the high data rate by using a scheme of four bits/symbol. For the highest rate of 9600 bit/s, two options for transmitting these symbols are given. The first is a fairly standard 16-point QAM signal state diagram and the second is a 32-point QAM constellation. The latter one would expect to be associated with a 5 bits/symbol modulation scheme, but in this case trellis coding is used (and this is, in fact, the example used to explain this technique in Chapter 3) which means that only four of these five bits are actual data bits contributing to real throughput. The two QAM options are shown in Figure 7.14. Lower data rate schemes are also permitted by V.32 at both 4800 and 2400 bit/s. V.32 is specified for both synchronous and asynchronous operation.

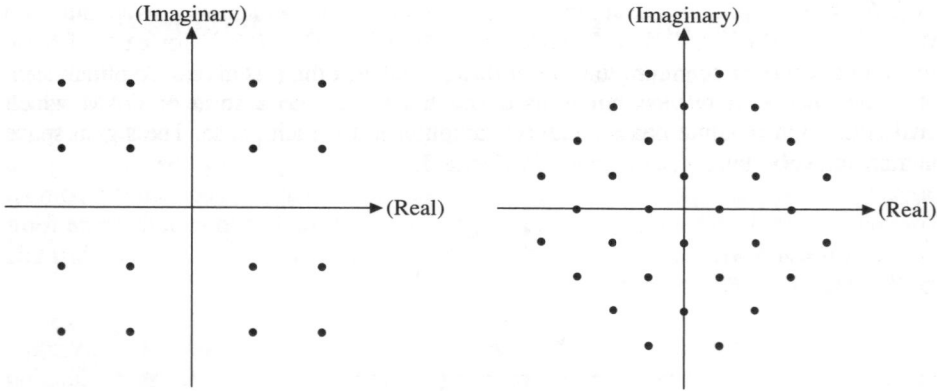

Figure 7.14   Signal state diagrams for 16 and 32 state QAM

V.33 Recommendation

This recommendation describes a modulation technique for providing 14400 bit/s full duplex over a four-wire leased line. This is achieved by a combination of phase and amplitude modulation. Trellis coding is employed (see description in Chapter 3). The scheme uses 7 bits/symbol, of which one bit is redundant. A lower data rate of 12000 bit/s is also included.

Wideband Modems in the V-Series

In order to make use of the old group-band circuits (12 FDM speech channels), a family of modems, known as wideband modems have been specified in Recommendations V.35, V.36 and V.37. These are generally used for synchronous transmission (although there are a few exceptions to this, such as for fax machines) and all three allow the option of an extra voice channel over the link. The data rates available differ between the three recommendations: V.35 specifies higher rates starting at 48 kbit/s with an option of 40.8 kbit/s; V.36 specifies higher rates starting at 48 and rising through 56 to 72 kbit/s; 96, 112, 128 and 144 kbit/s are all permitted by V.27. Amplitude modulation is specified in V.35 and V.36, and a combination of amplitude and phase modulation is recommended in V.37. As the population of group-band circuits has declined in favour of cheaper digital techniques, the use of wideband modems has ended. V.35 is used by PTTs to interface to 48K services: eg BT's Kilostream.

Error Correction and Data Compression for Modems (V.42)

With modem speeds becoming faster and faster, some of the techniques employed (eg, QAM) are theoretically more prone to errors in transmission than their slower speed counterparts. One countermeasure that can be taken in these circumstances is to increase the transmission power (which in QAM has the effect of making the signal space diagram more spread out, and thus making it easier to distinguish between the points). However, seldom is this a practical solution, and the alternative is to include some form of error protection with the data transmission. Some of the techniques for doing just this are described in more detail in Chapter 8.

There is a recommended approach from the CCITT for modems which employ some form of asynchronous to synchronous conversion (ie, those using coherent modulation techniques, but are still able to accept asynchronous data) such as V.22 and V.32. The CCITT specification is enshrined in V.42, in which two schemes are given. One is contained in the main body of the text and this owes its form to the link control procedure known as the high level data link control (HDLC). Link control processes, and in particular the HDLC method, are also dealt with in Chapter 8, but it is worth noting here that the new adaption of HDLC for modems is called the link access procedure for modems (LAPM).

The second correction scheme is contained within an annex to the main text and is based on a more character orientated protocol. Both these types of error correction have to be incorporated into the modem for it so be deemed V.42 compatible, but LAPM is the default mode. The details of the error correction process are exchanged in protocols after the physical link between the two DCEs has been established.

A further element of data processing is discussed in the document V.42 *bis*, which makes recommendations for data compression. This is based on a technique propounded by British Telecom, which is known as the Lempel-Ziv algorithm (and the coding method has thus acquired the title of BTLZ). The effect of data compression is to increase actual data throughput without the need to make the modulation rate any higher. Thus 19.2 kbit/s modems can be constructed by taking 9.6 kbit/s modems and applying data compression to them.

Modem Loop Testing using V.54

In order to carry out diagnostics on modems and lines, CCITT has recommended an approach to loop testing. Within V.54 four loops are identified (see Figure 7.9), and some V.24 interchange circuits are allocated for performing these tests. The operation of the loop testing procedures has already been covered under the description of the relevant circuits (140, 141 and 142) in this chapter.

Proprietary Modem Standards

Although proprietary modem techniques often present enhanced facilities, the disadvantages of non-compatibilty and approval for attachment to the public networks have often tended to dampen the impact these may have had on the modem market. However, in a few exceptional cases, such technology has become so popular that it has received the status of *de facto* acceptance. Two important examples of this phenomenon are the Hayes AT instruction set for intelligent modems and the family of microcom networking protocols (MNP). The former has found a niche through the ballooning PC market, and since it does not impinge on any of the CCITT areas of work (with the exception of dial-up facilities), faces little contention. The latter, although having gained a head start in terms of market acceptance, pre-empted work to be done by CCITT, which makes it politically a much more sensitive issue.

To complete this chapter, a brief summary of these *de facto standards is given.*

The Hayes AT Command Set

With the advent of PC technology, the possibility of cheap, intelligent machines which require communications facilities became very much a reality. From the ubiquitous success of the PC, the variety of applications being sought served to stimulate a new flexibility in modem technology. By incorporating microprocessors into their design, the modems themselves could be made to respond to many different circumstances and so the *intelligent modem* was born.

In order to provide the user with control over the modem facilities, software is supplied for the PC. By running this software, it is possible for the user to issue commands to the modem and ensure it is operating in the manner expected of it. By far the most popular and most widely accepted vehicle for this is the Hayes AT command set. Once the modem has been set in command mode by a special character sequence (usually three + characters — a *guard* time period before and after the sequence protects against a random occurrence having the same effect) instructions are sent to the modem preceded by the character string AT. From within the AT set of commands the modem's method of operation can be altered and a selection of the features available is given below:

—   Dial-up and answer a PSTN call.

—   Set modem transmission method, for example, V.22, V.23 etc.

—   Set synchronous or asynchronous working.

—   Set flow control, error control and compression.

—   Establish and select a profile for modem operation.

The last feature is a useful facility that would permit the user to establish a set-up configuration for an often used application, such as perhaps videotex access, and to select this profile with the minimum of fuss when required.

Microcom Networking Protocols (MNP)

Another step in the increasing sophistication of modem design was the realisation that the modem could take much more responsibility for link usage than just simply being a dumb transmitter. This has been confirmed by the adoption of V.42, trellis coding in V.32 and other developments which all use data pre-processing techniques to increase the overall efficiency of the transmission. The modem no longer just regurgitates the data sent to it, but is capable of making intelligent decisions of its own on how best to utilise the channel.

Prior to CCITT standardisation work, the activity surrounding the inclusion of link protocols in modems was concentrated around Massachusetts based Microcom Inc. who produced the family known as microcom networking protocols (MNP). These range from classes one to nine, the higher the number the greater the level of complexity. An important feature is that the higher classes have evolved from the lower ones, so that, for instance, an MNP class 5 modem could if necessary forget its enhanced features and talk down to a class 2 modem, by finding the lowest common denominator of their protocols.

Some of the salient features of MNP include the blocking of data for the purposes of link control, error control mechanisms incorporated within these blocks (to be able to run an acknowledgement based error correction scheme) and the compression techniques utilised by the classes MNP 5 and above. These are all subjects which are discussed in general in Chapter 8. However, for the time being we note here some of the 'landmarks' from amongst the MNP family:

— *MNP 2* Asynchronous transmission, 10-bit characters organised into fixed length packets.

— *MNP 4* Adaptive packet size selection ensures less overheads in poor transmission conditions. When many errors are occurring, smaller packets are sent so that, should retransmission be necessary, it will take less time. MNP 4 is incorporated into V.42 as an annex and is the recommended fall-back state for a V.42 modem if LAPM communication between the modems is not possible.

— *MNP 5:* This class features a real-time adaptive compression algorithm which is optimised by the data passing through the channel.

— *MNP 6:* A link negotiation scheme is used whereby the transmission starts off at a lower data rate and can be increased subsequently, given mutual agreement.

— *MNP 7:* Predictive compression gives even higher efficiency than MNP 5.

— *MNP 9:* MNP 7 (with slight modifications to the error control process) on a V.32 modem which gives a throughput three to four times quicker than would be expected of the modem's normal 9600 bit/s data rate.

During the discussions on V.42, various elements of MNP were mooted, particularly given the user base it had established. It was eventually decided that LAPM afforded more possibilities for future expansion, but that the ability to work using MNP 4 for fall-back purposes was a necessary part of compliance. However, although MNP algorithms were also suggested for V.42 *bis*, these were rejected in favour of BTLZ, which is aligned to LAPM. No MNP fall-back provision was incorporated into this document.

The Future of the Modem

Ever since digital circuits and digital access to the public network started gaining popularity, the death of the modem has been predicted by numerous pundits. So far it has refused to lie down and be buried and the increasing sophistication of these devices is clearly an indication of the continuing strength of the market. Interfaces have now been developed which perform for digital access what V.24 and its counterparts did for analogue devices and these are expected to play a role in the digital schemes of the future. They have already been incorporated into X.25 and ISDN, but the strength of the modem's position can be seen by the fact that even a temporary provision for V.24 interfaces has been made within X.25.

The only thing that is certain is that the predictions of modem's demise will continue to be premature so long as analogue (especially dial-up) continues to be a much cheaper option than digital. Also, the access to digital is still much less widespread than analogue. However, to date it has undoubtedly been the most enduring piece of communications technology.

8

Data communications techniques

Much attention has been given to the underlying technologies which make data communication possible. So far, much has been said regarding the communications networks which are available and the transmission techniques used to make channels suitable for data communications. Throughout these discussions it has been assumed that, on the whole, the data to be transmitted is essentially opaque. In the simple case of two DTEs having a conversation over a simple link, the data passing between the two has been their own business, having no bearing upon the data communication process.

In this chapter we will be turning this approach on its head. The channel will be assumed to be an invisible resource which the two DTEs make use of and we shall be looking at the considerations that apply to establishing a sensible conversation between the participants. The culmination of this will be to present an *accepted philosophy* for providing end devices with the ability to *interconnect* and *interwork* with one another. The purpose of this chapter is to build up to this by looking initially at techniques that are applied to the data itself which enhance the quality of an end-to-end communication or which introduce some sort of discipline into this process.

The chapter begins by discussing how to control the effects of data errors which occur during transmission. As Figure 8.1 illustrates, even a simple data communications system comprises a number of components, and each component can be a source of errors, some being more error prone than others. Protection against transmission errors must therefore be seen in perspective as only part of a solution to a total systems problem.

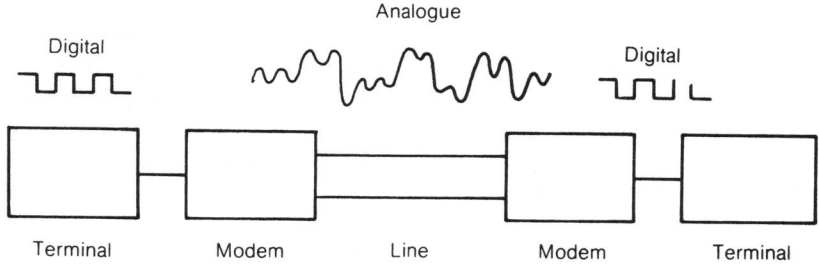

Figure 8.1   The elements in a data communications system

The recognition of the inevitability of errors in any part of a data communications system is the first step towards a remedy. The next steps are to evaluate the likely costs of errors and to examine the probability of their occurrence. A very high degree of error control can be applied to any part of a system, although it should be emphasised that the cost of guarding against errors must not be allowed to exceed the cost of allowing them to occur. Errors can be expected on data communications links and the nature of these and the available means or error control are discussed in this chapter.

Following the treatment of errors is a discussion on the need for means to control the flow of data. This is particularly relevant when either errors are being detected or one of the parties in the communication is slower at processing the data than the other one.

The introduction of procedures for protecting data against errors invariably presents overheads for transmission. The other side of the coin rests with the techniques used for data compression, in which much of the redundancy in the data is removed. Through such techniques much more efficient use of the communications channel can be made.

Having discussed how individual bits or bytes of data might be manipulated, the chapter continues by describing ways in which the data itself can be organised. Through an agreement of formatting, a common understanding between the end systems exists, enabling them to decide what part of the data represents information.

From all the techniques covered in the chapter, the threads are drawn together to show how they all belong within the process of link control. The supervision of the use of the channel by the communicating bodies is a fundamental ingredient in effective communications, and this can be achieved by establishing a consensus between the parties involved. In other words the communication is to be governed by *protocols* and to illustrate the role that these play, the most prominent link protocol, known as high level data link control(HDLC) is introduced. The discussion of protocols is then extrapolated into the wider communications scenario to conclude the chapter by presenting an ordered framework for data communications as a whole.

Error Control

Much attention has been given to this interesting subject in recent years and initial fear of transmission errors has been replaced to some extent by an enthusiasm to control them. There is, however, a need for consistency of approach to the problem. Residual error rates can now be reduced to infinitesimal proportions but it is unrealistic to design for a residual error rate of, say, 1×10^{14} in this area without protection to the same degree in other equally important parts of a system.

The level of protection deemed necessary must also be related to the nature of the applications and the penalties of producing faulty information. Clearly, errors in a system producing plain language transmission (where errors are largely self evident) need only a rudimentary form of error check, to give an acceptable level or error performance. On the other hand, an air traffic control system concerned with human safety can afford very few errors in its data transmission system. Such a system will have been designed using normal data transmission equipment and lines, which in themselves might have an unacceptable level of error performance. The systems designer, however, would have built into the system, error control techniques to reduce these errors to an acceptable level. Thus, some systems might tolerate an error rate of 1 bit in 10^3, others might necessitate 1 in 10^9 or better.

An error in a data transmission system means usually that where, say, a binary 1 has been sent, a binary 0 has been received. The other way in which errors can occur are where bits are lost, but this will have a catastrophic effect on the block being transmitted, easily recognised by the receive terminal.

Error rates are quoted usually as one error in so many bits (1 in 10^4 or 1 in 10,000 bits) but clearly the situation of 1 bit in error being followed by 9999 accurate bits is very unlikely, so that an error rate as quoted above really means that 'given a representative sample then the probability will be that 1 bit in 10^4 will be in error'. In computer systems such 1 bit errors are common but on data transmission lines errors also tend to occur in bursts.

General Sources of Error

The Operator

Although data communications links are prone to errors, they are by no means the worst offenders; sadly the people within a system are not beyond reproach. Terminal operators in an on line system are often the weakest link in an otherwise reliable installation and very high keying errors indeed are to be expected particularly, it seems, in the critical early days of operations. In batch processing systems, even the very best operators dedicated to a particular task will produce a high proportion of errors.

The Computer

Like anyone else, the computer professional may be fallible. Many residual program bugs are lurking now in computer systems seemingly ready to pounce at the most critical moment and then retire to await their next opportunity. The development of appropriate tools to enable the production of error-free programs is a long standing goal of software engineers. Computers themselves can make mistakes and the transient errors such as the dropped bit, caused perhaps by dust particles on a magnetic tape or disk are familiar to computer programmers and operators.

Transmission Errors

Background

Although the errors resulting from the above mechanisms are irritating, they are outside the scope of the data communication engineer. No amount of meticulous network planning will account for and rectify mistakes which can occur when operations are actually in progress. However, something can be done to alleviate discrepancies which might have been introduced during transmission.

The only statement that can be made with certainty about transmission errors is that they will occur. How often they will do so and what the distribution of errors will be on a particular channel is a forecast which the bravest communications engineer would not attempt. For instance, the factors affecting telephone circuit quality are numerous and varied, and, as we have seen in earlier chapters, the transmission of digital data along analogue paths raises additional problems which do not apply to the same extent when speech is transmitted along such paths.

Experience to date provides very firm evidence that digital transmission, together with stored program switching (as on System X), results in a massive improvement in transmission quality. For example, the noise amplification is virtually eliminated compared with analogue, as is impulsive noise directly caused by electromechanical exchange switching. Therefore, within the UK context, progressive improvements in quality may be expected with increasing penetration of digital technology.

Definition of Error Characteristics

Transmission errors in fact pose different problems for three different groups of people. The telecommunications engineers (and here modem manufacturers are included) must study the transmission problems likely to cause errors so that circuits can be engineered and modems designed to minimise their effects. In addition, the telecommunications engineers, through their PTTs, have the added responsibility of providing as much information as possible regarding the frequency and distribution of those conditions on their networks which are likely to cause errors and to recommend means of controlling them.

The designer of an error control system is primarily concerned with the control of errors with a view to providing the maximum protection against them with the minimum amount of redundant information. The prime interest is the distribution or pattern of errors as well as the frequency with which they are likely to occur; the designer will conduct laboratory and field experiments as well as using CCITT statistics in order to obtain the required information.

The systems analyst working on data communications systems apparently has the easiest transmission error control problem. Indeed, the problem may be ignored altogether in the belief that the level of protection which he needs against this type of error can be readily obtained. This is understandable with modern data communications systems for retransmission of blocks in error takes place automatically and the presence of transmission errors is rarely evident. the systems analyst could be forgiven for wondering what the fuss is all about and concentrating on other more pressing or exciting tasks. If the communications hardware has already been purchased and manufacturers' communications control software is being used, there may be very little the systems analyst could do about transmission errors anyway. However, a knowledge of data transmission error control may help a systems designer in a number of ways.

By considering carefully the effects of errors in a proposed system, the analyst can determine the proper degree of protection required. As a result he should be in a position to try and select an error control method which will give him the necessary residual error rate with the minimum redundancy. From then on, the systems analyst can regard errors as delays, which can increase costs or lengthen response times or both. These delays can be reduced considerably by careful attention to the block sizes and the error control procedures used. Later in this chapter, different types of error detection and correction codes will be discussed and the throughput of information in the presence of errors will also be considered. Before examining these subjects, it is useful to clarify what is meant by the various terms used to describe errors and to have some idea what causes them.

First of all let us consider the term error itself. This could not be easier, for the nature of computer data is such that the symbols 0 and 1 are mutually exclusive. If, therefore, a binary 1 is received when a binary 0 has been transmitted, an error has occurred. This is described as a single bit error if the bits on either side are received correctly. Errors also occur in groups and a two bit error group, for example is two consecutive erroneous bits with correct bits either side; an n bit error group is n consecutive erroneous bits with correct bits either side, and so on. Measurements of error groupings can show how common single bit error occurrences are.

Line disturbances also cause bursts of errors to occur. An error burst is defined by CCITT as a group of bits in which two successive erroneous bits are always separated by less than a given number (X) of correct bits; the definition going on to state that the last erroneous bit in a burst and the first erroneous bit in the following burst are accordingly separated by X or more correct bits. Such precise definition is necessary in international communications but can pose problems of understanding.

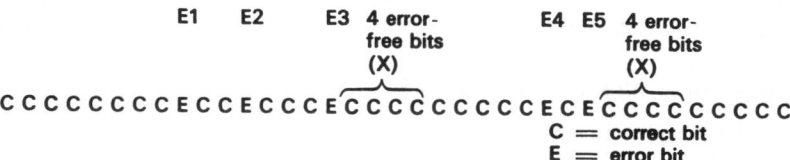

Figure 8.2 Analysis of a block for burst errors

Burst errors, in fact, are not as incomprehensible as they sound and are extremely important. If errors are to be detected, it is necessary to find out more about the way in which they are distributed. To say that an average error rate is, for example, 1 in 10^4 tells us very little, for errors will certainly not be conveniently slotted into a stream of data 1 every 10,000 bits. Error groupings are helpful but give no indication of the distances which separate the errors — a key factor in the design of effective error detection and correction codes. The analysis of data in terms of error bursts is very useful and, despite the wordy definition above, fairly simple. Consider the small block of data shown in Figure 8.2. For a 'guard' value of X = 4, this error pattern would be identified as two bursts of errors: one long burst stretching from E1 to E3, and a shorter burst between E4 and E5. E2 has simply been engulfed by the long burst and forms a part of it. Note in particular the non-uniformity of burst length. Figure 8.3 gives a typical graph of error burst size distribution. There are other definitions for error bursts which result in a constant length of burst, and the reader is invited to peruse some of the literature recommended in the bibliography. In particular, that of Wainberg and Wolf provides an easier way of judging the effectiveness of a burst error correcting code.

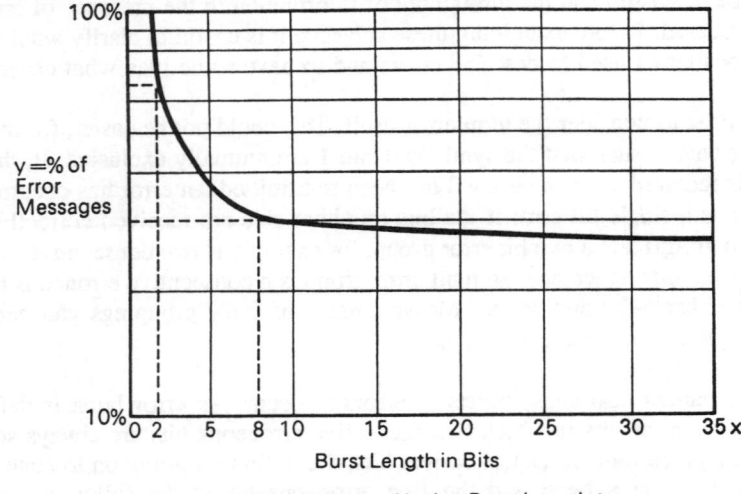

y =% of Messages Having Burst Length ≥x
Error Message Total = 23,589

Figure 8.3 Error burst size distribution (CCITT Blue Book, 37/22)

Having categorised errors and their pattern of occurrence we will briefly turn our attention to the mechanisms which are responsible for producing these patterns.

Impulsive Noise and Transmission Errors

Transmission errors are caused by a number of factors but impulsive noise is probably the major problem. The main source of impulsive noise is switched connections through automatic telephone exchanges.

The electromechanical switches in the exchanges cause vibrations which create movement on contact surfaces resulting in noise peaks or spikes with typical peak to peak values of 100 milli-volts. There is also some evidence to suggest that the quality of the final selectors in the called exchange exerts a considerable influence on the level of impulsive noise. This problem does not arise with digital switching. Noise power is normally expressed in dBm0 which refers to the ratio of the noise power at a point in the transmission path to the test level measured at the point (expressed in decibels). The maximum, number of noise peaks which occur during any period of 15 minutes on private circuits is published in the relevant PTT circuit specifications. For example, on a BT Network Keyline, a threshold level of 21 dBm0 is set which must not be exceeded more than 18 times in any period of 15 minutes (see Figure 8.4).

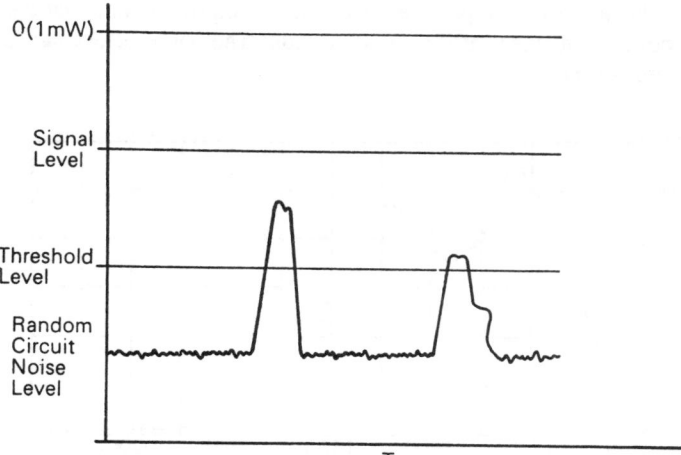

Figure 8.4 Noise peaks

Causes of Transmission Errors

Signal Power and Transmission Errors

As the most important factor affecting transmission error performance is the ratio of impulsive noise power to signal power, it would obviously be advantageous to increase the signal power of the modems used on telephone circuits. Unfortunately, there are limits which must be observed and these have been agreed internationally (CCITT Recommendation V.2). The reasons for this are, firstly, to avoid overhearing or crosstalk which can be caused in the local cable network by excessive signal power and, secondly, because the multi-channel carrier systems used on the networks have design tolerances which limit the power values of individual channels.

Calls connected on the PSTN are not only more subject to impulsive noise and interruptions than private circuits but the ratio of the level of these disturbances to the signal power is likely to be very much higher as a result of the greater attenuation of the signals. It follows from this that the error rate of calls over the PSTN will be considerably worse than on a private leased circuit.

Telephone Traffic and Transmission Errors

The incidence of errors on telephone circuits tends to follow the same pattern as that of the exchange traffic. The typical telephone traffic graph is shown in Figure 8.5.

This close correlation between busy hour traffic and data errors is an international phenomenon and can be seen clearly by comparing Figure 8.5 with Figure 8.6 which shows the distribution of erroneous blocks transmitted on a typical PSTN. The error peaks are due to the additional impulsive noise introduced by the automatic selectors in automatic telephone exchanges during busy periods. The same effect can be seen on private circuit connections.

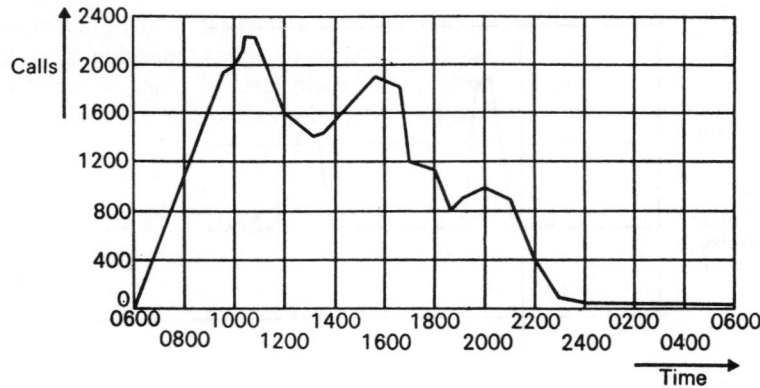

Figure 8.5 Telephone call distribution

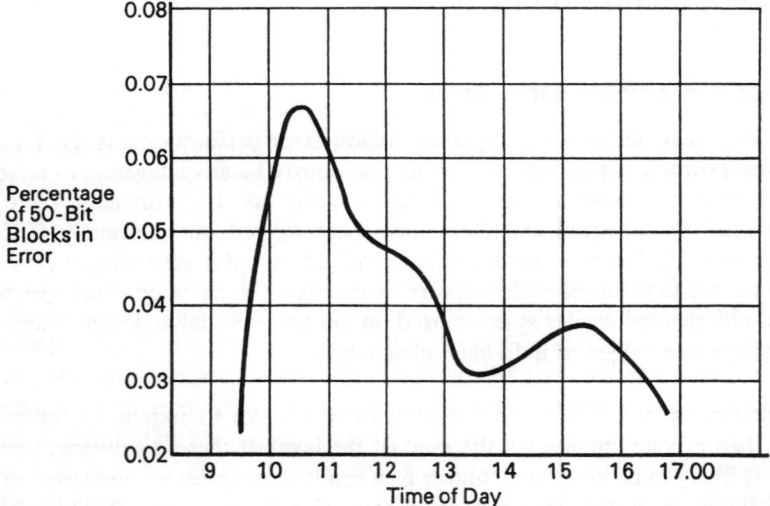

Figure 8.6 Time distribution of erroneous blocks on a typical PSTN

The results of tests performed throughout the world suggest that the error performance of data links in traffic off peak periods will yield better results than in the busy periods. This is particularly true of calls on the PSTN where there is also the added incentive to use off peak periods reduced call charges.

There may be a wide variation of bit error rates over different days of the week and again there are indications that this is probably due to variations in telephone traffic density.

Short Breaks in Transmission

A problem which occurs in all telephone networks is that of short interruptions in transmission during which the line signals may be lost completely. These have little or no effect on normal voice transmission but the consequences for data may be disastrous. In most error control mechanisms affected blocks would be easily identified and catered for, but where there are longer lapses it is possible for synchronisation to be lost and even for whole messages to go missing. At a certain level of noise, the modem stops sending carrier. The modem must then re-synchronise before sending more data.

Echoes

In speech conversations over long distance circuits, a person's voice may be returned or echoed back. These echoes are due to reflections which can occur whenever there is a change of impedance in the line such as a two- to four-wire conversion through hybrid transformers.

This talker echo effect is a nuisance when measured in tens of milliseconds, and above 500ms will inhibit speech altogether; some long distance circuits are fitted with echo suppressors to prevent echoes being returned but these can be disabled when the circuit is used for data transmission as it may be essential to have the return channel open for full duplex transmission.

Echoes have a different effect on data transmission. The problem is that data transmitted may be followed by a delayed replica of itself (listener echo) which may interface with the operation of the data receiver. Delays of fractions of a millisecond may be significant, and if the echoes are of sufficient amplitude, errors may be produced.

Other Hardware Components

As well as the switching apparatus and the line plant supplied by the network operator, we have also discussed (in previous chapters) some of the other components which are often used to perform data transfer. Generally speaking, equipment such as modems are obedient beasts and if a transmitted 1 is detected as a 0 (or vice versa) this is the result of a corruption during transmission, and misinterpretation by the receiver equipment of the incoming signal. Occasionally, however, the receive equipment will introduce errors of its own. These are mostly single error occurrences, except in the case of loss of synchronisation when one would expect a 50 percent error rate, or when carrier detection is completely out of phase and the whole received bit pattern is inverted (100 percent error rate).

Error Detection and Correction

Before discussing the techniques associated with detecting and correcting errors it is necessary to point out the difference in these two terms as they are sometimes confused. Quite simply, the process of ascertaining whether or not received information has been corrupted is error *detection* and this is a natural precursor to error *correction* mechanisms; it does not necessarily follow, however, that having detected an error in transmission a receiver will be able to correct it.

The methods used to perform these sorts of function generally fall into one or other of the following categories:

—   Information feedback.

—   Forward error correction (FEC).

—   Decision feedback, sometimes referred to as automatic retransmission on request or ARQ methods.

The first of these is generally used in asynchronous transmission, and the others in synchronous transmission.

FEC methods were, in fact, little used until recently. The subject matter is complex and mathematical, and heuristic approaches are not always fruitful. However, a short discussion based on a straightforward example is included for completeness. The curious and adventurous reader is referred to the titles in the bibliography.

However, ARQ methods, of which the basic acknowledge/not acknowledge (ACK/NAK) procedure is an example, are still by far the most common.

Information Feedback Systems

This method is widely employed for single bit error detection in low speed asynchronous data transmission. In Figure 8.7 the data entered on the keyboard does not directly activate the local screen, but is transmitted from the keyboard down the line to the receiving end which loops the signal back on the other half of the duplex pathway. When this returning signal reaches the originator it drives the screen and displays the character received. If this received character does not agree with that depressed on the keyboard an error is assumed to have occurred during transmission. The human operator is visually doing the error detection by comparing the known keyed input against the printed output fed back from the receiver to the originator. The technique is sometimes referred to as 'echoplexing'.

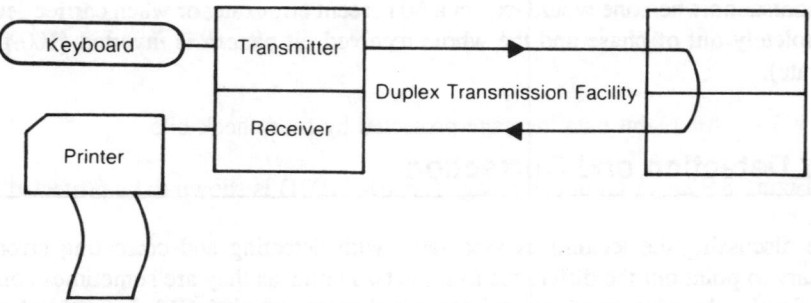

Figure 8.7 'Echoplexing' in asynchronous transmission

Forward Error Correction (FEC)

Unlike ARQ systems, forward error control (FEC) systems attempt to correct errors at the receiver without employing retransmission procedures. Sufficient redundancy must therefore be carried in the blocks so that, not only can the presence of an error be detected, but also its location with the block known. Once the position of an error is known, correction is achieved simply by investing the erroneous bit from 1 to 0 or vice versa. The technique relies upon the use of special coding schemes which entail carrying additional redundant bits in the block.

A Simple Example

Many complex coding schemes have been developed for forward error correction. A commonly cited example of these is the Hamming single bit error correcting code, which is one of the simpler methods used for error correction and, therefore, provides a useful introduction to the subject.

A Hamming code can be constructed for any group of data to give error correction of any one bit in the group which is detected as being in error; it can, therefore, be applied to a character or a block of information. The checking bits which are used occupy predetermined positions within the data field, the positions being all those with the value of 2^n. For example, in a protected data field of 15 bits, the checking bits occupy the bit positions 1, 2, 4 and 8 as shown in Figure 8.8.

Bit position	15	14	13	12	11	10	9	2^3 8	7	6	5	2^2 4	3	2^1 2	2^0 1
	I	I	I	I	I	I	I	X	I	I	I	X	I	X	X

I = Information bit
X = Hamming or check bits

Figure 8.8 Position of Hamming bits in a protected 15 bit data field

Bit position	15	14	13	12	11	10	9	8	7	6	5	4	3	2	1
	1	0	1	0	1	0	1	X	1	0	0	X	1	X	X

Figure 8.9 An 11-bit data message protected by four check bits

In Figure 8.9 an 11 bit data message (10101011001) is shown to be protected by four check bits.

Bit position	Binary value	
15	1 1 1 1	NB: Modulo 2 addition
13	1 1 0 1	(addition without carry)
11	1 0 1 1	produces even parity
9	1 0 0 1	
7	0 1 1 1	
3	0 0 1 1	
	0 1 0 0 ←	Hamming bits

Figure 8.10   Hamming bits produced from a modulo 2 addition

Bit positions 15, 13, 11, 9, 7 and 3 contain binary 1s and the binary values of these bit positions are added using modulo 2 arithmetic to produce the Hamming bits (see Figure 8.10).

Bit							X				X		X	X
position	15	14	13	12	11	10	9	8	7	6	5	4	3	2	1
Data field	1	0	1	0	1	0	1	0	1	0	0	1	1	0	0

X = Hamming bits

Figure 8.11   The full 15-bit field

The full 15 bit field is shown in Figure 8.11, the lowest order Hamming bits being inserted into the lowest order bit positions. At the receiver, the binary value of each bit position containing a 1 bit is added using modulo 2 arithmetic. If there have been no errors, the result should be 0 as shown in Figure 8.12.

Bit position	Binary value	
15	1 1 1 1	
13	1 1 0 1	
11	1 0 1 1	
9	1 0 0 1	
7	0 1 1 1	
4	0 1 0 0	
3	0 0 1 1	
	0 0 0 0 ←	modulo 2 result

Figure 8.12   No errors

Bit							X				X		X	X
position	15	14	13	12	11	10	9	8	7	6	5	4	3	2	1
Data field	1	0	1	0	0	0	1	0	1	0	0	1	1	0	0

↑
error

X = Hamming bits

Figure 8.13   With errors

If a single bit error occurs in transmission, the position of the error will be indicated by the modulo 2 sum. For example, assuming an error in the 11th bit position in Figure 8.11, the data would be received as shown in Figure 8.13.

At the receiver, the binary values of positions 15, 13, 11, 9, 7, 4 and 3 would then be added as shown in Figure 8.14. The modulo 2 sum in Figure 8.14 indicates that the 11th bit position is in error and the bit is then inverted from 0 to 1. If 2-bit errors occur, the position of the errors cannot be determined from the modulo 2 result, although, as this will not be 0, the presence of a double error will be indicated (it can be ascertained that this is a double error, which will ensure that this non-zero result is not misinterpreted as a single error position).

Bit position	Binary value
15	1 1 1 1
13	1 1 0 1
9	1 0 0 1
7	0 1 1 1
4	0 1 0 0
3	0 0 1 1
11	1 0 1 1 ← modulo 2 sum indicates bit position 11 in error

Figure 8.14   11th Bit position in error

The code can correct single bit errors and detect double errors but some multiple errors will escape detection.

Principles

The essence of all error correcting codes is the addition of extra bits over and above the information to be transmitted. Since these coding bits do not serve any information carrying purpose, they are often referred to as *redundant*, although this is not strictly true.

The basis behind adding the redundant bits is as follows. Take as an example the case where the number of information digits to be transmitted is k. The number of possible messages using this number of digits is thus 2^k. If a few coding bits, c, are added to the block so that the block length is now n (n=k+c) then the number of possible combinations of received bits is 2^n, which will be larger than the number of possible messages that could have been sent (which is still 2^k).

The trick of error correction schemes is to choose how to generate the extra coding digits. When added to the information bits there should still be only 2^k possible messages, but because of the increased number of combinations now available these are now very distinct from each other. These coded messages are known as code words. If a set of code words is such that for each one of them they differ in at least *3-bits* from any other code word then single error correction is possible. Looked at simplistically, in the case of a single error occurring, the received bit pattern will be 1-bit different from one code word, and at least 2-bits different from all the others. The decoder can thus make a reasonable assumption that the closest code word was the one that was originally transmitted.

The smallest number of bits difference between any of the code words is known as the Hamming distance, d, and from this the detection and correction properties of a code can be found. In general, for a code of distance d, the number of errors that can be detected is d-1 and corrected is (d-1)/2 (or the integer just below when the result is not a whole number). Thus, for correction to be possible the value of Hamming distance must be at least 3.

Advanced Techniques

The technique described above refers to a class of codes known as block codes, which implies the neat division of the data transmitted into a set number of bits; for any one block there is a known relationship between the constituent information bits and coding bits. Another example of this sort of coding approach are cyclic codes, of which Bose Chaudhuri Hocquenghem (BCH) codes are of the most powerful.

Sometimes it is the case that the data being transmitted is not binary in nature and may instead consist of symbols which are representative of binary clusters (dibits, tribits etc). For these instances, non-binary BCH codes have been developed as well as another family known as Reed-Solomon codes.

Finally, a completely alternative method of FEC can be seen in the family termed convolutional codes. When using such codes it is no longer possible to freeze a number of bits in time and identify the information and the appended redundancy. This is because they are generated by combining in some mathematical relationship both past and present elements of the information stream. The result is, therefore, always dependent upon what went before. Thus, by inspecting the sequences received and knowing what the initial mathematical derivation was, the most likely original information stream can be gleaned. Unlike block codes there is no defined decoding decision instance (when it is known that a block of data has arrived). For convolutional codes the decoding decision can be made after any number of received bits, although the longer the sequence, the more likely it is that the correct estimate will be made. The trellis coding employed in V.32 modems is one particular use of convolutional codes.

Comments on FEC Methods

Forward error correction is obviously a very useful technique, particularly in the case of simplex transmission where there is no return channel available to perform an acknowledgement based process. The same argument applies to channels where the link delay is large and thus the time spent waiting for confirmation is a large overhead projector (although this can also be overcome by using the windowing techniques described later in this chapter). There is also merit in a modest combination of FEC and acknowledgement protocols, which would cut down on the number of retransmissions required.

The coding schemes mentioned above only provide a small insight into the possibilities of FEC. FEC techniques have been proposed for dealing with random single errors per block, and even multiple random errors in a block; schemes for countering bursts of errors are also well documented. Knowing which to adopt relies largely upon anticipating the error performance of the channel itself and on judicious selection of the block size. Increasing the block size might mean less redundancy, but will also increase the likelihood of more than one error occurring in the block and the consequence of this is a more complicated coding algorithm to deal with multiple errors.

The general effect of FEC is to reduce the error rate of a channel but at the cost of a proportion of checking bits to information bits, which reduces the transfer rate. The transfer rate of an FEC method can be given as:

$$\frac{RK}{n}$$

where   R   =   the data signalling rate;

k   =   the number of information bits transmitted; and

n   =   the total number of bits transmitted,

although for convolutional codes the values for k and n are not so readily obvious.

ARQ Methods

The common feature of ARQ methods is the calculation by the transmitter of some function of the block's contents. The function is transmitted with the block and recalculated by the receiver. The calculated value is compared with the transmitted value and, if the two agree, then a positive acknowledgement (ACK) is sent back to the transmitter; otherwise a negative acknowledgement (NAK) is despatched, and this in turn triggers a retransmission of the block in error.

An unlimited number of functions can be calculated as a check value, but the ones which have found most favour are:

—   the block check count (BCC) as used in the basic mode protocols;

—   The CCITT cyclic polynomial as used in extensions to basic mode and which is standard in HDLC.

Parity Checking and the Block Check Count (BCC)

The standard practice in basic mode and similar character-oriented protocols is to provide checks at two levels.

First of all an eighth parity bit is added to each 7-bit information character. The convention in the basic mode standard is that this should be chosen so that the total number of 1 bits in the character is odd. This technique will detect errors in a single bit position, and is illustrated in Figure 8.15. It should be noted that although the single bit error is detected, it is not possible to determine its location.

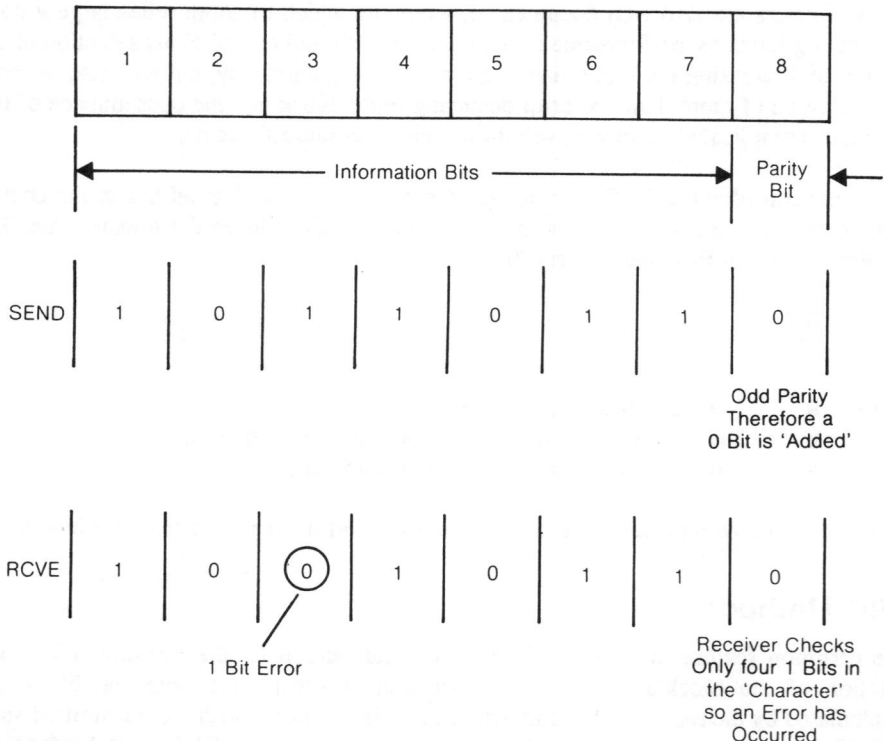

Figure 8.15 Character parity checking

In addition to the character or vertical parity check, it is also possible to define a single parity bit for each row of the block corresponding to specific bit positions of the constituent characters. This longitudinal parity check supplies the block check count (BCC), which is added to each transmitted block. Figure 8.16 shows how this is calculated. The BCC is composed of 7-bits plus a parity bit which is the parity of the BCC character not the summation of the parity bits in the block of text, and is accorded odd parity.

The longitudinal parity sense is arranged to be even. The opening start of text (STX) or start of heading (SOH) is excluded from the longitudinal parity summation but all succeeding characters including STX in a block started by SOH and the end of transmission block (ETB) or end of text (ETX) ending character are included. Synchronous idle (SYN) characters, if they occur, are excluded as a special case, because they could be introduced after formation of the block or text in some systems.

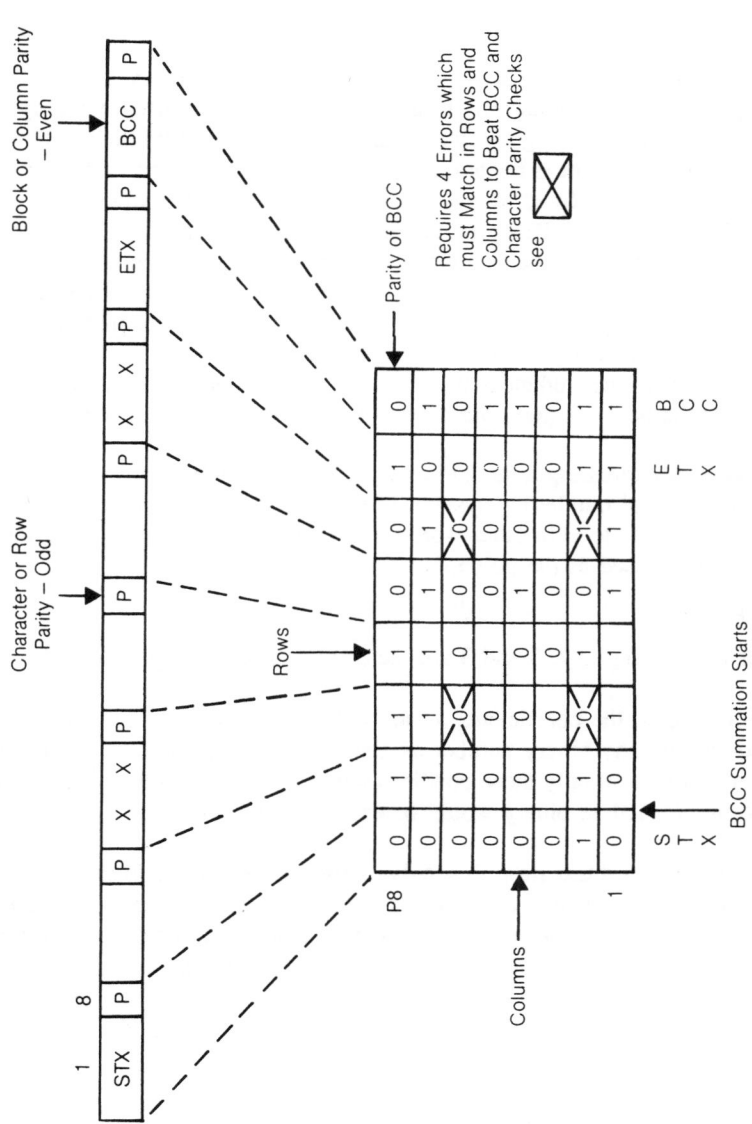

Figure 8.16 Character and block parity checks

This two co-ordinate parity check provides a powerful error detection capability. All block errors with an odd number of error bits can be detected as well as all 2-bit errors and some other even bit errors. Compensating errors are reduced but rectangular co-ordinate error patterns will remain undetected as indicated in Figure 8.16.

The Cyclic Polynomial Technique

Some of the most powerful codes in current use are known as cyclic codes or cyclic redundancy codes (CRC checks); these are not to be confused with cyclic error correcting codes. Cyclic coding involves a calculation at the transmitting station in which the block of data to be sent is treated as a pure binary number and is then divided by a predetermined number defined by a polynomial expression. The division is performed using modulo 2 arithmetic. This produces a remainder, forming the check digits which are transmitted at the end of the data block. At the receiving end the terminal repeats the division using the same predetermined number and dividing it into the received data including the check digits. If no errors have occurred during transmission the division will produce no remainder. Increased complexity of the cyclic code gives powerful detection but also greater redundancy with each block transmitted. The optimum codes are a compromise between good burst error detection and low redundancy.

The concept of expressing a number as a polynomial can be more readily understood by use of the decimal numbering system. In this, numbers are constructed representing units, tens, hundreds etc — in other words, successive powers of 10. Thus, the number 753 could be expressed as the polynomial:

$$7x^2 \quad + \quad 5x^1 \quad + \quad 3x^0$$

where x is 10. The digits 7, 5 and 3 are the coefficients of the polynomial.

In the binary case the coefficients of any polynomial can only be 1 or 0. Thus, it is possible to express any binary number as a series of powers of x (x is 2 in this case), with a term to represent only those positions where a 1 occurs in the binary number. As a simple example, 101 in binary would be expressed as $x^2 + 1$.

In order to demonstrate how this polynomial can be used, Figure 8.17 shows a combination of shift registers and a binary arithmetic function performing modulo 2 addition. This adder is the same as a logical exclusive-OR function (whose output is 1 when the two inputs differ, and is 0 when they are the same).

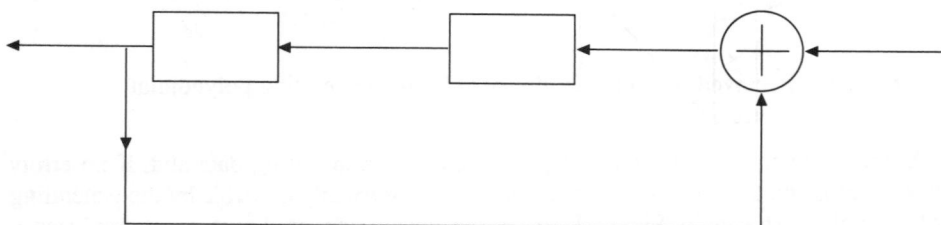

Figure 8.17 Shift register representation of the $x^2 + 1$ function

Any binary number (of any length) can be shifted into this arrangement and will, in some way, be operated upon by the function $x^2 + 1$ (since it is these two elements of the shift register sequence that are being added together). In actual fact, it is this process that is performed for the modulo 2 division. Thus, if the binary sequence 1001001010 (which could itself be written as a polynomial) were passed through the circuit in Figure 8.17, the result would be 1011110 with a remaining two 1s left in the shift registers. The patient reader may verify this by clocking through all the stages, but an easier way of corroborating this is to perform the equivalent of a long division, which is shown in Figure 8.18. Note that the main arithmetic process is a bit-wise modulo 2 addition with the generating polynomial and that a 1 is placed in the result at every point where such an addition is performed.

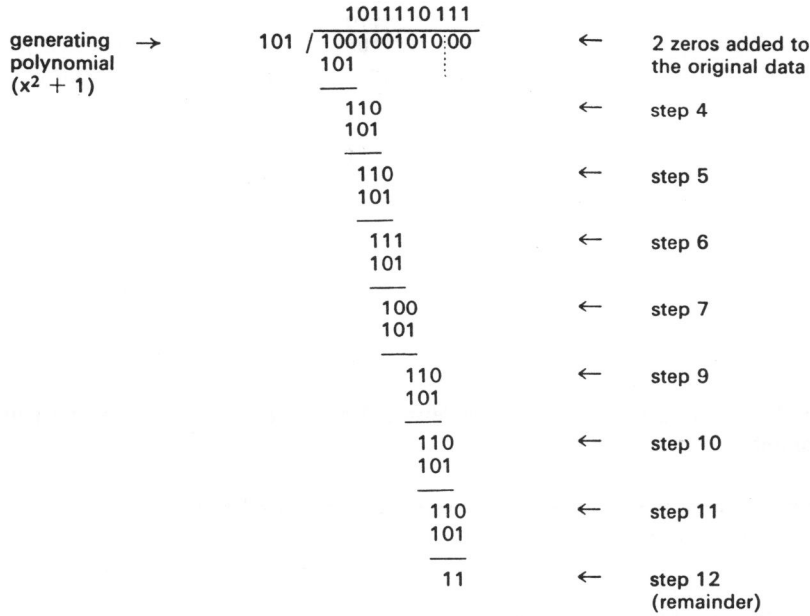

Figure 8.18 Received data exactly divisible by the generating polynomial

At the receiver, the calculation is performed on the incoming data and, if no errors have occurred on the line, the received data should be exactly divisible by the generating polynomial as shown in Figure 8.19. If the calculation at the receiver produces a remainder other than zero, this indicates that an error has been detected in the received data and the receiver requests a retransmission.

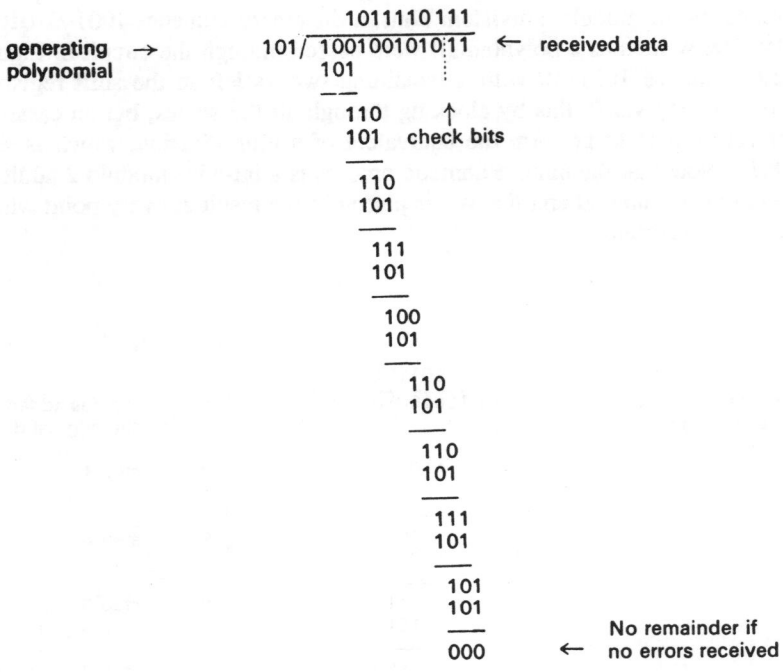

Figure 8.19 Division at the receive terminal of the received data by the generating polynomial

The polynomial which is generally used is that specified by CCITT Recommendation V.41. It is expressed algebraically as:

$$x^{16} + x^{12} + x^5 + 1$$

and a suitable circuit for encoding and decoding is shown in Figures 8.20 and 8.21. During the encoding process, gates A and B are held open, whilst C is closed. When all data has been clocked in, A and B are closed and C opened so that the CRC can be clocked out. At the receiver, the normal decoding process is performed with gate D closed and all the others open.

Through theoretical studies and subsequent practical experience this has been proved to be a very powerful error detection technique. For example, it has been found from computer simulation that when using a block size of 260 bits (including service bits and check bits) an improvement factor in the order of 50,000 is achieved. On a circuit with a mean error rate of 1 in 10^4 the residual error rate would, therefore, be in the order of

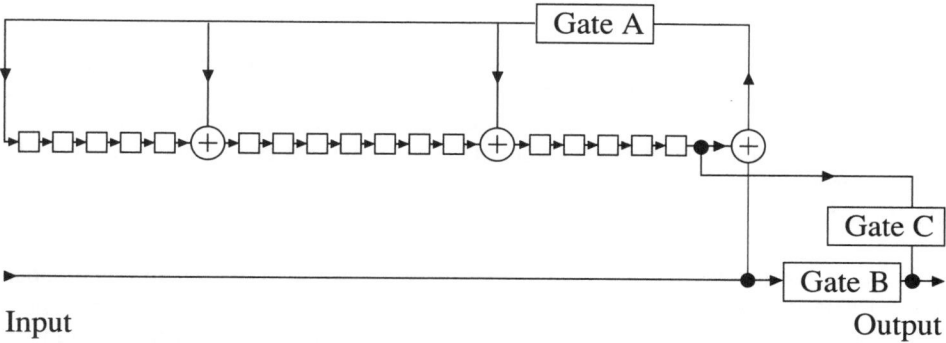

Figure 8.20 An encoding arrangement using the 16-bit polynomial $x^{16} + x^{12} + x^5 + 1$ (CCITT Recommendation V.41)

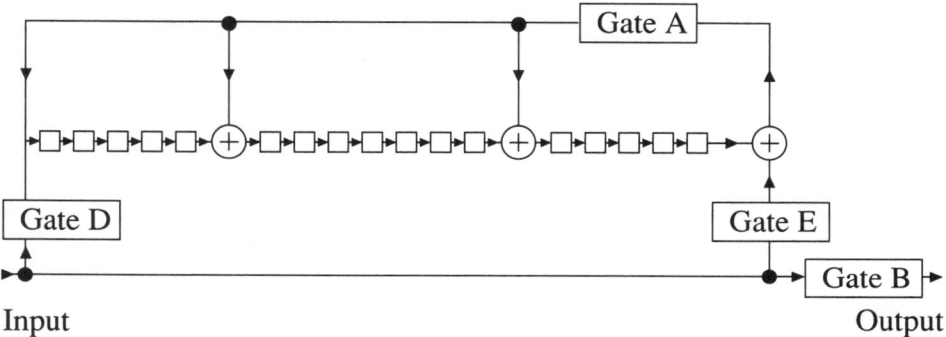

Figure 8.21 A decoding arrangement using the 16-bit polynomial $x^{16} + x^{12} + x^5 + 1$ (CCITT Recommendation V.41)

2 in 10^9, the 16 redundant bits comprising only 6.1 percent of each block. All odd numbers of errors within a block would be detected, as would any one error burst not exceeding 16 bits in length and a large percentage of other error patterns.

Efficiency Considerations and Variations and ARQ Techniques

Block Size

The size of the block used is often determined either by the operational requirements of a system or the hardware used, rather than the need to optimise the throughput of transmission links. In most cases, the blocks transmitted are of variable length although the upper limit is usually conditioned by the type of message or by the terminal characteristics such as buffer capacity. On alphanumeric display terminals, for example, the maximum block size is likely to be determined by the size of the buffer although the block sizes actually transmitted will vary considerably depending on the size of the input and output messages.

From an error control point of view, blocks should ideally be of fixed length, the actual length being determined by the probabilities of errors occurring and the nature of the error bursts likely to be encountered. There is, therefore, a conflict between the need for efficient transmission of information and other important design considerations. In on line systems, it is difficult to reconcile these differing requirements and compromise is necessary.

There are advantages in using large blocks. Firstly, as we shall see, where synchronous transmission is employed with half duplex control procedures, a significant period may elapse between blocks being transmitted. This period is constant whether blocks are long or short and the larger the block the less time will be wasted. Secondly, with a large block the proportion of redundant data to information may be smaller, for a given degree of protection, than with a smaller block.

There are also, however, disadvantages in using large blocks. First of all, these have to be retransmitted when necessary and this increases the local buffer storage requirement. However, with the declining costs of memory and intelligence, this factor is reducing in importance. Secondly, an increase in the size of the block transmitted produces a corresponding increase in the probability that the block will contain an error. This is compounded by the fact that a longer block will also take more time to retransmit and excludes other users from the link. Thus, finding an optimum block size is always a trade-off between throughput and expectation of channel error performance.

Throughput Efficiency

The throughput efficiency of a system in an error free situation is given by the equation:

Equation 1

$$E = \frac{\dfrac{B}{R}}{\dfrac{B}{R} + T}$$

where:
E	=	the throughput efficiency
B	=	the block length in bits
T	=	the total round trip delay
R	=	the input data rate (bit/s)

The losses due to parity bits or to control and synchronisation bits within blocks are ignored in this equation as these factors are independent of those given and their effect is best calculated and described separately. Also excluded is the effect of erroneous return messages, as the probability of these being in error is much smaller than in the relatively longer message blocks.

In considering throughput in the presence of errors, an additional consideration is the number of blocks in error relative to the total number of blocks sent. We can now extend equation 1 to allow consideration of throughput in the presence of errors so that:

Equation 2

$$E = \frac{\dfrac{B}{R}\,(1 - P(e))}{\dfrac{B}{R} + T}$$

where: P(e) = the block error probability

It will be seen from 1 that the duration of T is the critical factor in determining throughput efficiency. The total round trip delay T is the time delay between the end of transmission of one block (block n) to the beginning of transmission of the next (block n + 1). Some of the elements that T may comprise include:

— the loop propagation time of the line;

— receiver delay in recalculating the CRC;

— the effect of polling;

— modem turn-round times;

— modem propagation delays; and

— others of lesser significance.

Also to a large extent, the way in which the ARQ scheme is implemented can affect the delay T. There are various ways of organising the retransmission of erroneous blocks in ARQ systems and we now review these and their relative efficiency.

Idle-RQ

This is the method described in the discussion of basic mode protocols. Reference to Figure 8.22 shows that the requirement to transmit a block and then wait for a response before transmitting the next block (or retransmitting the same block) causes gaps (T) when the data blocks are not being transferred. For this reason it is sometimes referred to as idle-RQ.

Since the communications channel is usually the slowest part of a communication system, any inefficiency in its use impairs the performance of the whole system. If T is small compared with the total transmission time of a data block the reduction in efficiency may be quite small, but on wideband or long distance channels it will become increasingly inefficient as the block transmission time approaches the idle time.

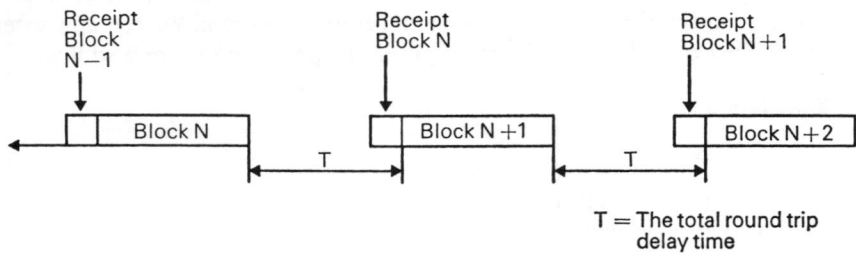

Figure 8.22 Simple half duplex error control (idle-RQ)

Consider now an idle-RQ system where a simultaneous 75 bit/s return channel is available and thus modem turn-round times can be ignored. In the first case, error free transmission is assumed so that the effect of errors on throughput efficiency may be seen in better perspective later.

A round-trip propagation delay of 25ms is assumed, as is a block length of 1000 bits (B), a forward channel data rate of 1200 bit/s (R) and a response over the 75 bit/s link of two characters (16 bits). In this case, the delay, T, is the sum of the propagation delay and the length of time to transmit 16 bits over the return channel, and can be calculated as 238ms (T). By using equation 1 it can be determined that the efficiency for this set of operational criteria is 77.77 percent.

Using the basic data from this example, we now assume that 1 block in 100 has to be retransmitted because of detected errors. Thus, using equation 2, an efficiency of 76.99 percent can be calculated, a drop of less than one percent.

These figures indicate that even if errors occur in a purely random fashion at a rate of no more than 1 bit in 20,000 at 1200 bit/s on a leased circuit, a maximum of 5 in 100,000 bit blocks would be in error. Due to the clustering effect of error bursts, the proportion of erroneous blocks would probably be a good deal less than this in practice. It will be seen from the simple example given that errors have a relatively minor effect on throughput efficiency with blocks of this size, when compared to the total round trip delay T. Nevertheless, although it is obvious from equation 1 that increasing the block size will improve the throughput efficiency in the absence of errors, the presence of errors does impose a restraint as the block error rate will increase with block size. There is, therefore, an optimum block size for any given bit error rate and total round trip delay T, which will give maximum throughput efficiency.

However, it will be remembered that there are other factors which influence block size. These can have quite significant consequences for the throughput efficiency of some idle-RQ systems, particularly where variable block lengths are used.

In the above examples a simultaneous return channel was used for message acknowledgement. Consider now data being transmitted over the PSTN at 2400 bit/s; assume that no simultaneous return channel is available and the modems, therefore, have to be turned round. Two reversals per block will be required, one to establish the return channel and one to re-establish the forward channel.

Using all the same figures as before (except the higher speed channel), and modems with a turn-round time of 75ms, the total delay, T, can be calculated as 182ms, less than previously, since the return channel is much faster. However, this decrease is to no avail, as the data rate also lessens the block transmission time, and the net effect is an efficiency quotient of 69.6 percent. Further increasing the signalling rate only exacerbates the problem. The same calculations for 4800 bit/s reveals efficiency down to 53.89 percent. However, it should be borne in mind that the overall throughput will still increase because of the higher signalling rate, albeit at reduced efficiency.

It can be clearly seen from this example that it becomes increasingly important to use longer blocks as the data signalling rate increases. However, the problem arises when the length of the block becomes greater as the block error probability also rises. Another problem is that the adaptive equalisers used on some high speed modems may increase the modem turn-round times and throughput efficiency may suffer. Full duplex modems do not have the turn-round problem, of course.

The above calculations have been presented primarily to demonstrate the inter-relationship of the more important factors which determine throughput efficiency. Costs have been ignored for the sake of clarity, but it will be seen that if the throughput efficiency is known, the cost per bit or cost per 1000 bits can be calculated for different circumstances.

Despite its limitations the idle-RQ technique is simple and reliable and has been employed in a large number of communications based systems. A major factor which influences the total round trip delay in the UK is modem turn-round time and this is avoided by using duplex facilities. There is a problem too on long distance circuits. While loop propagation times are normally small on circuits within a country the size of the UK, on intercontinental circuits, satellite links or circuits across larger territories (for example, the USA) may be extremely long and idle-RQ becomes less suitable. For example, let us assume a half duplex intercontinental connection via a satellite with a total propagation delay of 540ms with the other data as in the previous example; the throughput efficiency would then be an optimistic 37 percent on a circuit with no errors at all.

Continuous RQ

It will now be clear that much greater efficiency can be obtained with a procedure for transmitting in full duplex mode on duplex channels. Some indication of the effect on throughput efficiency can be gained by subtracting modem turn-round time from T in the example given in the previous section using the 2400 bit/s link, and the calculation repeated. When this is done the result is an efficiency of 93 percent. The class of continuous RQ techniques was developed for this purpose. In this approach, the data blocks are transmitted continuously over a duplex link with the objective of minimising idle time.

The procedure works as follows: A number of data blocks are transmitted in succession by the transmitter without waiting for individual block acknowledgements. At the receiver they are processed and acknowledged as in idle-RQ. However, by the time the transmitter gets an acknowledgement, it will generally have transmitted further data blocks, so that acknowledgements will lag behind the transmitted data blocks. This is illustrated in Figure 8.23.

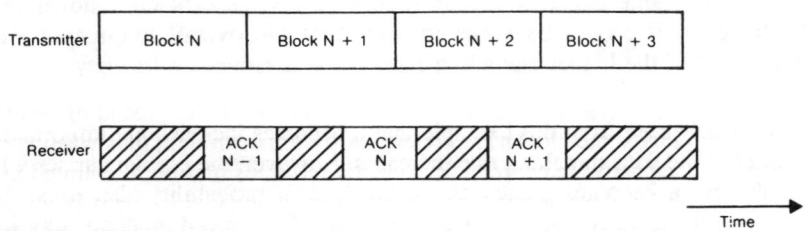

Figure 8.23   Continuous RQ

In this mode of operation a number of difficulties arise. First of all, at the transmitter each unacknowledged block transmitted needs to be stored in case retransmission is required. Secondly, at the receiver if a block is found to be in error, then a number of subsequent blocks may be received before the block in error can be retransmitted, and this may disturb the sequence of the data blocks. Therefore, in order to control sequence and to maintain correspondence between blocks and their respective acknowledgements, it is necessary to include in each block a unique block sequence number.

In continuous RQ there are two different techniques for handling retransmission; Go-Back-N, and Selective Retransmission:

> — *Go-Back-N*. In this scheme, when the receiver sends a NAK, indicating that the block with the corresponding sequence number must be retransmitted, the sender transmits that blocks, and then continues transmitting subsequent blocks, even though they may already have been transmitted. The procedure is illustrated in Figure 8.24.
>
> Following receipt of the erroneous N + 1 block the receiver ignores blocks N + 2, N + 3, N + 1, since they are out of sequence, and waits for the retransmitted block N + 1 to arrive. When block N + 1 has been transmitted, the transmitter continues with blocks N + 2, N + 3, and so on. Each end is able to keep track of the situation by reference to the block sequence numbers.

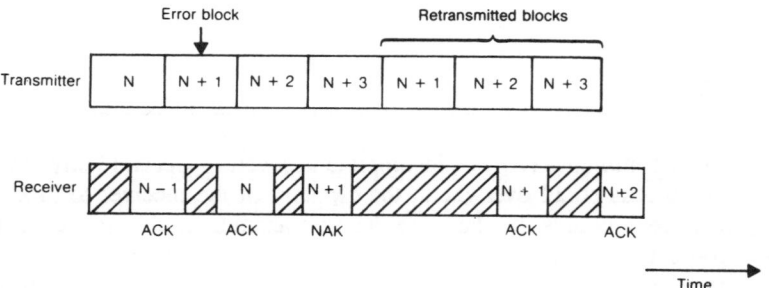

Figure 8.24 Go-Back-N

—   *Selective Retransmission.* A deficiency of the preceding approach is
that some blocks are unnecessarily retransmitted, thus wasting channel
capacity. In contrast for Selective Retransmission, illustrated in Figure
8.25, the transmitter only retransmits block N + 1, the block in error,
whilst the receiver accepts blocks N + 2, N + 3, even though N + 1
has to be retransmitted and will, therefore, be out of sequence when it
is received. After retransmitting block N + 1, the transmitter continues
with blocks N + 4, N + 5, etc.

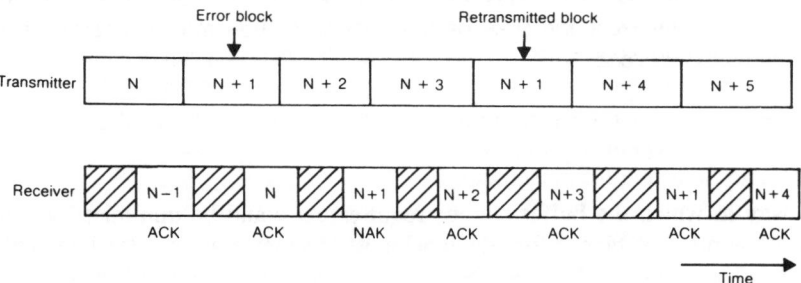

Figure 8.25 Selective Retransmission

Both approaches have their advantages and disadvantages. Selective Retransmission
makes more efficient use of channel capacity compared with G-Back-N, but entails a re-
sequencing operation at the receiver. Both methods introduce a buffering requirement
exceeding the single block storage requirement of idle-RQ. Go-Back-N requires buffering
at the transmitter to enable retransmission, and Selective Retransmission also requires it
at the receiver to enable the correct sequence to be re-established.

We have covered ARQ and its ramifications in some detail. This is firstly because the
basic ARQ principles are utilised in the HDLC protocol, and secondly, because it

provides a useful background for understanding HDLC which was designed to overcome the limitations inherent in character oriented protocols.

Flow Control

The discussion of continuous RQ was incomplete in several aspects. Firstly, unrestricted transmission of blocks could overwhelm the capacity of the receiver to process them. Secondly, if large amounts of data are being transmitted, the sequence numbers themselves could also become correspondingly large, and could quite easily exceed the space allocated to them in the block. These considerations introduce the requirement for an efficient flow control procedure. The chief functions of flow control are:

- to maintain the correct sequence of data end-to-end between transmitter and receiver;

- to ensure that transmission rates match the processing capacities of the two ends. This includes both processing capability and buffer storage provision;

- to supply an additional mechanism which will be required for one end to inform the other that it is congested. This tells the other side of the link that no further information, apart from perhaps control information, is to be transmitted until the congestion has cleared;

- to permit, in certain necessary circumstances, one end to regain control of the link. This introduces the requirement for a mechanism to interrupt the flow of information to transmit an urgent message or request;

- to optimise the utilisation of the link which is, of course, the final overall requirement.

One way in which the buffer storage requirement could be minimised would be to restrict the number of blocks for which acknowledgements are outstanding. This is the approach adopted in the Go-Back-N technique, N being the number of outstanding blocks.

A further refinement which keeps the sequence numbers down to a manageable size is to interpret them in a modulo arithmetic. This is equivalent to dividing the block number by the modulus number and using the remainder as the actual sequence number.

Thus, in an arithmetic equation using a modulus of 8, sequence numbers would continuously cycle round the values 0, 1, 2, 3, 4, 5, 6, 7. The restricted number of values does not constitute a limitation since in HDLC, where this is employed, the combination of sequence numbers with the modulus provides the actual basis for an efficient method of flow control. What is required on top of this is an agreement regarding how these numbered blocks are to be handled and some method of carrying this dialogue. These are the essential ingredients of a link control *protocol* and how these techniques are put into practice will be seen a little later in the chapter.

Data Compression Techniques

So far we have covered techniques which either add resilience or order (or both) to the communications process. In the course of all this discussion, control bits and bytes and blocks have been added to the information to be transferred without much concern as to the overall effect on the transmission process. In equation 1 above (and in 2) the efficiency of the information transfer was given ignoring these factors. However, the block of length B as used in this equation does contain overheads. A second value, I, can be introduced (which is usually less than or equal to B) and this represents the amount of pure information to be transferred. Thus, the efficiency of the information transmission process can be restated as:

Equation 3

$$E = \frac{\dfrac{I}{R}}{\dfrac{B}{R} + T}$$

In the usual case I is less than B, and as a result the calculated efficiency is reduced.

However, suppose techniques could be found whereby the size of the block B could be reduced without necessarily compromising the degree of information transfer. The result would be an increase in perceived efficiency, since in terms of the end systems, the conceptual value of I remains the same, but in actual communications terms B has been decreased. This is the function of data compression techniques.

Data compression makes use of actual redundancy that exists in the data in order to find ways of transmitting this using less bits. In many cases this requires some prior knowledge of the nature of the data, although this is not necessarily a drawback. The simplest techniques may seem remarkably obvious and not worthy of comment. However, it is surprising how easily these are overlooked, and they are often used in conjunction with some of the more complicated schemes. A brief description of common techniques is given here and for a fuller treatment the reader is referred to the bibliography.

Run Length Encoding

In many cases, the information to be transmitted consists of a document represented in ASCII or some other widely used character set such as EBCDIC. A particular tendency is for these documents to contain instances of character repetition. For example, tables of information may contain several occurrences of multiple space characters.

One way of compressing this information is to reserve a special character which is used to denote a *character run*. The way in which twenty spaces would be sent would then be in the following sequence — special character and space, followed by the binary value of twenty. Using this simple example a string of twenty bytes has been reduced to three.

Special cases of this technique exist where only one character is singled out as being repetitious and any use of the compression character denotes the occurrence of this one code. Thus, the compression is reduced to two (the special character and the repetition count) as there is no need to indicate the character being compressed. This technique has been used quite extensively with terminals that transmit a lot of null characters and is sometimes called *null suppression*.

One of the drawbacks of this technique is that compression only begins when there is a repeat length of at least four (or three for null suppression) characters. Below this threshold there is nothing to be gained by compressing. Thus, it is of no avail when the information consists of an often repeated pattern of characters. Another problem is that the amount of compression achieved is difficult to quantify without detailed knowledge of the document being transmitted; moreover, the degree of success achieved is closely dependent upon the document content. It also assumes the availability of a spare code to denote the compressed sequence.

Diatonic Codes and Pattern Substitution

Both of these techniques rely on the availability of special codes where compression has occurred. In the case of diatonic coding the occurrence of a pair character is denoted by one single special character and the more pairs that are to be substituted the more special codes are required. One example of a character pair suitable for such treatment might be carriage return and line feed; other potential candidates would again be dependent on the information to be sent.

In the English language, the most common character pairs can be identified by analysis of several texts drawn at random. However, suppose that the information consisted of the source code of some applications software. Due to the change to syntax and language use it has been ascertained that the common character pairings will be quite different in nature. Furthermore, different programming languages reveal their own particular biases and further analysis reveals that there is a strong relationship between the common pairs and the keywords found in the programming language. This suggests a refinement to the above technique, where not just pairs but whole groups of characters or phrases are replaced. In the example of the programming language, the most common keywords would be reduced to only one special character. This is less easy to accomplish with plain text but can still be used with some degree of success by substituting the words like 'the', 'and', 'this' and others. Even if this character pattern appeared in the middle of another word, for example 'and' in the word 'pandemonium' substitution would still be acceptable.

Statistical Encoding and Adaptive Techniques

The techniques discussed so far have relied heavily upon the type of information being transmitted and, in the main, have assumed some form of standard coding of characters in terms of the bits transmitted. A different technique, *statistical encoding*, although used extensively on character sets, has also proven versatile enough to be employed in applications such as the compression of the data transmitted by fax machines.

The technique is most readily explained by reference to a simple case, namely, the ASCII codes for alphanumeric characters. The makers of Scrabble were quick to realise that some letters occur with much greater frequency in the English language than others. Hence, there are twelve letter e's in the game and only one z supplied. This is a fair representation of the relative probability of occurrence, in actual fact, the measured value for the probability of a letter e is around 13. However, when this is translated into ASCII, it occupies seven (or eight with parity) bits. This is the same as infrequent letters such as z and q.

A much more effective representation would be to allow codes of multiple length, reserving the shorter codes for the common characters and the longer ones for those which are less frequently used. The best known example of this in operation is Morse code.

There are two well known algorithms for allocating event probabilities with variable length codes and these are known as Huffman encoding and Shannon-Fano codes. These two adopt slightly different approaches and the merits of each depend upon the circumstances. A rough rule of thumb is that Huffman encoding is more efficient when the spread of probabilities is not great; in statistical terms, this is described as a set of probabilities with low variance.

The applications of statistical encoding are not limited to character sets. For instance, a modified form of Huffman encoding is used to compress the data coming out of fax machines. The sort of output associated with faxes is a line-by-line scan of a page where the information consists of runs of black and white. The run lengths have a statistical nature to them and it is possible, therefore, to adopt a variable length code approach similar to the Huffman ASCII representations, but with special features to accommodate the more abstract nature of fax output.

One of the problems with statistical encoding, like other compression techniques, is the initial assumption of data content. In this case, the data to be sent has been analysed to establish the probabilities of the various constituent elements. It was stated before that in plain text the letter e is the most common and z the least; in between letters such as w and f occur with a lower than average degree of frequency.

However, suppose that the bulk of data transfer revolves around source code of software programmes. The result is that the estimations that were made for letters such as w and f, for example, will be wrong. This is due to their association with common programming language keywords such as *format, write, if* and *when*. Many other examples exist where the document content causes a skew of the actual probabilities. When combined with the processing overheads required to decode the compression, such a probability shift from the optimum can severely minimise the advantages of employing compression.

Since it is seldom possible to specify precisely what data will be transmitted, a slightly different approach can be adopted. This requires that both the communicating parties keep a track of the occurrences of each of the elements of data and continually recompute the respective probabilities. From this, the variable length codes are constantly re-assigned and hence the name *adaptive statistical encoding*. Obviously, the processing overheads involved are much greater than even straightforward Huffman or Shannon-Fano techniques. However, with the trend towards intelligent peripheral devices performing communications tasks, many practical applications of adaptive statistical encoding are now being implemented and, in many cases, this is completely invisible to the end users.

Structured Data Transfer

The techniques discussed so far have a common thread, which is that knowledge of the composition of the data has been an initial assumption. This is in marked contrast to the role of the electrical interchange circuits and signalling arrangements described in the preceding chapter. At this level the primary concern is the establishment, maintenance and termination of the physical link in terms of electrical signals representing streams of otherwise undefined binary bits, rather than conceptual structures such as characters, blocks and messages. However, having witnessed the power of such processes as error protection and data compression, the importance of defining these conceptual structures is fairly evident.

The practice of assembling bits into characters is universally accepted and needs little elaboration here. However, regardless of this, when faced with the requirement to construct and transmit larger messages, whether or not these consist of character strings, there are essentially two different approaches.

In both cases there is a fundamental process that occurs, identifying where a unit of information starts and where it finishes. In one case the information content is structured rigidly around the character format and is described as being *character significant*. In the second case, however, there is no such assumed discipline to the information (although it is often there) and, thus, the important unit of data reverts back to being the bit. This approach is most often referred to as being *bit significant*. These two techniques will be described by way of their best known implementations. Exemplifying the former is the basic mode of transfer, while the concept of bit significance will be seen to play an important part in high level data link control (HDLC). Whilst it is useful to appreciate the differences between these approaches, it is the techniques themselves which are of greater relevance, particularly in the case of HDLC.

Basic Mode

Basic mode is a character significant protocol based on the ISO standard 7-bit alphabet which has 10 characters (representing, for example, start or end of text), allocated to transmission control. The protocol was designed for half duplex operation between a transmitting station and one or more receivers on a physical point-to-point or point-to-multipoint data link.

The primary station is responsible for scheduling the data flow in the link by polling and selecting secondary stations, using specific address characters. This process authorises the secondary station to transmit data to and accept data from the primary station.

Transmission Control Characters

Data transmitted in basic mode is usually formatted into fixed length blocks, and section of each block are identified by the transmission control (TC) characters which have unique bit patterns. The 10 TC characters are listed below and their use as supervisory messages is shown in Figure 8.26.

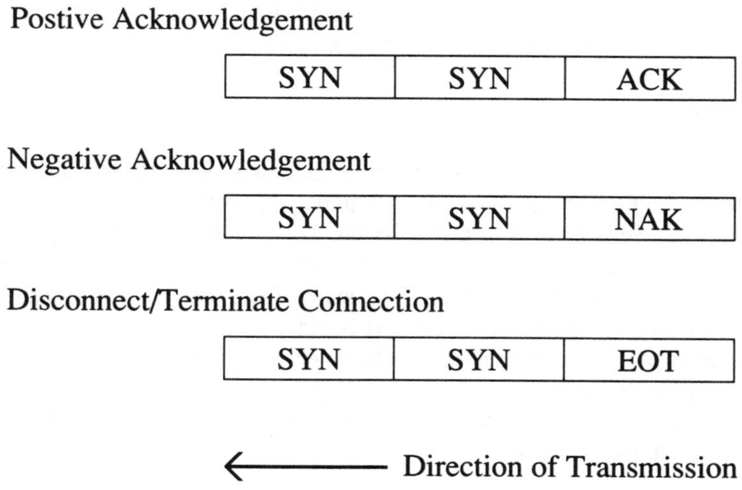

Postive Acknowledgement

SYN	SYN	ACK

Negative Acknowledgement

SYN	SYN	NAK

Disconnect/Terminate Connection

SYN	SYN	EOT

←——————— Direction of Transmission

Figure 8.26   Basic mode transmission control characters as supervisory messages

—   SOH (Start of Heading): indicates the start of the header (if present) of an information message or block.

—   STX (Start of Text): terminates a header or signals the start of a text string.

—   ETX (End of Text); signals the end of a text string.

—   EOT (End of Transmission): indicates the end of transmission of one or more text blocks and terminates the connection. It is also used as a polling character.

—   ENQ (Enquiry): requests for a response from a remote station.

—   ACK (Acknowledge): positive acknowledgement transmitted by a receiver in response to a message from the sender.

—   DLE (Data Link Escape): changes the meaning of other selected TC characters.

—   NAK (Negative Acknowledge): negative response transmitted by a receiver to a message from the sender.

—   SYN (Synchronous Idle): provides the means for a receiver to achieve or retain (idle condition) character synchronisation with a synchronous transmission control scheme.

—   ETB (End of Transmission Block): indicates the end of a block of data when a message is divided into a number of such blocks.

Basic Mode Messages

As all the blocks are sent over a synchronous link, it is vital to establish correct bit timing between the primary and secondary stations. This is done by sending at least two SYN characters at the beginning of transmission. These characters normally precede text blocks (or data) but may also be embedded in the data or sent continuously as *idle characters* when no data is being sent. The start of the text is then signalled by the transmission of an STX character. If the text is more than one block in length, then an ETB character is sent, otherwise an ETX character is sent (see Figure 8.27). Finally, a BCC (block character check) is transmitted to guard against the occurrence of errors in the data. The check starts with the STX character and ends with the particular end of block delimiter character being used.

The receiver is required to reply to the entire block either by acknowledging the block as being correctly received (using an ACK character), or received with errors (using a

NAK character). Assuming the block is correctly received, the positive acknowledgment reply would be as shown in Figure 8.26.

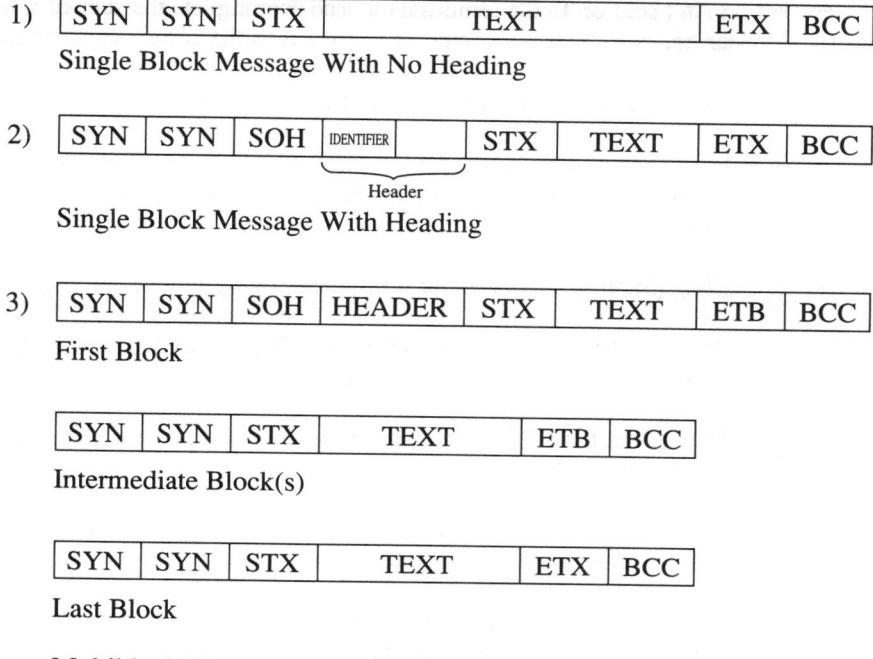

1) Single Block Message With No Heading

2) Single Block Message With Heading

3) First Block

Intermediate Block(s)

Last Block

Multiblock Message With Heading

Figure 8.27 Basic mode transmission control characters for data transmission

Code Transparency

This example shows that the line control method must be able to identify and act upon the unique sequences of bits which make up the transmission control characters. However, unpredictable data may well contain bit patterns which represent control characters, and so a DLE (data link escape) character is used to prefix any genuine control characters, for example DLE:STX. Naturally, DLE:STX could also occur in text, so as a simple rule to counteract this, whenever a DLE is detected in the text by the transmitter, it adds an extra DLE. This means that DLE:STX would be transmitted as DLE:DLE:STX. It then remains for the receiver to remove one DLE from each pair and to restore the text to normal. A single DLE prefix to another control character now identifies a true control sequence and transparent transmission is achieved.

Interleaving

In certain situations a number of stations can use a multiplexer to enable all the data to be transmitted over a single link. To ensure that data is passed over the link in an orderly manner, the multiplexer must interleave the data from each station. The multiplexer performs interleaving by giving each station on the network a specific time slot in which to transmit a character (or indeed a bit within bit oriented systems). Figure 8.28 shows four stations (A, B, C and D) which transmit over a single link. Every fourth character that passes over the link belongs to the same station. The demultiplexer at the receiving end has to take each character that it receives, and direct it to the correct destination.

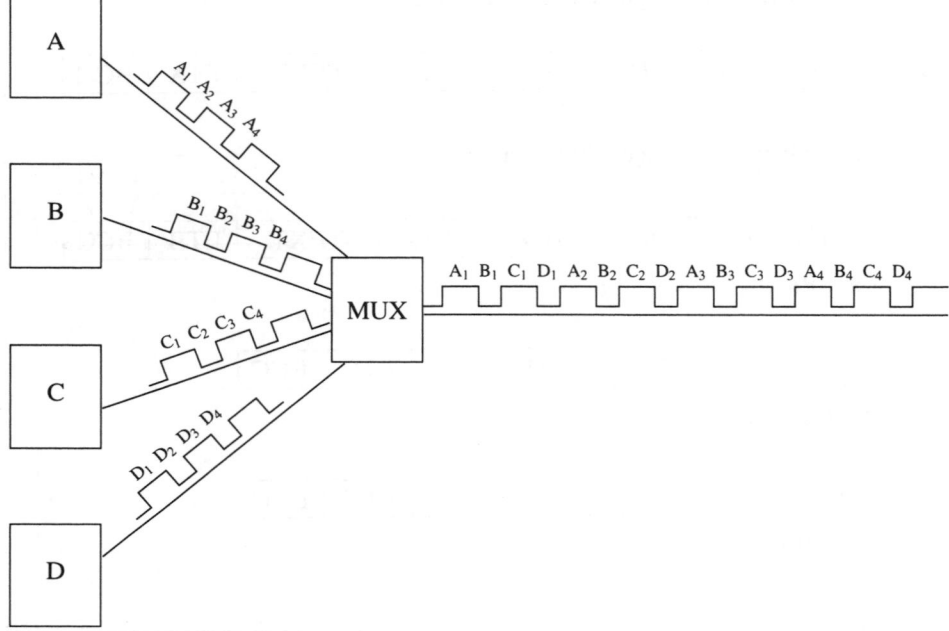

Figure 8.28 Data interleaving

Bit Oriented Protocols

The fundamental problem with basic mode operation is its lack of data transparency, which means that the system is inherently susceptible to certain bit patterns. Also, it insists on transmitting data over the communications link as a series of octets, even if the data itself does not naturally break down into 8-bit groups. Bit oriented protocols have been developed to transmit completely unstructured data where there is no division of user information into octets. This facilitates code transparency and also makes the data transmission more efficient.

High Level Data Link Control

Due to the drawbacks which are inherent in basic mode protocols, newer frame based and bit significant protocols were developed. These were based on the ISO developed HDLC and include SDLC (IBM's Synchronous Data Link Control) and ADCCP (Advanced Data Communications Control Procedure, developed by the American National Standards Institute, ANSI). HDLC is discussed in more detail in Chapter 10.

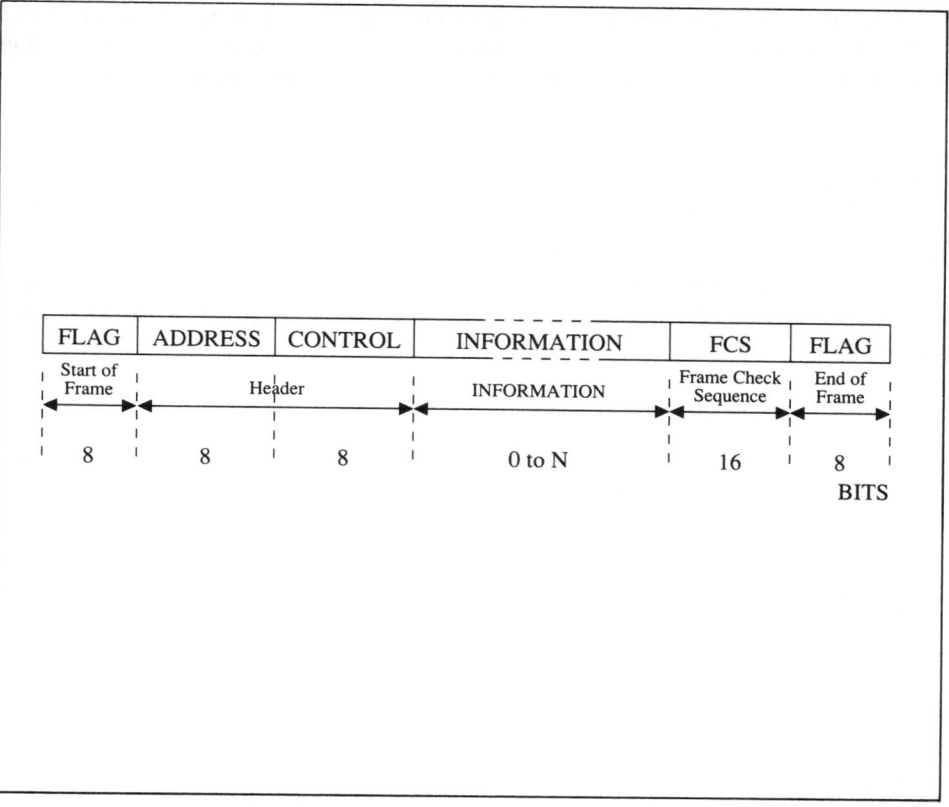

Figure 8.29 The HDLC frame

HDLC General Features

HDLC contains many features and benefits which are not present in earlier basic mode protocols. HDLC is able to:

— set up the link in either balanced or response mode as required, and close down the link with routine or emergency procedures;

— operate full duplex or half duplex exchanges of information, with flow control effected by the frame numbering and window size;

— handle single or multiple frame acknowledgement by using controlled retransmission of frames in error;

— operate over all types of link including point-to-point, point-to-multipoint, loop or ring;

— provide small control and code transparency overheads;

— provide a very powerful error detection system.

With such a range of attributes, it is not surprising that HDLC has been defined by ISO and has been incorporated into X.25 networks as an efficient and robust data link layer protocol for carrying packets across a link into packet handling networks.

Protocols and Architectures

So far, this chapter has focused on data link control procedures that provide the rules by which two or more devices may converse, or exchange information, over a data link in an efficient and reliable manner. In its most basic form, the protocol has to provide methods to:

—   identify the sender and receiver on a multipoint or dial-up network;

—   indicate the start and end of transmission;

—   detect errors and initiate some form of corrective action.

Since these link control procedures are so fundamental to the functioning of data networks, it is likely that they will continue to be the most important kind of protocol for quite some time.

However, the link control protocol represents only one layer of a hierarchy of protocols that, for example, allow transmission integrity over multiple links, or allow processes within one device to communicate with processes in another. By putting a set of protocols together into a hierarchy, an architecture is formed that provides a complete set of rules by which effective and reliable communications can occur.

Towards Open Systems Interconnection

One such protocol hierarchy, or architecture, has been defined in a model created by ISO. The aim of the model is to guide standards involving computers and communications, so that any computer or terminal connected to a data communications network will be able to communicate meaningfully with any other device on that network.

The ability to communicate in this way is called open systems interconnection (OSI). A set of protocols that permits OSI and the relationship of these protocols to one another is described as an open systems architecture.

The OSI structure calls for co-operation among systems from different manufacturers and with different designs. OSI therefore has to coordinate many activities, including:

—   interprocess communication (the synchronisation between OSI application processes — such as a person operating a data terminal, or a program executing inside a mainframe to access a database — and the exchange of information);

- data representation (the creation and maintenance of the methods used to reformat data exchanged between systems);

- data storage (the provision of storage media, file systems and database systems for providing access to, and managing stored data);

- process and resource management (the means by which application processes are initiated and controlled);

- integrity and security (the provision of information processing constraints that need to be assured through application processes);

- program support (the means of accessing the programs executed by an application process).

With manufacturers continuing with their proprietary protocols, these activities are difficult to coordinate. In addition, the continual advances in computing technology and communications are opening up a whole new range of opportunities and innovations that, until recently, would have been technically and economically unfeasible. These factors are making the goals of open systems difficult to achieve, since potential benefits of an open system can only be fully realised if diverse products, systems and services are able to intercommunicate easily.

9

Open Systems Interconnection (OSI)

This chapter is concerned with International Standards Organisation (ISO) Open Systems Interconnection reference model (RM/OSI). The chapter explains the purpose of Open Systems and how the development of the RM/OSI and corresponding standards, helps to overcome the problem of communicating data between different makes of computer system. The chapter continues with a description of the RM/OSI layers and the terms and concepts used to describe their function.

What are Open Systems?

The words *Open Systems* are loosely used to mean a system which is open to all others for the purpose of information exchange. A system may include one or more computers, associated software, peripherals, terminals, human users or operators, physical processes, physical transfer mechanisms and so on.

Open systems are intended to provide the capability to:

    —   *interchange* — portability to move application programs and data between systems with different underlying hardware and software;

    —   *interconnect* — to use standard communication highways to connect physically separated systems;

    —   *interwork* — to define standard mechanisms or protocols, for exchanging information between systems over the communication highways.

Open systems can be realised by using internationally agreed standards which address these areas of interchangeability, interconnection and interworking. Consequently open systems are concerned with standardising basic operating system services, data management and access, data structures and distributed applications, user interfaces and of course, telecommunications and network infrastructures.

Open Systems Interconnection (OSI) standards addresses the telecommunication and network infrastructure component of open systems. OSI standards enable different suppliers' systems to communicate *data* successfully. We will see how OSI addresses the issues of interconnection and interworking for data communication. To enable open systems communication the existence of a physical medium (ie, the bits of wire or cable) for connecting the systems, is assumed.

The Need for the OSI Reference Model (RM/OSI)

The Problem of Incompatibility

The requirement to enable the interconnection and interworking of different computer systems is fraught with the problems of incompatibility. Incompatibility comes in many guises, the inability to connect machines physically together because the plugs don't fit, the use of different character sets such as ASCII and EBCDIC, the use of different file structures such as sequential or random access — the list seems endless. Many suppliers offer one-off solutions to allow their systems to work, in some way, with others. These, however, only provide short-term solutions. A new requirement for connection to another computer system would require yet another proprietary solution. As is obvious, configurations can soon become very complex, expensive, and difficult to manage. As an exaggerated example of this problem Figure 9.1 shows the sort of configurations which could be required.

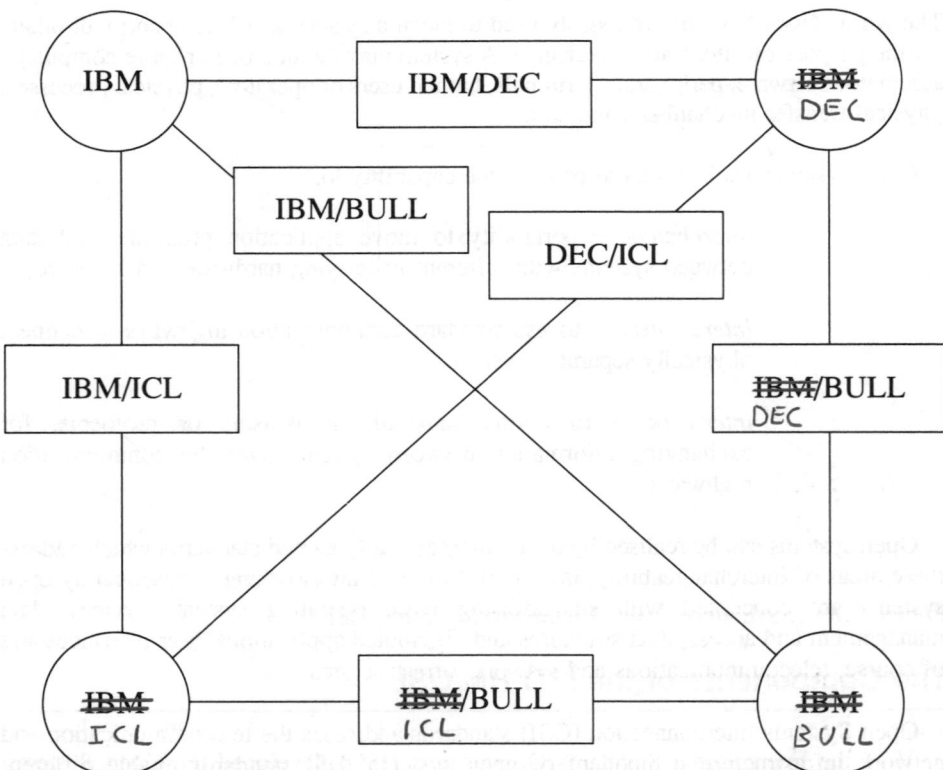

Figure 9.1 Potential interconnection and interworking requirement

Overcoming Incompatibility

The solution to the problem of incompatibility is simple in concept. Define a set of internationally agreed standards that would allow for the interconnection and interworking of different computer systems. Achieving it, however, is a little more difficult in reality. Consider, as a simple analogy, the use of national standards for connecting electrical appliances to the main power supply, for example, all appliances require a three pin plug to use the mains. However, this is only a UK national standard and your appliances would neither connect or work in Greece.

In using standardised computer system components an organisation will be free to purchase other standardised hardware and software which best meets its specific business needs. Organisations will no longer be tied to single supplier solutions. The RM/OSI provides the framework for these sets of data communication standards. As can be seen in Figure 9.2 the problem of incompatibility is simplified using a standardised mechanism for data exchange.

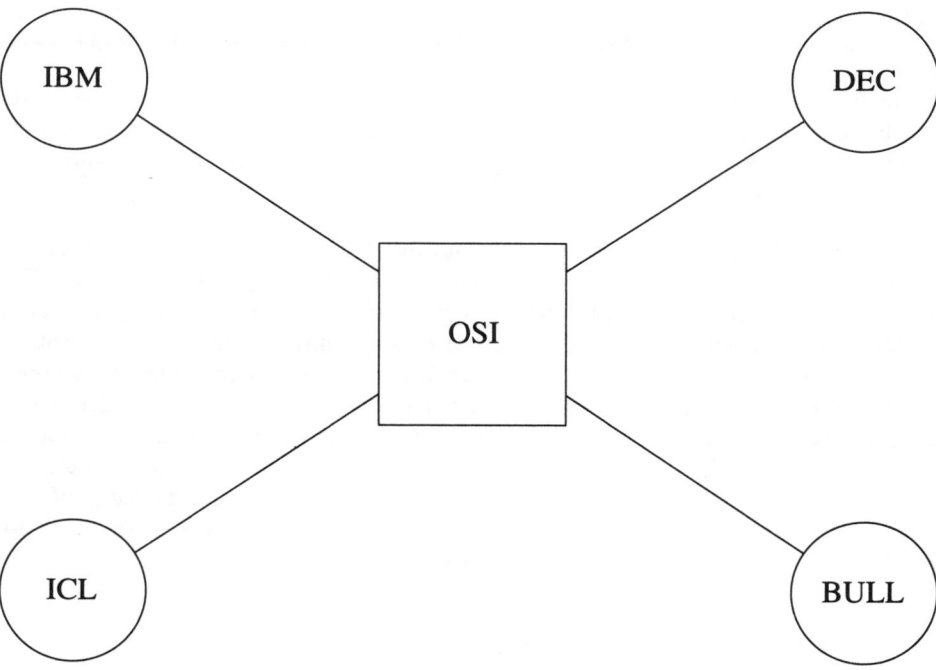

Figure 9.2   Interconnection and interworking using OSI

The Components of the ISO RM/OSI

In 1977 the International Standards Organisation (ISO) developed the Open Systems Interconnection Reference Model or RM/OSI for short. The RM/OSI seeks to ensure that any end system, regardless of make or model, will be able to exchange information (ie intelligent data) with any other end system.

ISO and the CCITT (Comité Consultatif International de Télégraphie et Téléphonie) have developed sets of protocol and service standards which are almost completely compatible. These standards fit into the RM/OSI framework which is made up of seven distinct layers. The seven layers of the RM/OSI encompass the standards for both *interconnection* and *interworking*. The framework, however, does not include the actual user applications which may require such data communication. Application choice is obviously up to the final end user, but the framework of standards must be able to cater for whatever application the user so desires. Examples of such user application groups include:

—   database systems;

—   office automation systems; and

—   process control systems.

The RM/OSI provides up to and including an application support layer of standards.

The RM/OSI does not define the actual physical medium (such as coaxial cable), over which the data exchange is to take place. However, some specific standards within the framework do identify the media over which they are to work. For example, ISO 8802.3I, Ethernet over unshielded twisted pair wires.

The RM/OSI seven layers can be split into two halves — those which provide for interconnection services — and those which provide for interworking services. The purpose of the seven layers is to provide a partitioning and separation of the sets of common services and functions, which are necessary in any system of communication. Each functional layer within the RM/OSI provides a set of services to the layer above and enhances the service provided by the layer below. One layer does not define one standard. Standards, in general, are driven by technological developments and not the other way around. For example, several different LAN technologies exist (Ethernet, Token Ring, Fibre Ring), and from these have been developed the standards of ISO 8802.3, 8802.5 and fibre distributed data interface (FDDI) respectively, all of which are standards within the data-link layer of the RM/OSI.

The Interconnection Layers

The interconnection group of standards makes up the bottom four layers of the RM/OSI. This group are basically tied to the network. The basic functions of these layers are summarised below.

Physical

This provides the functional, procedural and physical interfaces of communication links between equipment, for example, connecting cables to equipment, voltage magnitudes and handshake routines.

Data Link

This provides for the transfer and control of data across the physical medium, which includes error detection and correction procedures.

Network

The network provides routing, relaying of data, if required, in a manner which is independent of the actual network in use.

Transport

The functions provided here include error detection and correction, as well as multiplexing. Its basic function is to enhance the quality provided by the network layer below, if this is necessary.

The Interworking Layers

The interworking group includes the top three layers of the RM/OSI and basically provides the user applications with support services. The interworking standards, in the majority of cases, reside within the end systems themselves. Again, their functions are summarised below.

Session

This provides for the organisation and synchronisation of the exchange of the user applications data.

Presentation

This layer is concerned with how the information to be exchanged between user applications should be represented. This includes resolving character set differences, ASCII to EBCDIC, providing text compression and encryption services.

Application

The standards in this layer provide support for the user applications which wish to exchange information, for example, such information as the application parameters. Examples of a user applications requirements include file transfer and virtual terminal support.

Terms and Concepts

At this point, in order to gain a better understanding of the RM/OSI, it is important to become acquainted with some of the terms and concepts used to describe the RM/OSI. Figure 9.3 helps to clarify some of these terms and concepts.

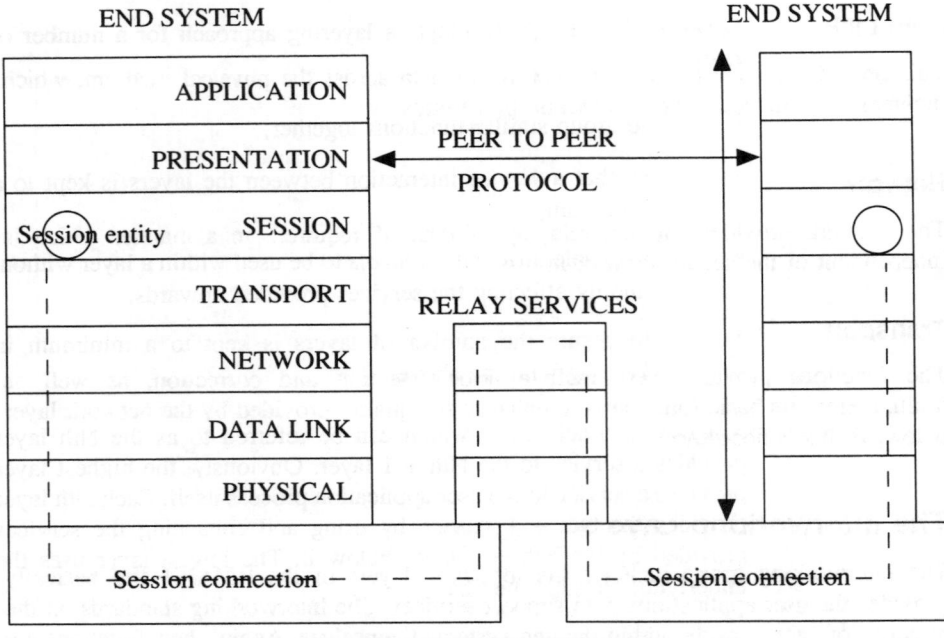

Figure 9.3   Terms and concepts used to describe the RM/OSI

Terms

— *An Entity*. An entity is a logical object which perms some well defined set of functions in associations with another entity at the same layer (a peer entity) in a different end system, for example, a protocol program or process.

— *Protocols and Services*. ISO defines two types of document for each standard residing within a layer, that of a *service* and a *protocol*.

• Protocol. This defines the sets of rules which govern how an entity of the same layer and importance, in different end systems, communicate with one another — (peer-to-peer). It also identifies the sets of commands that can be used between these entities (*protocol* commands).

• Services. This defines how a layer will communicate with the layers immediately above and below it, within the same system. The service of the layer below is used by the entities of that layer, consequently providing a service to the layer above. The service of a layer is used via *service primitives* commands.

- *Layering.* The RM/OSI adopts a layering approach for a number of reasons:

  - to group similar functions together,

  - so that a *service* interaction between the layers is kept to a minimum,

  - to enable different *protocols* to be used within a layer without unduly affecting the services provided upwards,

  - to ensure the number of layers is kept to a minimum in keeping with the above.

- *Nth Layers.* Each layer, which can be referred to as the Nth layer provides a service to the Nth + 1 layer. Obviously, the highest layer provides a service to the user application process itself. Each Nth layer provides the Nth + 1 service by using and enhancing the services provided by the Nth − 1 layer below it. The lowest layer uses the underlying physical medium.

- *Service Access Points.* The communication between the layers is achieved using a defined interface known generally as the *service access point* for that layer. As an example the Nth layer communicates with the Nth + 1 layer using the Nth service access point using a set of Nth layer service primitives.

- *Peer-to-peer protocols.* A peer protocol (a program/module) provides cooperation and communication with another peer protocol situated in a different computer system. Peer protocols are conceptually situated at the same layer and perform fundamentally the same functions but on different end systems.

- *Relay Entities.* Ideally the communication of two open systems would only require the use of two stacks of standard implementations, one on each end system. Sometimes it is not always possible for the implementation of these peer protocols to communicate directly, in this case a relay may be used. The relay entity provides a conversion service. As an example consider two end systems situated on totally different networks (X.25 and Ethernet). They would require a relay entity (a program/module/black box) to provide the conversion service (router) between the networks. Only the bottom three layers are present in a relay.

- *Addressing.* Each of a layer's entities has an address associated with it. Service access points are identified using this address and are consequently used by the next highest layer to which they may temporarily be connected. The remote end systems entity address, with which exchange is to take place, is referred to as an end-point identifier. A function is provided to map addresses from one layer to another in the same end system.

—   *Connectionless and Connection-oriented.* In connection oriented networks the sender of data requires that the receiver acknowledge receipt of the data. This provides an in-built system of flow control. Connectionless networks do not require such acknowledgement of receipt. In essence connection-oriented networks require a formal connection to be established between the end systems before data can be exchanged. Connectionless networks do not require such a connection.

The RM/OSI Layers

A more detailed description of the RM/OSI layers now follows. The descriptions are intended to describe the general functions and service to be provided by standards situated within these layers.

The Physical Layer

In essence this layer defines the means for putting data onto and taking data from the physical medium (wires, cables) which physically connect the end systems together. The unit of data exchanged through this layer is the bit. The physical layer shields the above layers from having to deal in bits.

The layer's function is to activate, maintain and de-activate physical links between the end systems. This physical link is also known as the physical connection. Such connections may either be of a point-to-point or multipoint nature. The layer also defines the electromagnetic and mechanical characteristics of the boundary (interface) between the physical medium (cable) and the physical layer itself. It provides the data link layer with these physical connections to the physical medium. Furthermore, the layer is able to provide a multiplexing service where several data link connections may operate over a single physical connection.

Many of the standards developed in this layer pre-date the RM/OSI model itself and were originally developed by the CCITT (for example, V series analogue, X series digital and I series ISDN recommendations). The standards within this layer, broadly define the following areas:

—   *mechanical* — this identifies the dimensions of the plugs/sockets and the functions allocated to each of the pins;

- *electrical* — this identifies what voltage levels can occur on the physical medium;

- *functional* — this defines what the voltage levels actually mean, say in binary terms 0/1;

- *procedural* — this defines the rules that apply to the various functions operating on the physical medium and the various sequences in which events may occur.

The physical medium (cable) itself does not form part of the physical layer.

The Data Link Layer

The main purpose of the data link layer is to provide a reliable data transmission service for use by the network layer. The data link layer shields the network layer from the characteristics of the physical medium. It also provides an error detection and, if possible, an error correction service for the network layer. The layer works in blocks of information (or frames as they are generally known). These blocks contain header and trailer information between which the data to be exchanged is sandwiched.

As well as being independent of the physical medium below it, the data link layer is independent of the type and nature of the data being passed down from the higher layers. To achieve such data independence the following services for the data link layer are defined:

- *connection establishment* — a data link connection is established between the communicating network entities so that an orderly and controlled flow of information can take place;

- *error detection* — detection of transmission errors occurring in the physical layer. This can be achieved using a simple parity mechanism, or a block sum check or a more powerful cyclic redundancy check (CRC) carried in the trailer part of the data link block/frame. The network layer is notified of any *detected* but uncorrectable errors;

- *error recovery* — the data link layer, in certain cases, is able to recover from error conditions without requesting help from the network layer;

- *flow control* — the control of flow of these blocks of information over the data link connection so that the buffer space of the remote network entity is not exceeded.

The data link layer provides use of the following mechanisms and functions to the network entities:

—   the use of the above services for the exchange of network data;

—   provision of *end-point identifiers* for use by remote entities;

—   provision and maintenance of the *sequence* (order) of the blocks of data being sent/received.

—   the use of *quality of service (QOS)* parameters to enable negotiation on the quality of the data link connection. QOS may be measured in the meantime between detected, but not corrected errors.

The Network Layer

At this layer a distinction can be drawn between packet-switched networks and circuit-switched networks. In the latter a complete *physical* connection between the two end systems is made and left open for the duration of the data exchange. In the former a *logical* network connection, through intermediate systems is used to direct packets of information along to their destination. Standards for the provision of both types are provided for in the network layer. An example of a packet-switched data network is X.25.

In general terms the purpose of the network layer is to provide the following services to the transport entities:

—   set up, maintain and clear down a network connection between the pair of transport entities wishing to exchange information;

—   provide, if required, a routing and/or relaying function between the transport entities, via intermediate systems;

—   perform multiplexing of several network connections on to one data link connection;

—   provide sequencing and flow control of the packets of information, at the intermediate system points along the data packet's route to its destination.

Obviously, the network layer is more active in packet-switched networks, actively supporting the transfer of data. The following mechanisms and functions are provided to the transport entities listed above:

—   provision of the above services for the exchange of transport data;

—   provision of end point identifiers, ie network connection addresses;

— enable the use of a quality of service (QOS) parameter to select the type of network connection required;

— enable the receiving transport entity to control the flow of the packets of information;

— notification of errors to the transport entity if they cannot be recovered by the network layer.

A network connection can be either a connection-oriented or connectionless type; X.25 is an example of a connection-oriented network, and Ethernet (a proprietary LAN protocol) an example of a connectionless network.

Connection over Sub-Networks

Consider the example where two end systems wish to communicate — one on an Ethernet LAN, and the other on a Token ring LAN, interconnected by an X.25 packet-switching network, ie the sub-network (see Figure 9.4).

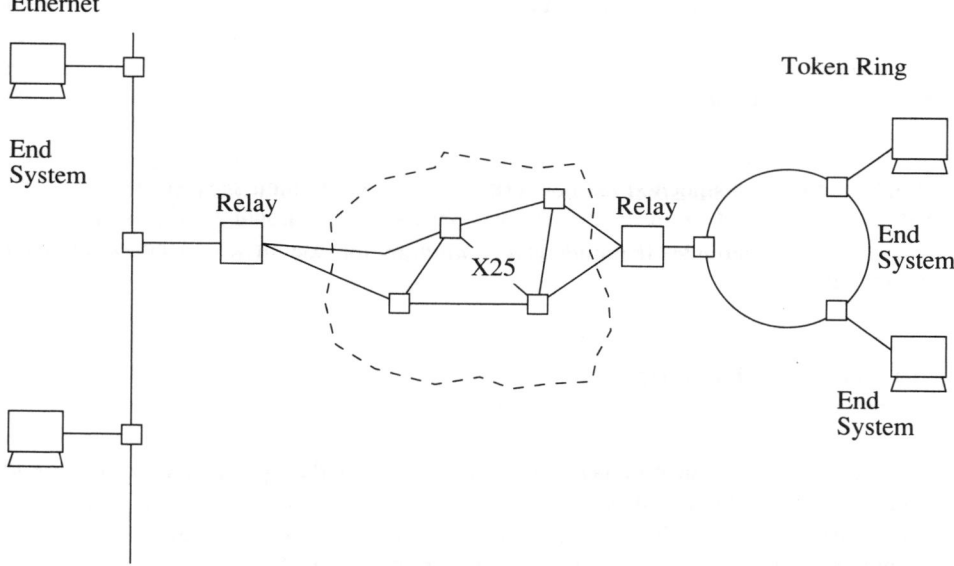

Figure 9.4 Connection over the sub-network

To recap, the network service provides the peer transport entities with total independence from the physical medium over which the transmission is to take place. The network layer must be capable of providing this service across interconnected networks of various kinds. The problem here for providing a network service from one end system to another, is two-fold. First, the underlying data link services provided by the Ethernet and Token ring LANs are different. Secondly, some kind of relaying function is required to carry the data over the X.25 networks.

The functions required within an end system to allow it to work over such sub-networks are termed *subnet access functions.* Each of the end systems wishing to communicate and the sub-network itself (or relay entities), will contain subnet access functions. This, in the OSI sense, can be regarded as sub-layers providing a service to higher sub-layers. Different sub-networks will have different subnet access functions and consequently provide different subnet services.

In using a connection-oriented network service, there are three phases of operation for the communication of data between the peer users of a layer. These are:

—   *connection establishment phase* — this entails establishing a logical connection between peer protocols subject to the QOS required by the user;

—   *data transfer phase* — this covers the use of the established logical connection to transfer the user of that layer's data. An expedited (fast) data transfer may be provided where the formal arrangement of a logical connection is not required;

—   *connection release phase* — this involves releasing or disconnecting the logical connection gracefully.

Implementing a connectionless network service is very much simpler as there is no need to have formal connect and release phases. The following discussion of the *transport, session* and *presentation* layers assumes the use of a connection-oriented network service.

The Transport Layer

The purpose of the transport layer is to hide the underlying network from the end computer system which wishes to communicate data. It provides such an end system with independence from the physical medium of the network. The transport layer also includes mechanisms for giving error free transportation of this data.

The users of the transport layer (ie TS users/session entities)) may request a particular class and quality of service (QOS). The transport layer is responsible for optimising the resources of the network to provide, if possible, the QOS requested. The transport layer is defined using two standards, that of a transport service definition and transport protocol. Their functions are briefly outlined in the following sections.

Transport Service Definition

The following services are provided by implementations of the transport service (TS) for use by the TS users. Such services include:

—   *Setting up a transport connection(s) (TC).* A TC is created between the two TS users wishing to communicate. More than one TC may exist between the same TS users (ie a multiplexing service).

—   *Selection of the quality of service (QOS) required.* The TS user is given the ability to negotiate and agree the QOS required. The use of a QOS parameter enables the TS user to specify the data rate, residual error rates and the transmission delay allowed.

—   *Provision of reliable transmission.* All errors, whether detected or undetected, by the lower layers, are to be corrected before the data is passed back to the TS user. The transport protocol decides on the class of transport service to be provided. Selection of this class is based on that requested by the TS user, through the QOS parameter, and on the underlying network type available, but is not necessarily provided.

—   *Provision of network independence.* The transport service provides independence from the underlying communications media which may be of several different types.

—   *Provision of transparent data transfer.* The data to be transferred across the TC is done so transparently, ie there is no constraint on the structure or content of the data being transferred. Transport Service Data Units (TSDUs) is the name given to the packets of data transferred between the TS users.

—   *Provision for flow control.* The receiving TS user has the ability to control the rate at which the sending TS user sends data.

—   *Mapping addresses.* A system of mapping TS users addresses through to the network service addresses is provided.

—   *Provision for the releasing of the TC.* Either of the TS users is allowed to release the TC.

The TC user uses the transport service via a number of transport service primitives (TSP) or commands, and their associated parameters. These TSPs in effect enable the TS users to communicate with the transport protocol. The following briefly describes the operation of each of the basic phases, establishment, data transfer and release of a transport connection (TC).

Transport Connection Establishment Service

There are four TSPs which are used by the TS user to establish a TC. These are:

-        T_CONNECT Request;
-        T_CONNECT Indication;
-        T_CONNECT Response;
-        T_CONNECT Confirm.

These TSPs carry it over a network connection NC (if it has already been established), certain parameters to the other TS user on the other end system. These parameters will include the QOS, expedited data, address information and user data parameters.

The Transport Protocol (TP)

The transport protocol (TP) within the transport layer is designed to:

-        operate between peer transport entities by using the underlying network service (NS). Protocol operation is achieved by passing a number of transport protocol data units (TPDUs) as user data in network service data units (NSDUs).

-        provide TS users with a consistent set of procedures when the underlying network is of varying quality. As such, five classes of procedure are defined for three different qualities of network. The class of procedures to be used between the transport entities is negotiable, this in part being requested through the QOS parameter during the connection establishment phase of operation between the TS users.

-        define the structure of the TPDUs which are the messages exchanged between peer transport entities.

-        help implement the TS to be provided to the TS users.

The Type and Quality of the Underlying Network

The class of the transport protocol is envisaged as operating over three types of network, types A, B and C. A being the most reliable and C the most unreliable. The types of network are distinguished by the way in which they are able to handle errors.

The transport protocol is able to deduce the type of network provided through the QOS parameter given by the network service. It can, therefore, make an appropriate choice on the class of transport protocol to be used, which may be one of five types from failures signalled by network layer without involving transport user, and recovery from TPDU errors.

Class	Network quality	Transport services provided
0	Type A	Flow control
1	Type B	Reliable transmission (error recovery)
2	Type A	Flow control Multiplexing Expedited data
3	Type A, B	Combination of class 1, 2
4	Type C	Class 3 and error detection

Creating a Transport Connection (TC)

After the TS user has initiated the connection establishment phase using the TSP T_CONNECT_Request, the transport protocol uses a number of TPDUs to:

- assign a network connection (NC);
- transfer TPDUs to establish the TC;
- transfer TSDUs from the TS user, across the TC;
- release the TC.

Figure 9.5 shows, in greater detail, the exchange of TPDUs and TSDUs to set up both an NC and TC.

The Session Layer

The *session* layer is responsible for the establishment and maintenance of a relationship between the final end users wishing to exchange information. Such end users may, in fact, be the application processes on the end computer systems, themselves.

The session layer provides a synchronisation and delimiting service, so that, two end users or application processes cannot transmit data simultaneously. For example, session layer services may be required to maintain the integrity of the data in a database. Here the session layer would prevent two users simultaneously attempting to update the same record. Again, there is a protocol and service defined for the session layer.

Session Service Definition

The purpose of the session service is to provide the SS user (presentation entity) with an organised and synchronised way of exchanging data with its peer SS user. The services provided by the session layer are grouped into what are known as *functional units*. These functional units may be combined and tailored to the particular characteristics and needs of the SS user. All implementations of the SS require at least the kernel to be in place. The basic functions of the session services are as follows:

— *Kernel* — session connection (SC) establishment, data transfer and session connection release.

— *Half duplex operation* — one way alternate data exchange between SS users.

— *Full duplex operation* — two way simultaneous data exchange between SS users.

— *Negotiated release* — the agreed release of the SC between the SS users.

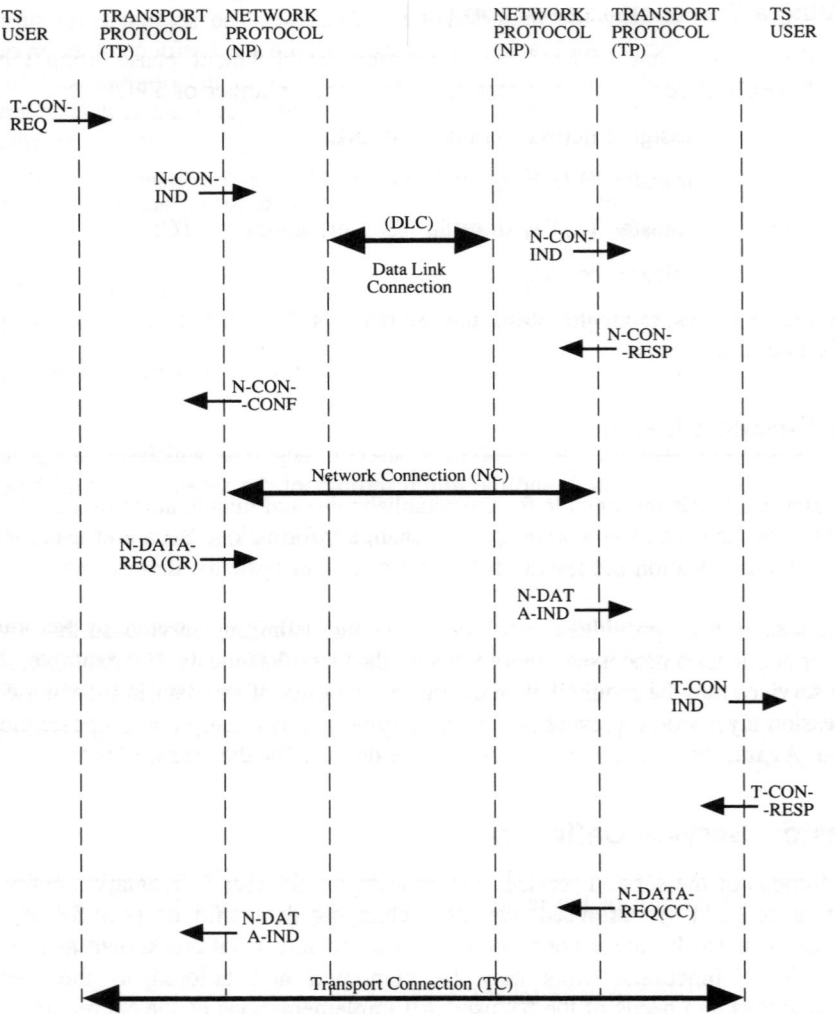

Figure 9.5   TPDU and TSDU exchanges to create a network and transport connection

—   *Synchronisation* — the synchronisation of the exchange of data.

—   *Resynchronisation* — the identification of synchronisation points for the restarting of data exchange should it be needed; either SS user may request resynchronisation.

—   *Activity management* — supports the activity management service, the major activity token being available when this functional unit is selected.

—   *Bracketing* — database management systems are important so that a transaction is not aborted halfway through to completion. This could destroy the integrity of the database. The session layer provides a facility for bracketing transactions, the successful completion of which will require a confirmation sent back to the original session entity. This is known as *bracketing* the messages to be sent in to activities. Should the transmission fail then the session entity sender retains the ability to re-transmit using the last bracketed message. The session layer also provides a service for the ordering and re-ordering of the messages to be sent.

—   *Expedited data transfer* — fast data exchange without a formal connection being established.

—   *Exceptions* — provides a service for the reporting of errors if the transport is unreliable.

—   *Negotiation* — enables SS users to negotiate which functional units are to be used, and the initial settings of the *tokens* described below.

—   *QOS* — enables the SS users to specify the quality of service (QOS) required for the SC.

—   *Quarantine* — is used for security reasons. The sending session entity can request that the receiving session entity does not release the data to its session user (presentation entity) until given permission to do so.

The session mechanisms that enable these functional units to be implemented are outlined below.

Tokens

Tokens are the attributes of a service connection. The ownership of a token allows the SS user exclusive use of the related service. There are four types of token available. These are:

—   data transfer token;

—   release token;

    —   synchronisation token;

    —   activity token.

Synchronisation Points

SS users may insert synchronisation points into the data that they are transmitting. These points are uniquely identified using serial numbers. Such synchronisation points may be one of two types:

    —   *major*, which separate complete dialogue units (for example, an entire file);

    —   *minor*, which define a data market within a dialogue unit (for example, records within a file).

Activities

These may consist of a number of dialogue units (for example, several files ready for transfer). A major synchronisation point would occur at the end of the activity.

Session Protocol

Fundamentally the same procedures and mechanisms are in place within the session protocol as for the transport protocol. The session protocol defines how session services primitives (SSP) are exchanged between the SS users as session protocol data units (SPDUs), the coding and structure of the SPDUs and how the underlying transport service (TS) is to be used to implement the session protocol by communicating its session service data units (SSDU).

The Presentation Layer

This layer is concerned with the syntax and data representation of the information to be exchanged between end users or application processes, presentation service users (PS Users). The presentation layer provides a conversion service between the PC users proprietary language format to a common language and back again. The layer provides independence of character formats and consequently provides machine independence. Basically, the data or information to be exchanged is first described in an abstract form. This describes the structure and type of the data and is known as an abstract syntax notation (ASN.1). ASN.1 may then be encoded into a transfer syntax data-stream for exchange. The transfer syntax is initially negotiated by the PS users. Encryption and data compression are examples of transfer syntaxes.

A *presentation context* defines a mapping service between ASN.1 and specific transfer syntaxes. The basic services provided by the presentation layer are much the same as those already described and include:

— *Connection establishment.* This involves setting up a presentation connection (PC) between the PS users. The establishment would include negotiating the set of functional session units described earlier and the presentation context to be used from a list of those that are currently active.

— *Connection release.*

— *Context management.* The PS users are allowed to define new contexts or delete from those in the active list.

— *Information exchange.* Full and half duplex operation are catered for by the presentation layer.

— *Dialogue control.* This is a similar service as that provided by the session layers. It includes token management, synchronisation, resynchronisation and activity management.

In practice, the presentation layer is often implemented as part of the application or session layers. A service definition and protocol have also been defined for this layer.

The Application Layer

In a truly distributed system environment a variety of application processes will require data communication with one another. These application processes may require the use of a number of different services to transfer files and exchange their messages and commands. It is for this reason that there is more than one *service* definition and *protocol* standard defined at the application layer.

The application layer is the point where the application processes (user program processes) requiring communication, interface or interact with other OSI protocols and services. The application layer standards are concerned with the transmission of the semantics of the data (ie the meaning of the data (for example, a file/data structure), as well as the data itself. The lower layers are concerned only with the passing of data. It is at this point in the RM/OSI that data communication transfer is treated as a virtual terminal session, a file transfer or a mail service.

A number of such application layer standard developments have been undertaken including file transfer, access and management (FTAM), virtual terminal protocol (VTP), message handling system, (MHS), and job transfer and manipulation (JTM). Unfortunately there is some confusion in this area because the original modelling concepts and terms have been modified to encompass new ideas. The following describes the two models and the terms and concepts used to describe them.

Model One

In the original model two groups of standards were defined. These were:

— *common application service elements (CASE)*. These standards provide functions for the general interfacing of applications, that is, the establishment and release of application connections, or associations as they are known in the OSI environment. These services are generally required by all application processes (AS Users).

— *specific application service elements (SASE)*. These provide the services and protocols required for specific applications services, that is file transfer access and management (FTAM), virtual terminal (VT), job transfer and manipulation (JTM), and message handling system (MHS).

Model Two

The revised model of the application layer and the terms used to describe it are now outlined. The revised model, instead of separating the two services into common and specific functions, defines a general application layer structure. Figure 9.6 helps to clarify the terms and concepts employed. Communication between application processes can now be represented by an association between application entities (AE). The application process communicates with the application layer via a user element (UE) which has an AE associated with it. Application service elements (ASE) describe sub-divisions of an AE. ASEs describe the types of work which the application process wishes to be performed.

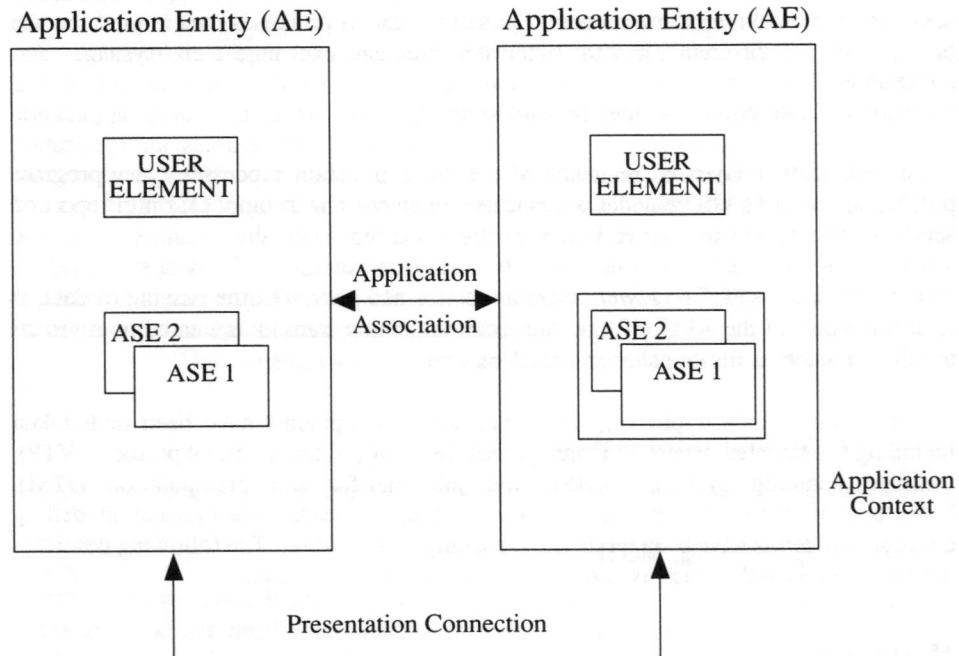

Figure 9.6 ASE, AE, and UE

ASEs may be of two types:

— *Reusable*. These can be thought of as analogous to the functions of CASE. They provide general services; examples include ACSE, CCR, ROSE and RTSE.

— *Application specific*. These are analogous to SASE because they are intended to define specific service requirements. Examples include FTAM, JTM, VT and MHS. Such application specific ASEs may again be further sub-divided into aspects of the service they provide. An *application context* describes a combination of ASEs which are required by an application process to achieve its work objective.

Reusable ASEs

— *ACSE — association control service element*. This service allows an application process to *begin, end* and *halt* presentation connections (PC).

— *CCR — commitment, concurrency and recovery services*. These allow an application process to have control over the *atomic actions* of a connection. Application processes may be divided into discrete atomic actions. Atomic actions can be viewed as a sequence of operations which the application process is to undertake, between itself, on the one hand, and the other application process on the end system.

— *ROSE — remote operations service elements*. Some application processes are inherently interactive. ROSE enables an application process to request that a remote application performs an operation. ROSE provides a service to allow the remote application to report on the success or failure of the requested operation.

— *RTSE —reliable transfer service elements*. This service is provided to recover from communication and end system failure and minimises the number of retransmissions needed for recovery.

Application Specific ASEs

The following specific services ASEs reside within the application layer of the RM/OSI.

Electronic Messaging, MOTIS and X.400

MOTIS and X.400 are, in essence, concerned with the standardisation of electronic messaging. Electronic messaging services and systems have been available for many years now in a variety of different forms, ie telex, teletex, public electronic mail, (for example, Telecom Gold), *private inter-company electronic mail* and, of course, *facsimile*.

Facsimile is probably now the most popular form of electronic document transfer but, as such, does not fall directly under the electronic messaging banner.

Current messaging services have provided organisations with the ability to convey hard copy messages and documents quickly throughout the world. However, these current services do have their drawbacks. These include slow speed, potentially high error rates, limited geographic coverage, interworking problems and the inability to modify and change the messages sent electronically, once received.

By standardising the component parts of message handling systems (MHS) and the improvement in carrier technology, it is now possible to overcome many of these problems. MHS are simply store-and-forward type electronic mail systems.

X.400 is a generic term for a series of eight recommendations from the CCITT, which define an architecture for a MHS. An implementation of X.400 would allow computer application processes, or people, to communicate by submitting and receiving messages consisting of structured or unstructured data. Such implementation would provide for the interworking of different office systems equipment. These recommendations cover the following areas.

An MHS network model (X.400)

MHS components comprise:

— *Users.* These include dumb terminals, local host computers or a third party electronic mail facility connected through a public data network.

— *UA — user agents.* These provide the users local interface to a message transfer service (MTS). Several UAs may be associated with one user and one MTA.

— *MTA — message transfer agents.* These are the message exchanges that route the messages to their final destinations.

MHS services comprise:

— *MTS — message transfer service.* This is a general service which is operated between the MTA and the UAs to provide an application independent, store-and-forward message transfer service. In general it provides the enveloping service of the message to be transferred. Envelope information could include the originator and recipients address, general information about the contents of the message, an indication as to how urgent the message is and any special services required for this message. Special services provided might include the conversion of the document to another format.

—   *IPM — interpersonal messaging service.* This is the only user type service to have been defined so far. Each user service is designed to perform a specific user task or application; IPM is one such specific user task. IPM provides users with mail/messaging facilities and the ability to interchange office documents. Other user services might be defined for a user's particular application requirements and need not be internationally agreed. One such user service could be defined for carrying EDI trade documents.

This model also details the protocol layering for an MHS under the following headings (see Figure 9.7).

- an overview of the IPM and MTS (X.401), describes the basic service elements and optional user facilities of the MHS;

- the rules for converting information types, for example, telex to ASCII (X.408);

- the presentation layer transfer syntax used by an MHS, (X.409);

- details the facilities provided by ROSE and RTSE which are sub-layers of the MHS architecture;

- specifies the service and protocol for the message transfer layer (MTL) within the MHS model;

- specifies the content protocol for the IPM service and includes the format for memo headers and multi-part body types;

- details the specialised means of access for the teletex terminals which allows them access to the functions of the IPM and MTS.

—   *MOTIS — message oriented text interchange systems.* This is the ISO name for both the message handling environment and the messaging standards which are a superset of X.400.

Consider an MHS example, the originating users UA submits its message to its MTA. The message is enveloped with the information required to enable the message transfer system to deliver it to the recipients UA and consequently the recipient user. The transfer may require the use of several MTAs.

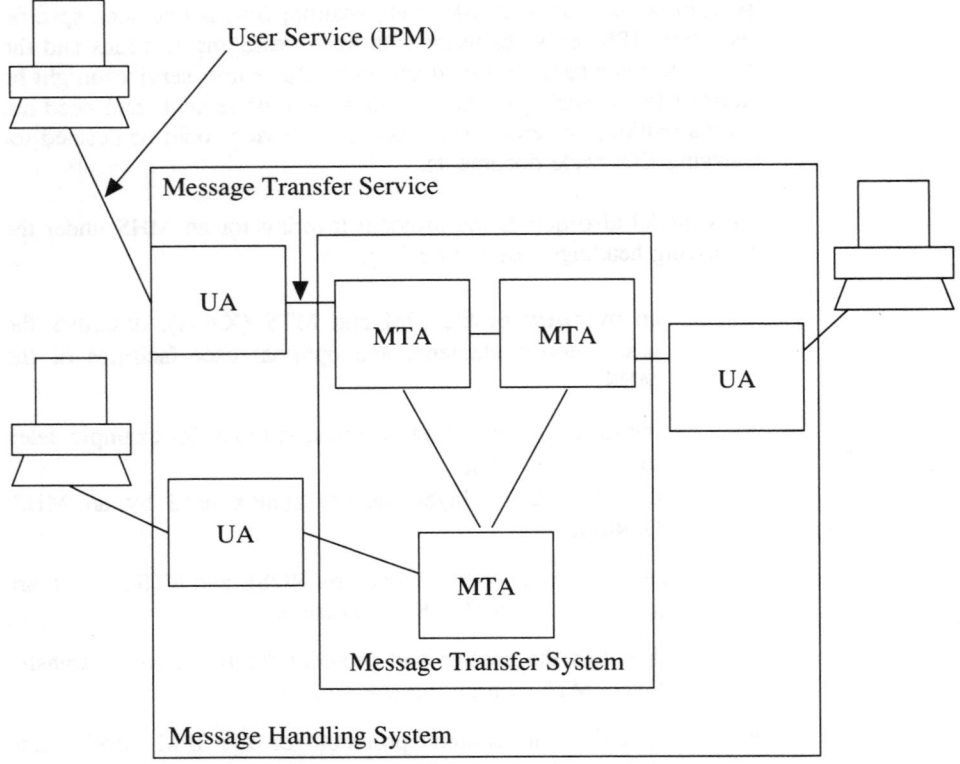

User Service (IPM)

Message Transfer Service

UA

MTA — MTA

UA

UA

MTA

Message Transfer System

Message Handling System

Figure 9.7 The basic MHS model

File Transfer Access and Management (FTAM)

FTAM is an ISO standard designed to provide the ability to transfer, access and manage files which may be stored on different end systems. File servers and the distributed databases of remote clients, are examples of applications which may require the use of such a service. The sort of file types that may be supported by FTAM include: *sequential, random access* and *single key indexed sequential files*. It is the intention that FTAM will eventually be able to support hierarchical, network and relational file types and, consequently, their database counterparts.

FTAM can be broken down into five parts:

—   *Virtual filestore.* This in itself is concerned with three areas, the filestore model, the actions of the filestore and the sets of attributes of the files.

— *Filestore model.* The filestore model defines a description of the real files which is totally independent of the way it is implemented on the end system. The model describes a real file in abstract terms. For example, linking a unique file name *attribute* to a tree structured list of the data units contained within the file. A plan is provided which enables the virtual filestore to be mapped on to the real filestore implementation.

— *Actions of the filestore.* This defines the actions which can be performed on the file, its attributes or data units. Two groups of actions exist:

— actions on complete files (creation, select, change read attribute, open, close, delete),

— actions on the contents of the file (locate, read insert, replace, extend, erase).

— *File service.* This defines the set of service primitives and their associated actions upon the filestore model. These primitives are used and exchanged by application entities (AE) during an association.

— *File protocol.* This defines the actual interactions which may take place between AEs in order to use the elements of the file service.

Virtual Terminal (VT)

In an open systems environment the ability to use a variety of different terminals with different computer systems is necessary. Consequently, a requirement exists for a virtual terminal standard. Two such standards exist, a protocol (VTP) and a service (VTS). The standards alow for terminal-to-terminal, terminal-to-process and process-to-process interworking. The major application for VT is seen as the terminal-to-process work, such as in a data entry type applications. The other interactions being achieved by other application support services. For example, terminal-to-terminal using MHS interpersonal messaging (IPM) for messaging and FTAM for bulk file transfer between terminals, and open distributed processing (ODP) being used for process-to-process communications.

Fundamental to the understanding of VT is the conceptual communication area (CCA). This contains a wide ranging set of generic terminal functions including clear screen, background colour and foreground colour. The specific terminal can be mapped on to these generic CCA functions. Consequently, the CCA allows a variety of terminal types to be used.

Because there are so many different applications types (terminals and processes), they are likely to require varying levels of the VT service. *Service classes* are used to define these ranges of functions for the groups of application types.

Classes defined so far are:

— *Basic VT class*. This class provides functions for simple PCs to be mapped on to the CCA, for example, colour modes, full character repertoire and text editing features.

— *Forms class*. This class provides functions for data entry type application processes within, say, administration departments. Here it is possible to predefine the screens to be used for the entry of data. It provides a user definition option.

JTM (Job Transfer and Manipulation)

A user may want to enter jobs remotely for processing at a later time or date. JTM service historically derives from remote job entry (RJE) type applications. Batch processing enabled jobs to be submitted from a remote terminal and the results to be printed at a remote printer. The era of interactive systems has seen this mode of operation decline, but a requirement for such an operation still comes from within the academic community.

The JTM standard allows an *initiator entity* to produce a work specification which can be sent to another end system for execution. The work specification obviously specifies what operations are to be performed and when. JTM provides functions other than this simple job entry. These include:

— enquiries as to the status of a job;

— operator manipulation and modification of the job;

— cancellation of jobs;

— operator to operator communication service;

— system message transfer;

— simple user message transfer.

Functional Profiles

Many standards are now in existence within each one of the seven layers. The problem for many organisations is to define which of these standards is most suitable for their particular business environment. A *functional profile* defines such sets of standards.

The seven layers of the RM/OSI make up the horizontal slices; functional profiles are vertical slices of the RM/OSI. Examples of functional profiles include: government OSI profile (GOSIP) defined by the CCTA, technical office protocols (TOP) originally defined by Boeing and manufacturing automation protocol (MAP) defined by General Motors.

The functional profile document highlights what the user organisation would like to see included in implementations of the RM/OSI (the protocols stack). Having defined a functional profile the next stage is to gain endorsement by the vendors themselves. This is why only the very largest organisations (with large purchasing power) have the capability to achieve such endorsement.

The profile work which is driven by the user organisation is not just a list of standards to use, but includes procurement advice and guidance on associated issues. In general, a functional profile will consist of a general specification which will include:

— references to international standards and recommendations;

— details of the relationship between the base standards to be selected;

— identification of the base standards and the specific technical problems which they are required to satisfy.

Because many international standards within the RM/OSI are still immature, references are still made to *de facto* or national standards, where necessary, as interim solutions.

User Services

In essence *user services* build upon and use the application support services described above. The following user services simply add value to and provide semantics (meaning) to the user data that is to be exchanged. As an example, consider office document architecture (ODA) which defines a way of formally structuring all types of documents you may find in an office environment, for example, memos, letters, reports. The FTAM support service could provide the means by which to transfer the documents described by ODA between end-systems. Electronic data interchange (EDI), ODA and standardised generalised mark-up language (SGML) define three such standards that provide so-called user services.

Electronic Data Interchange (EDI)

Within many areas of inter-company trading the generation and movement of traditional paper-based documents (invoices, purchase orders, credit notes, etc) is financially expensive. EDI or *paperless trading* helps reduce such overheads. EDI provides a means of electronically transporting trade data (trade documents) directly between the computer systems of the two companies wishing to trade. The trade data that is to be transferred is described in a standardised language and format, so that any computer system with a conformant parser can access and process the trade data received. EDI defines this standardised language and format.

EDI Architecture

A general EDI reference model has been defined. It has been designed as a hierarchical structure consisting of the following components:

- *File/functional group.* This defines a collection or batch of trade documents of the same type, for example, all the invoices for one particular company.

- *Message.* This is a division of a trade document into a number of messages, for example, a purchase order may consist of a *header* (consisting of customer and supplier details), the *details* (listing the ordered products and quantities) and the *trailer* (giving total quantities and values).

- *Segment.* Each message can be divided into segments, for example, one segment might consist of the supplier's name, address and reference number.

- *Data elements.* This is the actual vocabulary of EDI. The actual elements are the items to be conveyed between the trading partners, for example, expiry dates, product codes, etc. Each segment might consist of several data elements.

The EDI reference model defines these components for the construction of an EDI trade document. The other parts required of an EDI standard include:

- an EDI vocabulary and syntax definition;

- a set of *message design rules* to allow the above components of the reference model to be grouped together, to form the messages and trade documents;

- a transmission format which includes start and stop control to allow the trade documents to be stored and transmitted in a file format (perhaps using the service of FTAM or MHS).

- a set of standard definitions for the user data and system control characters.

EDI Standards

Unfortunately, work by a number of different bodies has led to the development of several specific EDI standards. Electronic data interchange for adminstration, commerce and transport (EDIFACT), forms the basis for a truly international EDI standard. However, there are two industry EDI standards TRADACOMS and ODETTE. Furthermore, a number of national EDI standards are also prevalent in Germany,

Holland, Belgium, South Africa, Sweden, Australia and France. The EDIFACT standards work completed so far may be summarised as follows:

- *vocabulary* is trade data element directory (TDED);

- *syntax rules* comprise electronic data interchange for administration, commerce and transport (EDIFACT); there is an implementation guide to EDIFACT;

- *EDIFACT message design guidelines* and *EDIFACT standard messages*.

Value Added Data Services (VADS)

Because of the cost and expertise required to prove interconnect and interworking solutions, EDI services are provided by a third party or value added data service (VADS) supplier organisation. The VADS supplier will be able to provide several proprietary communication interfaces to an organisation via leased line and/or dial-up. In essence VADS provide a gateway or protocol conversion service for their clients. Interactive and store and forward types of trade data transfer are usually available, the former giving almost instantaneous transfer of the trade data between trading organisations. The latter, which is the more common service, uses a VADS computer to act as a remote mailbox for the storage of trade data.

Some of the major VADS suppliers include:

- *INS — international network services.* INS is a merger of ICL VADS and GEISCO. The services are provided using the Mercury network in the UK and GEISCO's international network. INS has brand named its services to particular sectors. These are:

    - TRADNET, using the TRADOCOMS standard, is the most widely used within the UK. It has been aimed at the consumer goods industry.

    - TRADNET International, aimed at the international trade market.

    - MOTORNET providing an EDI service to the motor industry.

    - DISH (data interchange for shipping) designed to speed up information flows between major exporters.

    - BROKERNET to provide information services to the insurance industry.

    - DRUGNET for the health care sector such as doctors, pharmaceutical industry and government departments.

    - PHARMNET for the pharmaceutical industry.

— *IBM UK*. IBM's EDI service is an in-house standard and is based on their information exchange service. The service is provided to the transport, shipping, insurance, construction, retail distribution and manufacturing businesses, over IBM's private international managed network service (MANS).

— *Istel*. Istel provides the EDICT service on their private *Infotrac* carrier network. The EDI service is provided to the manufacturing, travel, health, finance, retail and distribution business. Other VADS include DEC, Travinet (a subsidiary of Thomas Cook and the Midland Bank) and BT.

Office Document Architecture (ODA)

The generation and manipulation of office documents, whether they be letters, memos, graphs or reports, has increased rapidly over the last decade. This is in some way due to the advance of technology which has brought about cheap word processing systems, desktop publishing, and mini-based office systems. As a consequence the volume, sophistication and quality of such office documents has increased. Again, there is a problem of incompatibility. There are many different suppliers of proprietary office document products, which perform the tasks of creating, copying, merging, storing and printing of office documents. Many are unable to exchange and use the documents created by another supplier's systems. Such office document products have been unable to interwork due to the lack of standardisation in the way office document structure and content is defined.

The ODA standard defines the way in which any office document is structured in terms of chapters, paragraphs, titles, etc; the content allowed and how this information should be presented for transmission between two systems. The contents of a document described using ODA can include text, graphics and image components.

Office Document Interchange Format (ODIF)

The ODIF standard defines how the ODA document is encoded for transmission between end systems. ODIF defines a number of different interchange streams. One such stream contains general information about the whole document (document profile). This can be used by the end system to see if it can support the document to be exchanged.

As an example consider the transmission of a written document created by a word processor package on system A to system B, where a graphics package is to be used to add a diagram. On system A, an ODA converter would translate the document into an ODIF format which would then be transferred to system B. This transmission could be achieved using perhaps OSI support services or proprietary protocols. The ODIF format received by system B would apply a reverse conversion to the local ODA format.

ISO ODA Standard

The ODA standard is currently divided into seven component parts:

- introduction and general principles;

- document structures;

- document profile;

- ODIF;

- character content architectures;

- raster graphic content architectures;

- geometric graphic content architectures.

Conceptual View of an Office Document described by ODA

The following is a conceptual view of any office document described by ODA.

```
DOCUMENT --- PROFILE
          --- BODY --- STRUCTURE
                  | -- CONTENT
```

Document Profiles

The document profile component contains a general description of the office document. This can be transferred to another system independently of the body for, say, interrogation purposes. The general description consists of:

- *characteristics,* which describes the application that generated the document, the document structure, the content architecture used (character, raster, CGM) and page size, etc;

- *management,* identifies the author, the title, creation, date, history, distribution list, etc;

Document Body

- *document structures* — ODA considers the structure of a document as one of two types:

  - *logical,* breaks the document down into a hierarchical tree structure of logical objects. These objects top down may consist of chapters, sections, paragraphs, sentences, all the way down to individual units of text;

- *layout*, also breaks the document down into a hierarchical tree structure, but the layout objects are concerned with how the information is presented on the final document in terms of position. These objects listed from top down consist of page, header, frame, block, footer, etc.

ODA attributes describe the characteristics and relationships between the objects of layout and logical structures. The following are examples of attributes:

— *logical attributes* which include the protection attributes that identify whether the logical object such as a text block may be modified or not;

— *layout attributes* include the dimensions which will be needed to identify the size of the layout object, such as a page;

— *layout style attributes* may define the link between logical and layout structure; an example is the specification that says all new chapters must start on a new page;

— *presentation attributes:* these attributes are closely associated with the content of the document, for example, emboldening, font selection.

— *content architectures:* currently there are three types of content defined for ODA.

— *character elements* which are text based elements based on the eight-bit encoding scheme;

— *raster graphics* which are two-dimensional images formed by the grouping of rectangular black and white picture elements. This is incompatible with CCITT group 2 and 4 facsimile bit mapped images;

— *computer graphics* which allow for the description of two-dimensional graphics from graphics characters. It uses the computer graphic metafile (CGM) encoding as defined in ISO 8632.

Document Classes

Many organisations use common structured documents such as memos and letters over and over again, ODA allows the definition of document classes to represent these commonly used documents (Figure 9.8). The common structure and attributes are put into what is called a generic definition, while a specific instance of that document is known as a specific definition.

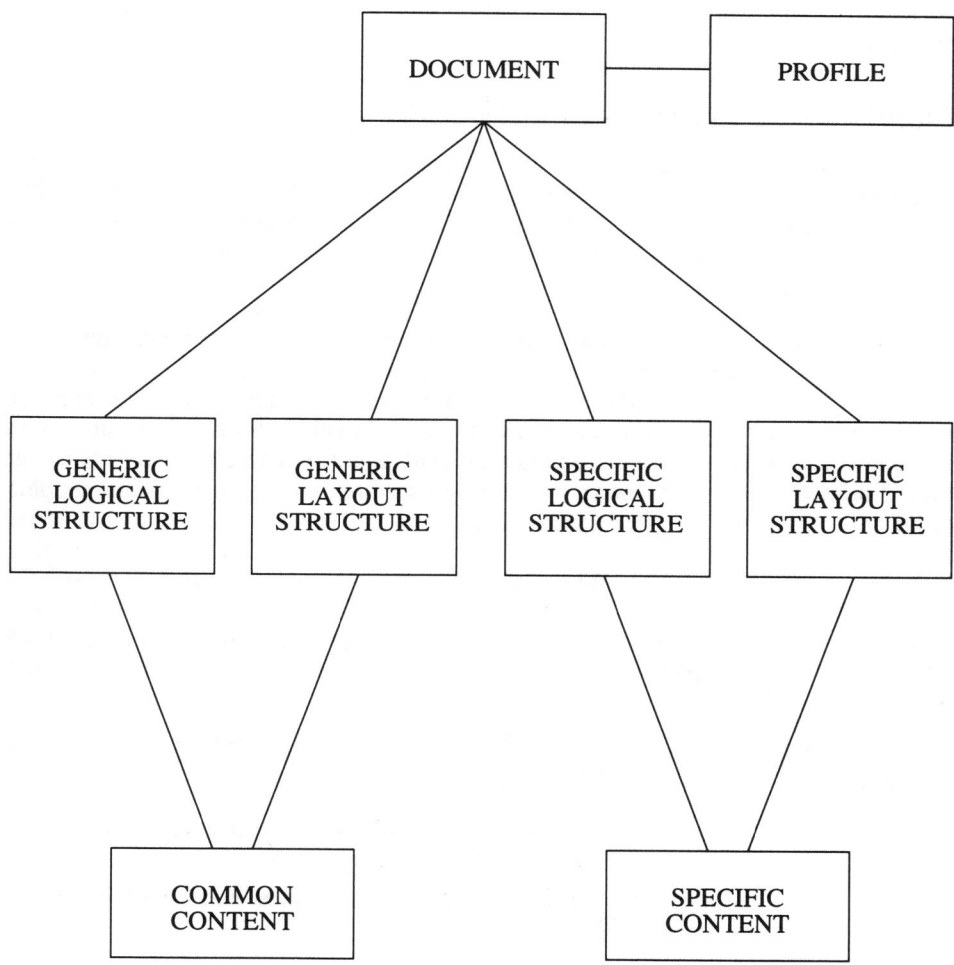

Figure 9.8 Illustration of an office document described by ODA

Document Processing Model

All office documents are created, edited, imaged or printed on to a particular device. The ODA standard describes these three processes editing, layout and imaging and how they operate on the structure and content of an office document.

Document Application Profiles

As with OSI, ODA is a very rich standard and not all elements or components of the standard are pertinent for all types of application. Consequently user groups are defining document application profiles or subsets of the ODA standard. Some of the major document application profiles include MMI and PM1 developed by the CCITT, NIST/TOP and ONA (BT open network architecture).

Standardised Generalised Mark-up Language (SGML)

SGML has many similarities with ODA in that both are concerned with standardising the description of the *structure* and *content* of office documents. The difference is basically historical, ODA has grown out of the word processing and office systems environment whereas SGML has originated from the techniques used in text formatting found on mainframe systems within the publishing industry.

SGML defines a formal abstract syntax for marking-up (tagging) a document in a general form. Marking-up refers to the process of breaking down a document into different elements and tagging them with associated attributes. The tags may be interpreted in a number of ways depending on the environment for which the document is intended, ie word processing, desktop publishing, etc. For example, the document maybe printed in draft form or viewed in a typeset form on a high resolution graphic terminal.

SGML is very flexible and documents can include text, computer graphics, scanned images, tables, mathematical formulae and spreadsheets. The intelligent manipulation and processing of the contents, however, is limited to text-based characters only. Such flexibility means that potential applications include computer assisted electronic and database publishing.

The ISO SGML standard is a *natural English* text standard. In SGML all documents held in an electronic form are made up of two types of information:

— the content, which is the actual information of the document;

— the mark-up, which specifies the attributes and structure of the document contents.

SGML Mark-up

The tagging of particular elements and contents of a document is known as marking-up that document. The tags are used to define the beginning and end of an element, for example, chapters and paragraphs. Certain elements may in themselves form sub-elements of another element. In this way it is easy to define a flexible structure for documents that could perhaps be viewed in several different ways. As an example, consider the following use of delimiter type tags.

```
<chapter>
<introduction>
<paragraph>
This is a paragraph
<end paragraph>
<end introduction>
<body>
This is another paragraph
<end body>
<end chapter>
```

These marked-up sentences would define a document with the following logical structure.

CHAPTER --- INTRODUCTION --- PARAGRAPH
|_____ BODY

An entire document could be marked-up in this particular way.

As in ODA both a logical and layout structure can be defined by this mark-up language. A concrete syntax can be used to tie the SGML abstract syntax for use within a particular type of environment.

SGML defines four types of mark-up tags:

— *descriptive*, defines the delimiters used to tag elements (parts of a document) and the attributes associated with those delimiters, as shown above;

— *entity reference*, may define constants which are to be substituted, when they are processed, for predefined text or text strings;

— *mark-up declaration*, allows the definition of new mark-up tags for SGML by their user;

— *processing instructions*, defines processing instructions to be used when processed, however, this reduces the independence of SGML.

Content Types

Documents described by SGML can contain text, image, data, voice, but the standard defines only two data types for describing such contents, those of character and bit.

Document Type Definitions

These perform a similar function to an ODA's document profiles. Their purpose is to define a set of mark-up declarations which in turn specify the elements, attributes and entities to be used within the document. It allows a system to interrogate the document type definition and identify the document's structure and content without transferring the whole document.

Global Application Services

Global application services identify those services that are potentially required by all the layers of the RM/OSI. They form, in themselves, sets of standards. Consider, for example, the need for a security service within a communications network. Security is required at the *physical* layer to prevent access to the media or tapping of the wire. Security is also required at the *application* layer to secure access and use of application support services such as file transfer, (ie to stop unauthorised users transferring or accessing private files). The ISO are currently addressing the need for global application services. The following acts as a brief introduction.

Naming and Addressing

In the exchange of information between open systems using OSI there is a need for peer-to-peer protocols to operate. There is consequently a need to know the identity of the entity (implementation of a peer protocol) with which the data exchange is to take place. The structure and content of entity and network object names and addresses is being standardised.

OSI Directory Service

The OSI directory service is a special distributed database facility to hold all of the standard name and address information about the network objects. The scope of the work on directory services covers the *naming and addressing* information contained within it and the means by which a user can variously access it, add to it, delete from it and modify it. The directory could hold terminal addresses, program addresses, in fact, any network related addressed.

OSI Management

To ensure that the resources and components which make up a network are used as effectively as possible, knowledge of their usage, user(s), fault conditions, throughput etc, is required. In fact, management of a communications network is concerned with managing:

- *faults* — testing/detection/isolation/notification/recovery;

- *performance* — congestion/delay/reliability/utilisation;

- *security* — files/records/access;

- *accounting* — billing/charging/usage;

- *configuration* — addition/deletion/initialise/termination of resources.

The management framework being proposed for OSI systems is to provide the necessary sensors to gather such information and, consequently, enable effective management. The conceptual management framework proposed specifies three levels of OSI management:

- *protocol management* — management of the events within a particular protocol at a particular layer in the RM/OSI; such events are defined within the protocol and service definition of that layer;

- *layer management* — the management of interactions and operations between the protocols of a layer;

- *systems management* — the management of events within a complete seven layer stack.

Security

There is a growing need for security within data communication networks as the number of private and public telecommunications services increases. Threats to such networks include:

- the unauthorised modification of data;

- masquerading as a friendly entity;

- the denial of service and message repetition.

As a result of such threats the data being communicated between end systems, may be destroyed, corrupted, modified, stolen, removed, lost or disclosed to an unfriendly party. As an addition to the RM/OSI, a security architecture has been defined which addresses some of the problems. The following security service requirements have been identified:

- *authentication* — provides confidence in the other user on the other end system;

- *access control* — protection against unauthorised use of network resources such as files and software;

- *data confidentiality* — protection against data disclosure;

- *data integrity* — protection during modification;

- *non-repudiation* — provides proof of sending and receiving data.

Implementation of the above services would require the use of mechanisms such as encipherment, encryption, passwords, tokens, cards and cryptographic check values. Other useful mechanisms include event detection and audit trails.

10

X.25 Packet-Switched Networks

This chapter covers X.25 and HDLC. X.25 is a network access protocol used with packet-switched networks. HDLC is a link level protocol used extensively in wide area networking. It is used at link level in X.25.

CCITT Recommendation X.25 is a very stable network interfacing standard. It is fundamentally unchanged since 1980. The overwhelming majority of terminals and hosts support it and many LAN products use it in *routers* and *gateways* to wide area networks. It has been incorporated into the OSI model as ISO 8208 (see Chapter 9).

In the early 1980s, the packet-switching networks market was dominated by a few suppliers. In the last five years, demand has increased, especially for private networks and many more communications suppliers have brought out X.25 products. Additional facilities were added to the Recommendation in 1984 and 1988. A large number of facilities are optional and not all have been incorporated into the different suppliers' products. Many networks are still running with the 1980 version of the standard, especially public data networks; British Telecom's PSS for example. It should be noted that Datagrams were dropped in the 1984 version and that no public data network has ever offered this facility. Users need to be clear about which facilities, mandatory and optional, they require. For this reason many users have produced detailed statements on the communication service they require, for example, the Government OSI Profile (GOSIP) which has been written to define levels of conformance required from products procured for British government departments.

The chapter is completed by a description of the high-level data link control (HDLC) protocol, which is used extensively in all major manufacturers' products. One mode, the *asynchronous balanced mode*, is used as X.25's level 2 protocol.

Overview of X.25

The X.25 Recommendation defines the interface between data terminal equipment (DTE) and data circuit terminating equipment (DCE). The definition of the DCE is, strictly speaking, the equipment that converts DTE signals into a form in which they can be transmitted over a physical circuit. However, in the sense used in X.25 it refers to the access node or the packet-switching exchange to which the DTE is connected, but in this interpretation the DCE also includes the modem on the subscriber's premises. Figures 10.1 and 10.2 illustrate the arrangements.

It is essential to recognise that X.25 is a network access protocol. It makes no assumptions about the way in which the network functions, other than that the packets involved in an interaction between two DTEs are delivered in the order in which they enter the network. Therefore, providing the two DTEs employ equivalent implementations of X.25, they should be able to intercommunicate satisfactorily, and the network should appear transparent to them.

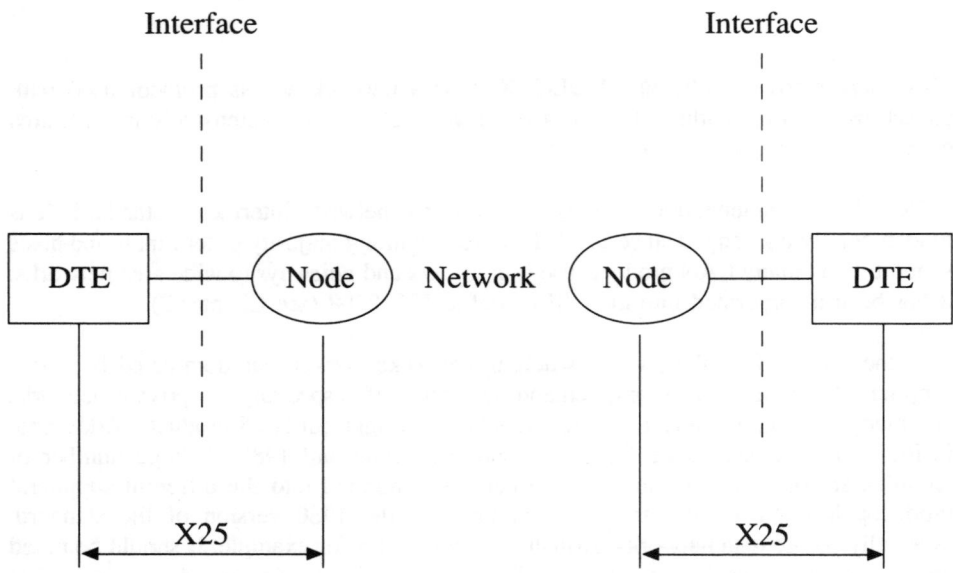

Figure 10.1 Applicability of X.25

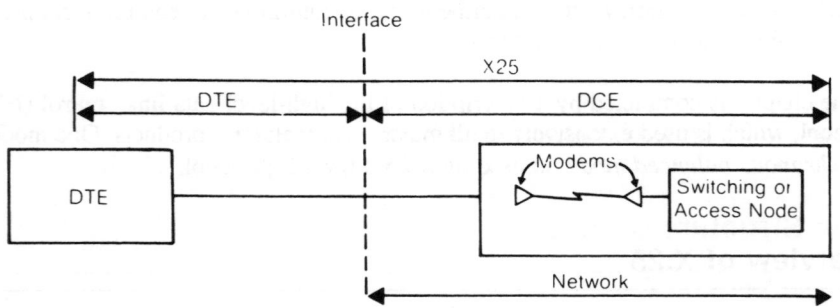

Figure 10.2 The X.25 Interface to a packet-switched network

X.25 applies only to DTEs connected by a synchronous transmission circuit and which have sufficient intelligence to implement the X.25 protocols. During the development of the standard it was realised that there would be a requirement to connect simple terminals with limited intelligence, accessing via asynchronous leased circuits and PSTN.

A set of associated standards, X.8, X.28 and X.29 were developed to meet these requirements. Central to X.25 is the concept of the virtual circuit and a corresponding virtual call which enables the network access circuit capacity to be split into logically separate channels, the packets corresponding to a number of different calls multiplexed onto the same, physical circuits. This capacity sharing extends across the network and then onto the access circuits of the destination (or destinations) to which the virtual calls are addressed.

The X.25 specification merely defines the technique for accessing the network, and there are other situations for which it has been implemented, which are not included in the official Recommendation. These include:

—   as a node-to-node protocol within the network. A number of PTTs have implemented networks in this way;

—   the highest level of X.25 (level 3) used as an end-to-end protocol between two computers or DTEs.

These are illustrated in Figure 10.3.

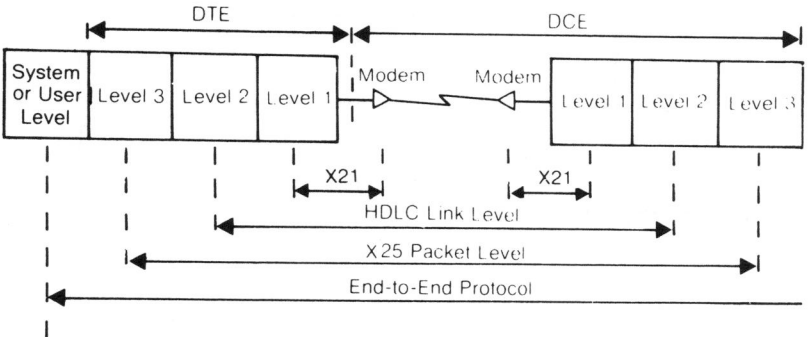

Figure 10.3   Protocol relationships in X.25

The Structure of X.25

X.25 comprises three distinct and independent levels, and the procedures at one level utilise the functions of the level immediately below, but are independent of the manner in which the latter are implemented. There is a close correspondence between these three levels and levels one, two and three of the seven layered architecture. Indeed, the structure of the ISO seven layered model, particularly in the lower layers, has been strongly influenced by X.25. (See Figure 10.4).

The three levels in X.25 are:

— the physical interface;
— the link control or link access protocol;
— the packet level protocol.

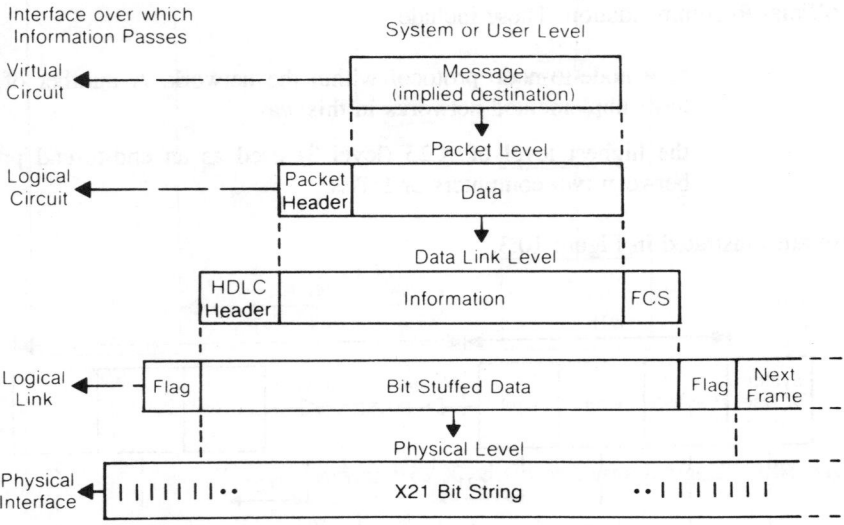

Figure 10.4   Hierarchical information flow

X.25 specifies the protocols for exchanging information between corresponding levels in the DTE and the DCE. Figures 10.3, 10.4 and 10.5 illustrate the relationships between the levels. A packet is transported in the information field of an HDLC frame, and each level accepts information from a higher level and adds header or trailer information before passing it to a lower level. The *packet* as such remains uninterpreted except at the packet level (level 3). Figure 10.5 illustrates how the logical structure of X.25 maps onto the physical configuration.

Level 1 — Physical interface

For the physical interface X.25 utilises the X.21 standard or X.21 *bis*. The important characteristics of this level are that it provides a bit-serial, synchronous, full duplex, point-to-point circuit for digital transmission. The physical arrangements are depicted in Figure 10.6.

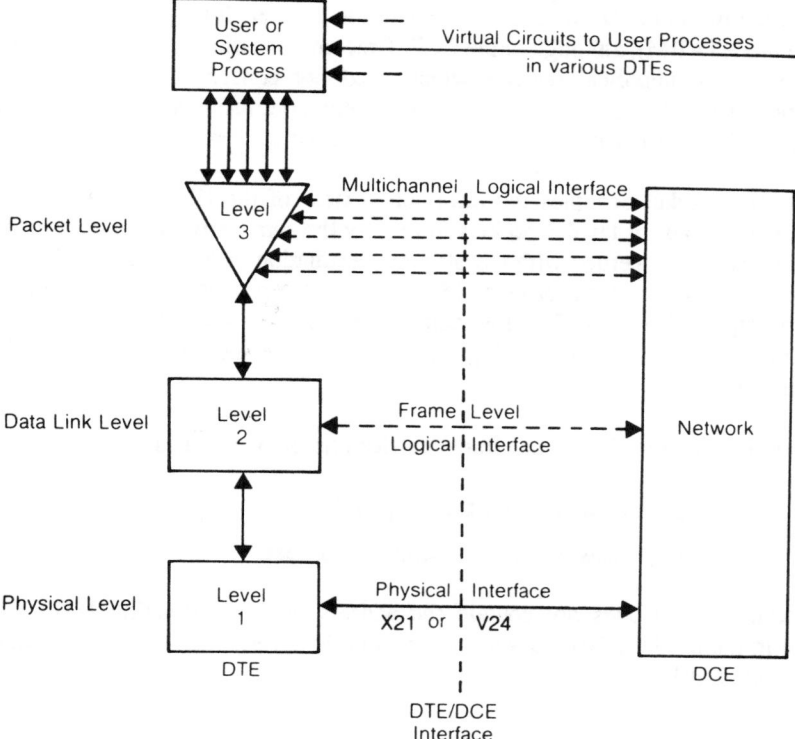

Figure 10.5 Logical structure of the X.25 interface

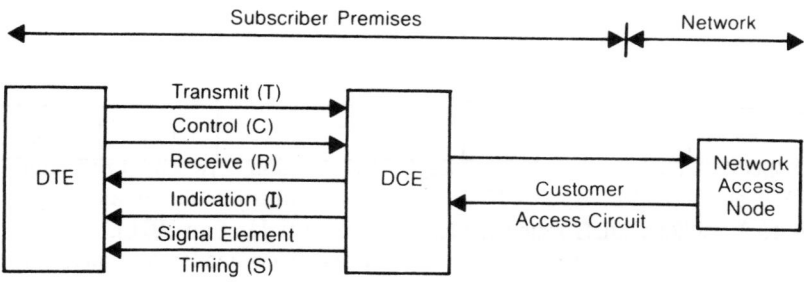

Figure 10.6 X.21 Physical interface

Level 2 — Link Control

The aim of level 2 is to provide an error free transport mechanism for packets generated at level 3 across the DTE-PSE link. It uses the frame structure and procedures of the high level data link control (HDLC) protocol to achieve error control and flow control.

A fuller description of HDLC is given later in the chapter. It utilises the asynchronous balanced mode (ABM), although in the CCITT recommendation for X.25, this is referred to as LAPB, for historical reasons which need not concern us here. The two are compatible, and under ABM either station may send commands at any time and initiate responses without receiving permission from the other station.

The level 3 packet occupies the information field of the HDLC frame (see Figure 10.7). Level 2 treats all level 3 packets in exactly the same way; it does not inspect the packet, and is not concerned with the ultimate destination or logical channel number of the packet. The only concern of level 2 is to convey the packet across the terminal-PSE link, uncorrupted by errors. The flow control performed at level 2 applies to the DTE-PSE links and extends across all logical channels since the latter have no individual identities at level 2.

The address field in the HDLC frame carries one of two values:

—   *commands* contain the recipient's address; and

—   *responses* contain the sender's address.

The address field does not contain the address of the destination DTE, which is specified in the header of the packet contained in the information field of the frame, but is not required at level 2.

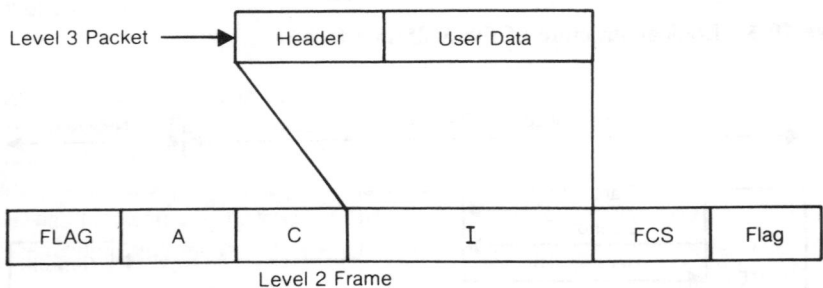

Figure 10.7   X.25 HDLC frame format

Level 3 — Packet Level

The packet level of X.25 defines the packet types and transitions which occur when various types of packet are transmitted or received. This level also performs a multiplexing function by converting the single channel provided by level 2 into a number of logical channels, and makes provision for independently controlling the flow of information on each logical channel. Error recovery, which may entail clearing and re-initialising the channel, can be applied either to a single channel or to all of them. The method of flow control and error recovery is very similar to the method employed in level 2, and there is also a similarity between some of the commands and responses of the two levels.

In summary, X.25 level 3 provides the following facilities:

- multiplexing of logical channels onto a single data link;

- error and flow control across the local interface between DTE and DCE (not end-to-end);

- guaranteed packet-sequencing;

- interrupt facilities;

- error recovery by reset and flushing out of packets;

- virtual calls end-to-end between DTEs;

- permanent virtual circuits between DTEs;

- virtual circuits which perform end-to-end addressing;

- packet size conversion between DTEs.

Packet Formats

There are a number of distinct packet types and formats defined to support information transmission and a variety of control functions. The schematic in Figure 10.8 identifies the main types and the significance of the packet fields.

The maximum length of the data field can be specified at call establishment time or at subscription time, depending upon the particular network. In all versions the default is 128 bytes and the maximum is 4096 bytes. If the maximum packet size differs at each end, thus resulting in a lack of correspondence not only between packet sizes, but also between the counts of packets transmitted and those received, the protocol can assume responsibility for packet fragmentation and reassembly. The more data indicator (M) is utilised for this purpose.

The qualifier bit (Q) is used for transmission between an X.25 DTE and the packet assembly/disassembly facility (PAD) supporting asynchronous DTE access, and distinguishes between packets destined for the remote DTE and those packets which are directed to the PAD to modify its mode of operation.

The address field comprises 14 digits. The first twelve are for use by the network administration, and on public networks, are used to address each individual DTE which is connected to the network and to identify different countries and individual networks within countries. The two remaining digits are for optional use by the subscriber, and could, for example, address a specific application or process within the DTE. This optional address field will pass transparently through the network.

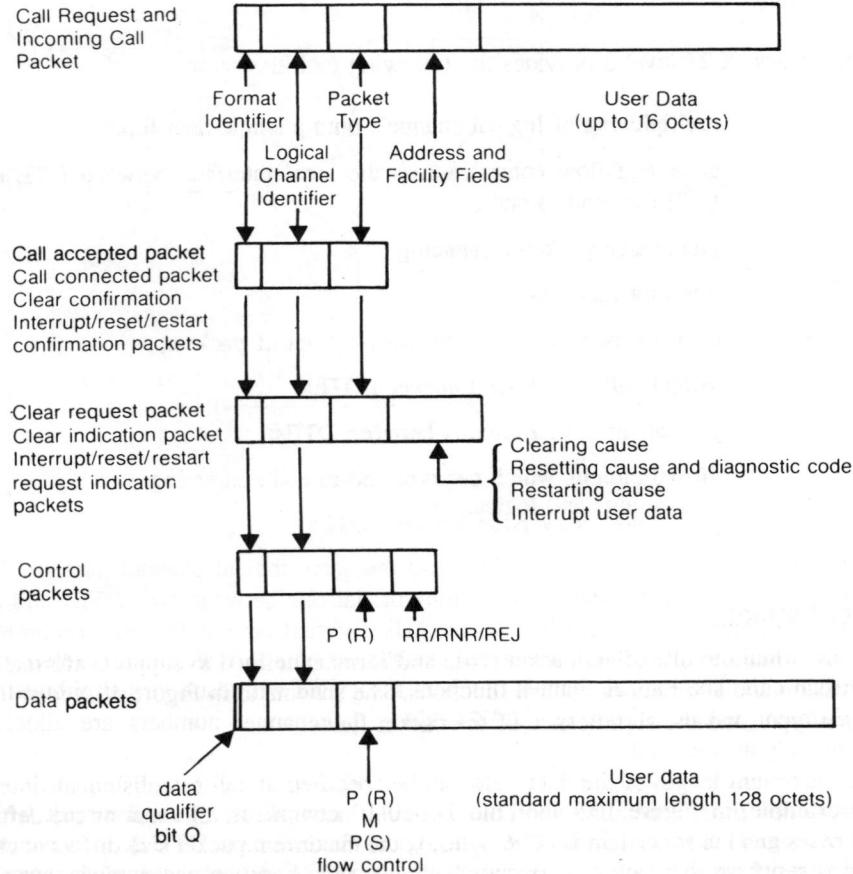

Figure 10.8 X.25 Packet formats

The facilities field is used by the DTE during the call establishment phase to specify amongst other things, the network facilities required. This is where optional user facilities are provided by the network.

Logical Channels and Virtual Circuits

An X.25 terminal can support several virtual calls simultaneously, by means of packet-interleaving. Each call is allocated a logical channel number, and all packets related to that call carry this channel number. The logical channel number is, in fact, in two parts:

— a logical channel group number less than 15; and

— a logical channel number less than 255, giving a total of 4096 possible channels.

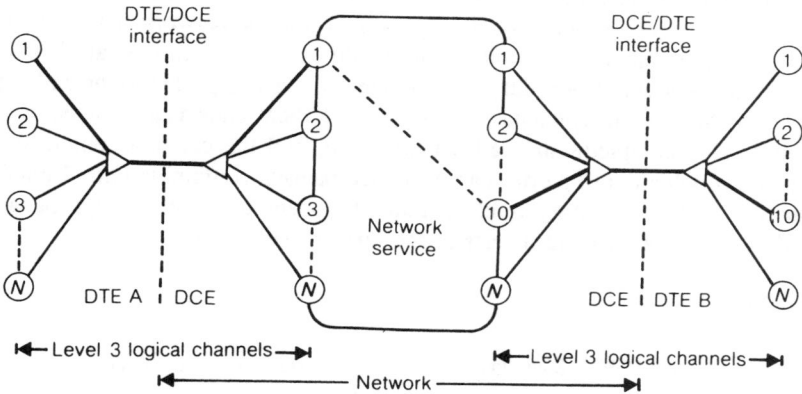

Figure 10.9 Virtual circuits and logical channels

Both the DTE and its supporting DCE use the same logical channel number. The logical channels define two-way associations or liaisons between two DTEs and are illustrated in Figure 10.9. These liaisons are called virtual circuits or, more accurately, switched virtual circuits (SVC), and a call established on a virtual circuit is referred to as a virtual call. The logical channel numbers associated with each virtual circuit may have different values in the two DTEs, since the channel numbers are allocated independently at each end of the link.

For example, in Figure 10.9 the path 1-1-10-10 comprises a virtual circuit which associates logical channel 1 in DTE A with logical channel 10 in DTE B. However, the logical channel numbers are usually chosen from opposite ends of the number range, to avoid collisions.

The major difference between a switched physical circuit and a switched virtual circuit is that the latter is only allocated circuit capacity or bandwidth when packets are being transmitted. Another type of circuit is the *permanent virtual circuit,* which is a permanent association. It is allocated a permanent logical channel at each DTE, and is analogous to a conventional leased circuit. They are set up at network configuration time.

Datagrams, dropped in the 1980 version, were mentioned in the introduction to this chapter. Certain logical channels would be reserved for datagram use and held permanently in the data transfer state. A datagram is a single self contained packet, sent to the network with no set up or clear procedures.

As we have already observed, the X.25 specification merely defines a set of procedures for accessing the network. This allows some freedom of choice for internal switching and routing techniques employed within the network.

Virtual Call Establishment

The procedure for establishing and terminating a virtual call is illustrated in Figure 10.10. A virtual call is set up by sending a call request packet to the network. This contains the address of the called party and details of any facilities required on the call. It may also contain a limited amount of user data. The network routes this packet to the called party, who respond with a call accepted packet; the network then sends a call connected packet to the caller. This completes the call set up phase, and the call then enters the data transfer phase in which both parties can exchange packets containing data. Either party can clear the call by sending a clear request packet and, on receipt of a clear confirmation packet, a terminal reverts to the idle or ready state.

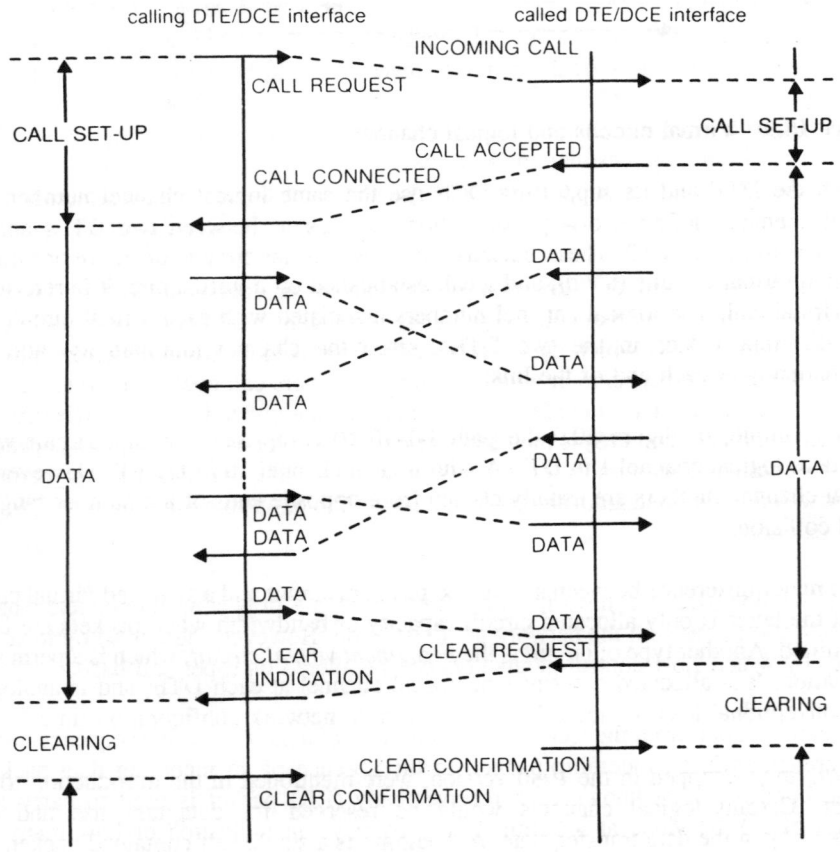

Figure 10.10   Call establishment, data, and call clear phases of a virtual call

Both the calling and called DTE addresses are required to set up a virtual call, and, although the called DTE does not strictly need to know its own address, and this could be stripped off by its serving DCE, inclusion of the address does provide an opportunity for the network to check for incorrect addressing. Each DCE also arranges for incoming packets to be allocated the correct local channel number before onward transmission to the DTE.

Permanent virtual circuits are treated as a special type of virtual call; there is no call set up phase and the logical channel is permanently in the data transfer state. The clear indication packet may contain the reason for the clearing, for example, DTE out of order, network congestion, number busy, remote or local error.

Flow Control at Level 3

Since the primary purpose of level 2 is to provide a reliable transport mechanism using error prone circuits, the virtual circuits at level 3 should be comparatively error free. Therefore the emphasis at level 3 is on controlling congestion. Flow control is generally interpreted as a local function, and the acceptance of data packets by the serving DCE does not imply that the destination DTE has agreed to accept them, but merely that the network has sufficient storage to hold the packets and await their acceptance by the remote DTE.

An independent window flow control mechanism analogous to that of level 2 is applied on each logical channel. Every data packet transmitted contains a packet send sequence number P(S) which cycles repeatedly from 0 to 7. Some networks may also provide the extended scheme in which sequence numbers can cycle from 0 to 127. The first data packet sent across the DTE-DCE interface in a given direction carries the send sequence number 0, and P(S) is then incremented by one for each subsequent packet sent in that direction. The number of packets or window size that can be sent before an authorisation is required from the receiver is negotiable, up to a maximum of 7 (or 127 when extended).

Both window size and maximum data field length are flow control parameters agreed at subscription time with the network administration. However, there is an optional user facility by which flow control parameters can be altered on a per call basis.

An authorisation from the receiver takes the form of a packet receive sequence number P(R) contained in an incoming packet. In the absence of an incoming data packet, a special control packet known as a receive ready (RR) packet is used to carry P(R). Receipt of a P(R) value at a terminal only conveys an indication of how many data packets the network is prepared to accept, and does not necessarily imply that previous packets have been received correctly to use the (R) value as a form of delivery (D) indication.

The principal flow control commands are:

— receive ready (RR);

— receive not ready (RNR);

— reject (REJ).

These operate in a similar way to their counterparts at level 2.

Provision is made for priority interrupt packets to carry data from DTE to DTE, and these are not bound by the normal flow procedures. An interrupt packet may overtake earlier packets and will be acknowledged by an interrupt confirmed packet.

The principal error recovery mechanisms are *reset* and *restart*, and these can be initiated from either end. They will usually be invoked as a result of either a sequence error detected at level 3 or an error detected at a higher level, possibly in a higher level protocol. An error at a lower level which perhaps resulted in a link reset at level 2 does not automatically have an effect at level 3. In such an event, a restart at level 3 might be necessary, but this would have to be explicitly requested.

Reset re-initialises a specified virtual circuit or permanent virtual circuit in both directions and flushes out any data or interrupt packets in transit.

The *restart* function operates in a similar manner to reset but is applied to all virtual circuits and permanent virtual circuits. Paralleling the call establishment procedure, the two functions are initiated by issuing a request, which becomes an incoming indication at the remote DTE; and it is then followed by a confirmation returned to the originating DTE.

Packet-Switched Network Facilities

In addition to the basic facilities such as the virtual call, X.25 can be used to support the provision of a range of other facilities and services. Depending upon the network administration, some of these may be present amongst the basic offerings, but in general they are supplied as options. The user must specify which optional facilities he requires, and they will generally incur additional charges.

Over 20 such options have been defined, and they include:

— *fast select* allows up to 128 bytes of data to be carried in a call request packet. Data can also be carried by a call accepted packet and in a clear request packet;

— *closed user group* allows a user to communicate with other terminals in the same closed user group, but prevents communications with other terminals; the overall effect is to create the security of a private network within a public network;

— *reverse charging* and *acceptance* are two complementary facilities allowing the request and acceptance of reverse charge calls;

— *non-standard default packet sizes* allows selection of other than the normal default maximum packet size of 18 octets;

— *flow control parameter negotiation* allows selection of maximum packet size and window size across the network interface;

— *hunt group* are calls with the hunt group address distributed across a designated grouping of DTE-DCE interfaces;

— *call redirection* allows incoming calls to be redirected, if the DTE is out of order or busy, to an alternative — this can be a list which is tried in order.

X.32

CCITT Recommendation X.32 specifies the interface between a packet DTE and a packet network through public switched telephone network (PSTN), integrated services digital network (ISDN) or circuit switched public data network (CSPDN). The call can be initiated from either the terminal or the network. Procedures for identifying the caller are specified and duplex or half duplex operation is used. Once the call is established, X.25 procedures are used.

The PAD and the X.3, X.28, X.29 Recommendation

The PAD facility was developed to enable unintelligent DTEs operating in asynchronous mode to access other DTEs connected through a full X.25 interface. It is now used extensively by PCs with asynchronous emulation. PAD stands for packet assembly disassembly and it is associated with, and located at, the packet switching exchange or access node.

The PAD Functions

The primary function of the PAD is to accept the serial character strings generated by a start/stop DTE, and to convert these into the packet formats specified by X.25 level 3 on behalf of the DTE. The packets are then despatched across the network to the packet switching exchange serving the destination X.25 DTE to which they are delivered under the normal X.25 arrangements.

Packets flowing in the reverse direction, on reaching the exchange serving the start/stop DTE, are converted into character strings acceptable to the DTE. A separate set of procedures contained in Recommendations X.3, X.28 and X.29 govern the operation of the PAD and specify the interfaces between the PAD and the start/stop DTE on the one hand, and the X.25 DTE on the other. The inter-relationships are shown in Figure 10.11 and the Recommendations are described in the following sections.

X.3

This Recommendation defines the basic functions of the PAD. The start/stop DTE not only has limited intelligence, but terminals vary widely in their characteristics; also the remote DTE may only recognise and be able to converse sensibly with the terminals having specified characteristics.

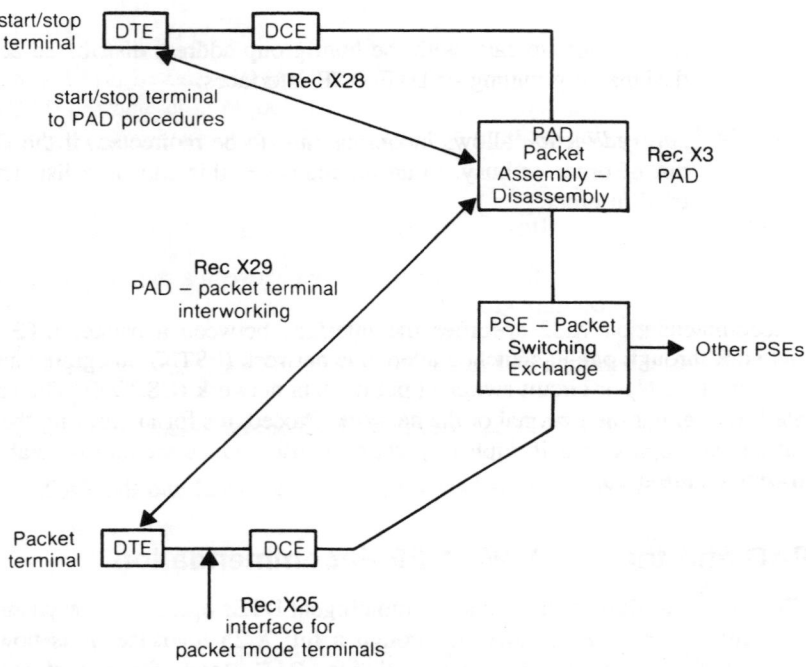

Figure 10.11   X.3, X.28, X.29 and X.25 relationships

The PAD is, therefore, provided with a set of parameters, the values of which can be set either from the terminal or the remote DTE which is being accessed. So far, more than 20 parameters have been defined and they relate to such features as:

—   whether echo-checking is required;

—   selection of the data forwarding characters (these are the characters which must be inserted by the terminal to define packet boundaries, and to instruct the PAD when to assemble and forward a packet;

- auxiliary device control signals to activate a locally attached device at the remote terminal such as a cassette or diskette drive;

- specification of terminal format effectors such as line feed or carried return;

- padding delay requirements to allow the terminal time to effect a line feed (applicable mainly to mechanical terminals).

Thus the PAD can, in effect, supply some of the terminal handling functions, and also enables a *foreign* terminal to simulate the characteristics of a terminal recognised by the remote packet mode DTE.

Initially, the parameters can be set to one of a number of standard profiles, but their values can be altered or negotiated between the start/stop terminal and the PAD or the remote X.25 DTE at logging-on time or during a call. The procedures are specified in X.28 and X.29.

X.28 — Interface between the Start/Stop Mode DTE and the PAD

The Recommendation specifies the protocol to be used between a start/stop terminal and the PAD. It lays down procedures for:

- establishing a call to the PAD from the start/stop terminal;

- setting the PAD parameters to the required values;

- exchanging data between a start/stop terminal and the PAD.

Access to the PAD

Access to the PAD can be:

- via the public switched telephone network;
- over analogue leased lines.

Where digital data services exist, access will also be possible:

- over digital leased lines, using a Recommendation X.20 or X.20 *bis* interface; or

- over the circuit-switched data network, using a Recommendation X.20 or X.20 *bis* interface.

The Recommendation X.20 *bis* interface is compatible with the Recommendation V.24/RS232 interface, thus allowing existing terminals to use the new digital data services.

Where digital data services exist they can be expected to provide a more reliable link than analogue circuits, and with increasing multi-frequency dialling penetration, there should be substantial improvements in call establishment time for dialled calls (less than two seconds).

PAD Initialisation

Having accessed the PAD, the start/stop terminal sends a service request signal. This enables the PAD to detect the data rate and code used by the terminal, and to set the PAD parameters to an appropriate initial setting. Provision is made for a password to be required at this stage, to prevent unauthorised usage. Procedures are given whereby the start/stop terminal user can change and/or read out the values of the PAD parameters.

Virtual Call Set-Up to the Destination Packet Terminal

The start/stop terminal user indicates to the PAD the address of the packet terminal to be called, together with the facilities required on the call. The PAD sets up the virtual call, and then prepares to carry out the packet assembly/disassembly function during the data transfer phase.

During the data transfer phase the start/stop terminal can recall the PAD (ie escape from the data transfer phase and enter into a control phase) in order to alter the PAD parameters, reset or request the status of the virtual call, send an interrupt to the packet terminal, or clear the call.

X.29 — Interface between a Packet Mode DTE and the PAD

This protocol is essentially the same as the X.25 protocol used for connection to other packet mode terminals, and call establishment and flow control follow normal X.25 procedures. However, it has extra procedures to accommodate the presence of a PAD. When a packet mode terminal is called by a non packet mode terminal the intervening PAD sets the value of particular bits in the first octet of the call user data field of the call request packet to indicate that PAD assistance is being given. In general these bits quality the use of the following three octets in identifying any additional protocol to be used on the call. The first four octets together form the protocol identifier field, and are followed in the call request packet by up to twelve octets of user data.

In the reverse direction, the packet mode terminal may communicate with the PAD by means of qualified data packets (in which the Q bit discussed under X.25 is set to one). The information in such packets is identified by the Q bit value as being for use by the PAD, and not for passing to the non packet mode terminal. This procedure allows the packet mode terminal to read, and in certain cases, set the values of the PAD parameters described under Recommendation X.3.

A particular coding of bits eight and seven of the first octet of the protocol identifier field allows the identification of non-CCITT protocols to be used on a virtual call. This opens the way for an organisation such as the International Organisation for standardisation (ISO) to standardise terminal-to-terminal protocols and their individual identification using this mechanism.

The coding of the above bits is:

	bits 8 and 7	=	0 0 for CCITT use
—		=	0 1 for national use
—		=	1 0 reserved for international user bodies
—		=	1 1 for DTE-DCE use

In addition to the facility for reading and setting the PAD parameters, two other types of message can be sent from the packet mode DTE to the PAD. It can cause a *break* signal to be sent to the start/stop terminal, and it can request the call to be cleared down.

High-Level Data Link Control (HDLC)

High-level data link control (HDLC) is really a misnomer, because it is a link level protocol and should not be confused with the high-level protocols. The essential features of HDLC such as the frame structure and its nested construction or onion skin architecture had their origins in the work of a British Standards Institution (BSI) committee in 1968. ISO had been studying protocols since 1962, and a major result of this work was the basic mode protocol. The BSI ideas were taken up within ISO, and the study of data link control procedures contained in association with the evolution of a model for overall communication systems architecture and resulted in the standardisation by the ISO of a more sophisticated method of data link control, known as high-level data link control procedures (HDLCP).

The HDLC standards are much sounder (as standards) than the earlier basic mode standards. They were evolved using *prospective standardisation*, the planning of standards before many conflicting systems became established in the market. Taken together, they specify a powerful and sophisticated group of facilities, which are very widely used. Although they have great versatility to cover a wide spectrum of applications, at their core lies a relatively simple set of procedures; for the great majority of systems these will suffice, and will thus readily permit interworking.

Overview of HDLC

The HDLC family has the following characteristics.

HDLC was originally designed for two-way simultaneous operation between a *primary* station and one or more *secondary* stations on a physical point-to-point or multipoint data link. No provision is made for secondaries to communicate with each other.

The primary station is responsible for scheduling the data flow in the link by authorising secondaries to transmit. This may be a one-for-one response basis or a long term delegation which applies until terminated by the primary. All transmissions are in *frames*. Transmission is bit-oriented which means that any sequence of any number of bits can be transmitted in a frame. Each frame has an address field, a link control field, and space for an information field. All frames carry a frame check sequence (FCS) using the CCITT cyclic polynomial sequence for error checking. When the frame contains only link control information, the information field has zero length.

In frames from the primary (*command frames*) the address identifies the station or stations which are authorised to receive the frame. In frames from the secondaries (*response frames*) the address identifies the secondary of origin.

The control field provides for two independent numbers (modulo 8) for sequencing the forward and reverse flow of frames in the link.

There are three components to HDLC:

- the frame structure;

- the elements of procedure;

- the classes of procedure.

The first specifies the common frame structure, including the error checking and bit sequence transparency mechanisms and the size and position of the address and link control fields.

The *elements of procedure* specify the commands, the responses and the sequencing information which can be coded into the control field for link control and error recovery purposes.

The *classes of procedure* define various modes of operation of a link (master-master, master-slave, etc); each class uses an appropriate selection of the commands and responses defined in elements of procedure.

HDLC Variants

The introduction of link protocols using the standardised frame format is increasing and all of the major manufacturers support it, either as HDLC or under their own acronym:

- SDLC, IBM synchronous data link control;

- ADCCP, American national standards institute (ANSI), advanced data communications control procedure.

HDLC Frame Structure

Figure 10.12 shows the uniform structure. Each flag sequence identifies both the start of one frame and the end of the previous frame. More than one flag sequence can be used if a longer time interval is needed between frames. If the transmitter has no information to send, it may transmit either continuous binary 1s or flags. It may also abort the transmission of a frame (to save time when retransmission is called for) by sending at least seven continuous 1s.

Each frame begins with an address octet and a control octet, followed by a data field (which may be omitted) and closes with a 16-bit frame check sequence. The address field has 256 combinations but no rigid rules are specified for allocating addresses to secondaries. The contents of the control field are defined in the elements of procedure standard which is discussed later in this chapter. A mechanism is provided for extending the length of the address and control fields, but they are always a multiple of octets.

Flag (F)	Address (A)	Control (C)	Information Field	Frame Check Sequence (FCS)	Flag (F)
	8 bits	8 bits	Variable	16 bits	

Checked Information

Figure 10.12 HDLC frame structure

The frame check sequence is applied to all fields of the frame excluding the flag sequence. The information field can contain any number of bits, subject to some upper limit set by the system, and in any code. As shown in Figure 10.12, the address and control fields normally consist of one octet but they can both be extended to two if required.

Code Transparency

The flag pattern could occur in the combined A, C, information and FCS fields and Figure 10.13 shows how a transparency is achieved. If the transmitter finds a sequence of five adjacent 1-bits, an extra 0-bit is inserted before transmitting the next real data bit. To keep the system simple an 0-bit is inserted even if the data bit following the five adjacent 1-bits is already 0. The receiver can thus be sure that an 0 which follows five adjacent 1-bits has been artificially inserted and the process, known as *bit stuffing*, ensures that the receiver can distinguish the true frame delimiters.

Bit-oriented operation is obtained by passing the incoming bits (including the stuffed 0s), in sequence, past an 8-bit *window*. As each new data bit is received, the window is opened to see if the last 8-bits are a flag sequence. If not, the window is closed until the next bit is received. When a flag is framed in the window the receiver accepts it as a frame delimiter. If the following octet is not another flag the receiver assumes that a frame has commenced and it removes any stuffed zeros from this and the ensuring bit-stream until another flag appears in the window, at which point it assumes that the previous 16-bits (unstuffed) are the frame check sequence. Hence, the frame length between flags can be of any number of bits equal to or greater than 32. Frames with less than 32-bits between flags are discarded. The inclusion of the frame check sequence prior to the end of frame delimiter implies a short FCS buffer in the receiver, so that the traditional cyclic check summation can be correctly phased, but this does not pose a problem. It will be evident that, compared with the procedure employed in the character-oriented protocols, this is far tidier and is a well structured method for securing code transparency.

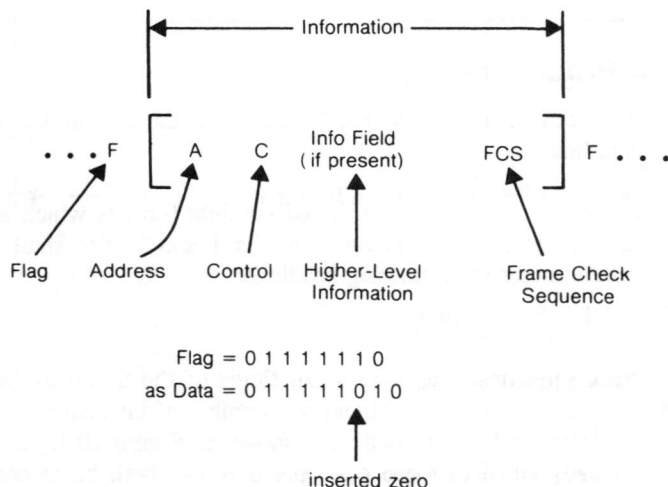

Flag = 0 1 1 1 1 1 1 0
as Data = 0 1 1 1 1 1 0 1 0

inserted zero

Figure 10.13   Bit stuffing and transparency

Synchronisation

In a two way simultaneous character-oriented system (for example, basic mode), the loss of character synchronisation can be disastrous because it becomes impossible to identify the delimiters which separate data, commands and responses. Recovery is a painful process involving time-outs and total re-synchronisation in both directions. The bit-oriented HDLC system is superior because frame synchronisation is restored whenever a flag is detected. The loss of a flag, the simulation a flag, or the loss of octet synchronisation during a frame, is recovered, quite simply by the error control system and is indistinguishable from a simple frame error.

Implementation

Figure 10.14 shows how the *nested architecture* facilitates a structured approach to constructing and dismantling a frame. However, most of these functions are implemented by custom built integrated circuits.

The receiver follows a logical sequence of operations on the incoming data. The first step is to identify the flags which mark the beginning and end of a frame. All the data between the flags is subjected to an inserted 0-bit removal process, in order to recover the original data. The frame check sequence (FCS) is applied and this permits transmission errors in the address, control or information fields, or the FCS itself to be detected. (If a flag is corrupted, the receiver will assume, erroneously, that the 16-bits preceding the next flag it detects are the FCS, and this will yield an invalid check result.) If the frame is error free, the address and control fields are examined and can be checked for procedural errors. Only if the frame is valid will the receiver proceed to deal with the information bits.

Elements of Procedure

The elements of procedure are concerned largely with the use of the control field controlling the data link.

The basic single octet control field has three possible formats which are shown in Figure 10.15. The format type is contained in bits 1 and 2. We shall describe the elements of procedure under the following headings:

—   sequence numbers;

—   operational modes;

—   information frames;

—   numbered supervisory frames;

—   unnumbered command and response frames.

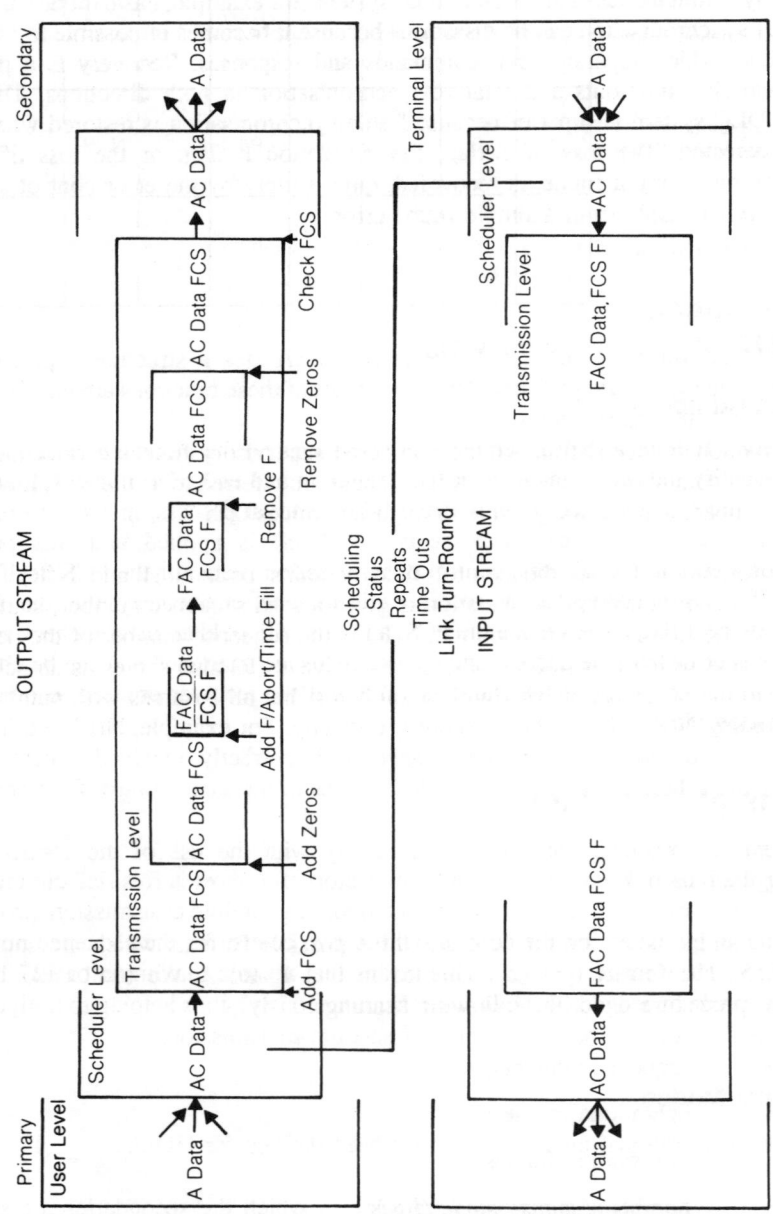

Figure 10.14   Architectural concepts

	1	2	3	4	5	6	7	8
Information Frame	0		N (S)		P/F		N (R)	
Numbered Supervisory Frame	1	0	S		P/F		N (R)	
Unnumbered Frame	1	1	M		P/F		M	

Figure 10.15   Basic control octet coding

Sequence Numbers

Both information frames (I-frames) and numbered supervisory frames carry either one or two sequence numbers in the control field. These are referred to as the N(S) and N(R) sequence numbers and normally cycle through the values 0 to 7.

In an information frame, the control field contains both N(S) and N(R). Every information frame transmitted by a station is given a send sequence number, starting at N(S) = 0 for the first I-frame transmitted. N(R) is the sequence number of the next information frame which that station expects to receive. N(R) also serves as an acknowledgement to the other station, the implication being that all I-frames with numbers up to and including N(R−1) have been received correctly. For example, N(R) = 6 implies that all frames up to and including 5 have been correctly received. There is no relationship at all between the N(R) and N(S) values in the same control field and they cycle asynchronously.

The range of 0 to 7 has been found satisfactory for most terrestrial circuits and applications. However, in longer distance links involving satellite transmission, provision has been made, by extending the control field to two octets, for the sequence numbers to cycle through the range 0 − 127. This means that up to a maximum of 127 frames can remain unacknowledged (ie, be *in-flight* simultaneously), thus helping to mitigate the approximate half-second round trip delay of satellite transmission.

Operational Modes

The following operational modes have so far been defined for HDLC:

— *normal response mode (NRM)*, in which the secondaries are strictly disciplined to transmit only when specifically instructed to do so. This is aimed at the control of multipoint links;

— *asynchronous response mode (ARM)*, in which a secondary is free to transmit information until the mode is changed;

— *asynchronous balanced mode (ABM)*, in which a secondary temporarily assumes an equivalent role to the primary until the mode is changed;

— *normal response mode extended (NRME)*, extended version of above;

— *asynchronous response mode extended (ARME)*, extended version of above;

— *asynchronous balanced mode extended (ABME)*, extended version of above.

Information Frames

The first type of frame, the information format frame, is used for transmitting information. As we have seen, an important property is the *piggy-backing* of acknowledgements for multiple frames onto a single information frame, rather than the one-for-one ACK and NAK procedure of character-oriented protocols.

Numbered Supervisory Frames

Supervisory format frames are used to acknowledge correct reception of all information frames up to and including the one numbered N(R)—1 and to convey control signals. Supervisory frames never contain an information field, and therefore do not require an N(S) number. An N(R) number is present, and has the same significance as for an information frame.

Bits 3 and 4 of the control field are coded to provide four controls:

— *receive ready (RR)*, is a combined acknowledge and proceed signal. N(R) identifies the next required frame;

— *receive not ready (RNR)*, is also an acknowledgement specifying the next required frame, but with a warning that the station is temporarily unable to receive it;

— *reject (REJ)*, indicates that a frame has been received in error. N(R) identifies the number of the frame at which retransmission should start;

— *selective reject (SREJ)*, indicates that a specific frame has been received in error and requests the retransmission of that frame only. This mechanism introduces more complex sequencing problems and is only intended for sophisticated systems, such as satellite links.

Each of the above can serve either as commands or as responses.

Unnumbered Command and Response Frames

Unnumbered frames contain no sequence numbers and the five bits 3, 4, 6, 7 and 8 can be coded to provide 32 supervisory functions. Unnumbered frames cannot contain an information field (I-field), but this cannot be sequence number checked unless a separate numbering system is included in the I-field by the user. Unnumbered commands and responses are used for link housekeeping functions such as mode setting, status signalling and problem reporting.

- *set normal response mode (SNRM)*, sets the link into a normal response mode, but also performs link reset functions;

- *set asynchronous response mode (SARM)*, sets the link into an asynchronous response mode and performs a link reset function;

- *set asynchronous balanced mode (SABM)*, sets the link into an asynchronous balanced mode, but also performs a link reset function;

- *reset (RSET)*, performs a pure link reset function without changing the link operational mode;

- *SNRME*, is an extended version of SNRM;

- *SARME*, is an extended version of SARM;

- *SABME*, is an extended version of SABM;

- *unnumbered acknowledgement (UA)*, reports that the secondary has accepted the SNRM, SARM or SABM command;

- *unnumbered information (UI)*, signals an information frame without a sequence number;

- *unnumbered poll (UP)*, requests transmission from the addressed secondary without specifying the restart point;

- *command reject (CMDR)*, reports that the secondary is unable to obey a command from the primary either because it does not include the command in its repertoire, the command is illogical, the frame is too long, or the sequence numbering is wrong. The rejection frame is extended by 3 octets of information which repeat the rejection control field, declare the send and receive counts and indicate specific reasons for rejection;

- *frame reject (FRMR)*, has the same effect as CMDR but FRMR can be used either as a command or a response;

- *exchange identity (XID)*, allows for address identity to be exchanged. This may be extended later to include agreement of optional elements to be used;

- *disconnect (DISC)*, restores the link to a disconnected mode. SNRM, SARM or SABM must be sent to restart;

- *request disconnect (RD)*, requests disconnection;

- *disconnect mode (DM)*, reports that the secondary has obeyed the disconnected command;

- *set initialisation mode (SIM)*, a command which initialises the link. Use may be extended later to embrace more sophisticated link initialisation functions such as program load;

- *request initialisation mode (RIM)*, requests initialisation of the link;

- *test* accompanies a special test message frame.

The Poll Final (P/F) Bit

All frames contain a P/F bit in bit position 5 of the control field. It is referred to as the P bit in command frames (ie frames sent by the primary) and the F bit in response (ie frames sent by the secondary).

Originally the P/F bit was designed to allow the primary to instruct a secondary to respond and, in the reverse direction, for the secondary to indicate the final frame in a sequence. This procedure, therefore, has an obvious application in multipoint operation using the normal response (NRM) mode.

However, its use has now been extended to distinguish repeated sequences from original sequences for error control, known as *check pointing*. Discipline is exercised by following the rule that in asynchronous response mode or asynchronous balanced mode, a response with F set to 1 must be sent when a command is received with P set to 1. A second command with P set to 1 should not be sent until a response with F set to 1 has been received.

Flow Control

The procedures described above provide the basis for an efficient and resilient method of flow control and error control. It is beyond the scope of this publication to give a detailed and comprehensive account of how this operates; a brief description of some of the principles must suffice. In addition to the N(S) and N(R) sequence numbers carried in the frames, the two ends also maintain sequence counts or state variables.

Send State Variable V(S')

The send state variable denotes the sequence number of the next I-frame in sequence to be transmitted. V(S) can take on a value of 0 to 7. The value of V(S) is further increased by one following each successive I-frame transmission, but is not allowed to exceed the N(R) of the last received frame by more than the permitted maximum number of outstanding frames (k). We have seen that under normal conditions k has the value of 7.

Receive State Variable V(R)

The receive state variable indicates the sequence number of the next I-frame in sequence to be received. This V(R) can take a value between 0 and 7. The value of V(R) is increased by the receipt of an error free, I-frame, in sequence, whose N(S) equals V(R).

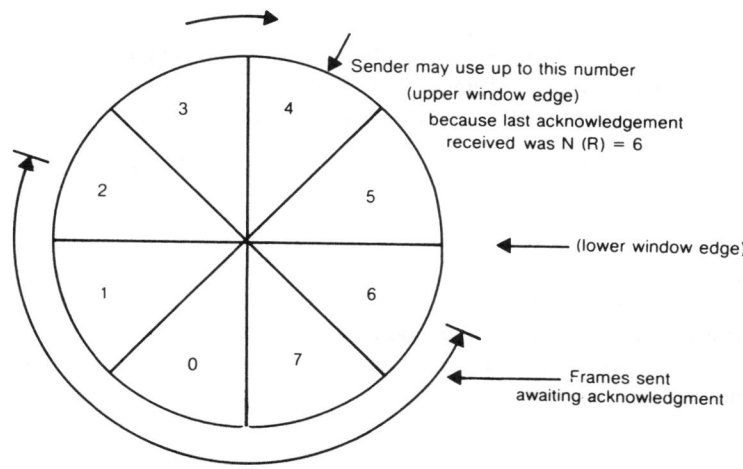

Figure 10.16   The window mechanism

When a station has an information frame to transmit (ie one not already transmitted, or one having to be retransmitted) it will transmit it with an N(S) equal to its current V(S), and an N(R) equal to its current V(R). At the end of the transmission of the I-frame, it will increment V(S) by one. If V(S) is equal to the last value of N(R) received plus k, the station will not transmit any new information frames.

The number of frames (k) which can be outstanding before an acknowledgement is received is sometimes referred to as the *window*, through which the values of N(R) recycle. The window is progressively *closed* by the transmission of each information frame, until after the seventh it is shut, and so no more information frames can be sent in the channel, until an acknowledgement is received. Figure 10.16 illustrates this mechanism. This shows that:

—   the last acknowledgement value N(R) received was 6, thus acknowledging satisfactory receipt of frames numbered up to 5;

—   frames numbered 6, 7, 0, 1 and 2 have been transmitted but are awaiting acknowledgement;

—   two slots are still available and frames with sequence numbers 3 and 4 can be transmitted without any further acknowledgement being received.

Several commands and responses are provided to initialise the link and to indicate whether the station is able to receive frames. For example, if temporarily embarrassed for space to receive frames, a response not ready (RNR) frame may be sent. This indicates that no more information frames should be sent. To enable acknowledgement of received frames, this also carries an N(R) number. The instigator removes the hold condition by the transmission of a response ready (RR) frame with the appropriate address.

Classes of Procedure

The frame structure standard is quite rigid and fully definitive with the one simple exception of address allocation. However, the elements of procedure standard contains many operational modes and command/response options and it is not possible to guarantee compatibility between systems by using this standard alone. When this was recognised, the ISO decided to specify two classes of procedure standards, one for multipoint link control and one for point-to-point operation. These are sometimes referred to respectively as the unbalanced and balanced classes of procedure.

Unbalanced Class

In the unbalanced class of procedure, one station (designated the primary) has total responsibility for control of the link. The link may be point-to-point or multipoint (supporting one primary station and a number of secondary stations). Only the primary station can issue commands and these always contain the address of the secondary for whom the command is destined. Responses sent by a secondary always contain the address of the secondary.

Two modes of operation are defined: the normal response mode in which a secondary can only transmit after receiving permission to do so from the primary; and the asynchronous response mode in which a secondary may transmit without receiving explicit permission from the primary. Only one secondary at any time may be in the asynchronous response mode.

The standard gives a basic repertoire of commands and responses for this class of procedure, together with 10 optional functions which can either extend or restrict the basic repertoire to alter the characteristics of the procedure. The optional features supported must be stated in any implementation of the standard. Typical of the additional optional functions described is the command and response for initialising a remote station.

Balanced Class

The characteristics of the balanced class of procedures is that the stations at either end of the data link are combined stations which perform as both primary and secondary stations. This standard was developed in consultation with CCITT, since this is the class of procedure that applies to the DTE-DCE interface on a public data network, where equal control of the link is required at either end. The operational mode is known as the asynchronous balanced mode (ABM); either station may send commands at any time and initiate responses without receiving permission from the other station. Commands contain the recipient's address. Responses contain the sender's address.

The standard specifies:

- the addressing scheme;

- the basic repertoire of commands and responses for this class;

- twelve optional functions to modify the basic repertoire (achieved mainly by the addition or deletion of commands and responses in the basic repertoire);

- operating procedures.

A welcome inclusion of this standard deals with conformance, ie the extent to which an implementation conforms to the standard. The 12 options are rigorously numbered, and a description of the class of operation must indicate the options provided, for example, class BA 2, 3 is the balanced asynchronous mode class of procedures with optional features 2 and 3 (for improved performance and single frame retransmission) implemented.

A station conforms to the balanced class of procedures (with optional functions) if it implements all commands and responses in the basic repertoire as modified by the optional functions.

11

Integrated Services Digital Network (ISDN)

The Origins of ISDN

In order to cope with the ever increasing need to transmit inherently binary-based information between computer controlled systems, a widespread trend towards increasing digitisation has arisen within modern communications. This trend is even occurring in the field of voice communications, where it is convenient to convert analogue voice signals into a digital form which makes them suitable for digital transmission across a digital link.

There are many powerful reasons for using fully digital telecommunications. For example, the processing and switching of digital signals is faster, more reliable and more economical than analogue processing and switching (see Table 11.1), so call set up times are reduced. The circuits which perform the digital processing are also capable of detecting and correcting errors, giving virtually error free transmission. Consequently, the number of retransmissions within a call is dramatically reduced, and so transmission becomes more economical. The resulting calls over digital links are clear and virtually noise free, irrespective of the geographical distance between the transmitter and receiver.

Once the digital links over existing networks have been established by the caller, there are many existing applications which may be used (see Table 11.2). Some of these applications require their own dedicated network due to the different transmission characteristics needed to carry digitised speech, data, moving images or drawings. However, using many different user terminal types (such as telephones and telex machines) to access many different services is very uneconomical, and the advantages of having a single network to provide many services from a single terminal are vast. It is these advantages which are fundamental to the concept of an integrated services digital network (ISDN).

The following definition of ISDN has been offered by one writer:

Basically, ISDN can be thought of as a huge information pipe, capable of providing all forms of communications and information services (voice, data, image, signalling) over existing twisted pair wiring. The large bandwidth it provides can be used for virtually anything and everything, with the user deciding how the bandwidth is to be dynamically allocated. It is an information utility accessible from a wall outlet, much like today's electrical utility, with a variety of devices able to be simply plugged and unplugged. It allows mixing and matching equipment from various vendors without concern for compatibility. It will be compatible, due to emerging worldwide standards.

Table 11.1 Comparison of PSTN and ISDN

	Comparison
Maximum Data Rate	PSTN 9.6 kbits duplex
	ISDN 64 kbits duplex
Call Set-up Time	PSTN 5 − 25 secs
	ISDN 1 − 2 secs
Error Rate	PSTN 1 in 1000
	ISDN 1 in 100,000
Signalling	PSTN slow and primitive
	ISDN fast and advanced
Additional Services	PSTN None
	ISDN Many

Table 11.2 Existing networks and services

Network	Service
Public switched telephone	Telephone
Telex	Telex
Circuit switched digital	Teletex
Public packet/circuit data	Data
TV	Television
Radio broadcasting	Sound

In Europe, these standards are set by the European Telecommunications Standards Institute (ETSI) and are derived from recommendations formulated by the CCITT. Their definition of ISDN is neatly summarised by Recommendation I.120 of the 1988 Blue Book. This states that:

The main feature of the ISDN concept is the support of a wide range of voice and non-voice applications in the same network. A key element of service integration for an ISDN is the provision of a range of services using a limited set of connection types and multipurpose user network interface arrangements.

From these two definitions it can be seen that from both the users' and the regulatory bodies' point of view, ISDN is part of the total integration and convergence of voice and data services within information technology and specifically within telecommunications.

The history of ISDN goes right back to the early part of the 1960s, when the development of pulse code modulation (PCM) transmission systems heralded the start of the transition from analogue to digital transmission. Early PCM systems were used to increase the capacity of existing short distance cables, and it was soon realised that these systems could also provide the basis for an integrated network using digital transmission, signalling and switching on a global scale.

Concurrent with the migration from analogue to digital transmission, was the development of digital switching and common channel signalling. The combined result of these developments was the concept of the integrated digital network (IDN). The IDN was to consist of digital stored program control (SPC) local exchanges which would be interconnected by high speed digital transmission circuits on to which a postal, telegraph and telephone authority (PTT) could add extra digital services.

To complement the IDN, digital customer access would be required in order to give fully digital end-to-end communication. This was to be achieved by introducing integrated digital access services which would extend the digital link from the serving digital exchange to the subscriber's premises and would enable subscribers to access the IDN digitally (see Figure 11.1). The services offered by the IDN were to include a number of ISDN type services and would allow subscribers to connect telephony, text, data and low speed image devices to networks such as the public switched telephone network (PSTN), ~~the public switched telephone network (PSTN)~~, the public switched data network (PSDN), or the telex network (see Figure 11.2).

The developments in ISDN have evolved over a number of years, and will continue to do so for many years yet. It is recognised by PTTs and equipment suppliers that during this evolutionary period, ISDN will have to be offered as a short term overlay to selected customers located around ISDN compatible exchanges until such time as a global ISDN has emerged. The rate at which ISDN roll-out has occurred to date and is continuing, is very much dependent on a country's attitude to ISDN, and also to a large extent on the availability of standards, equipment and applications.

The CCITT Recommendations

The CCITT Blue Book Series (Volume III — Fascicles III.7, III.8 and III.9) published after the 1988 Plenary Assembly, contains the overall guidelines and I-series Recommendations for ISDN produced by Study Group XVIII. Also involved in the ISDN studies were Study Group II (who looked at ISDN numbering issues in recommendations E.110-E.333), Study Group VII (X-series Recommendations for data communications), and Study Group XI (digital switching, common channel signalling and customer access signalling). A summary of the different groups of I-series recommendations is given in Appendix 11.1 at the end of this chapter.

IDA, IDN and ISDN

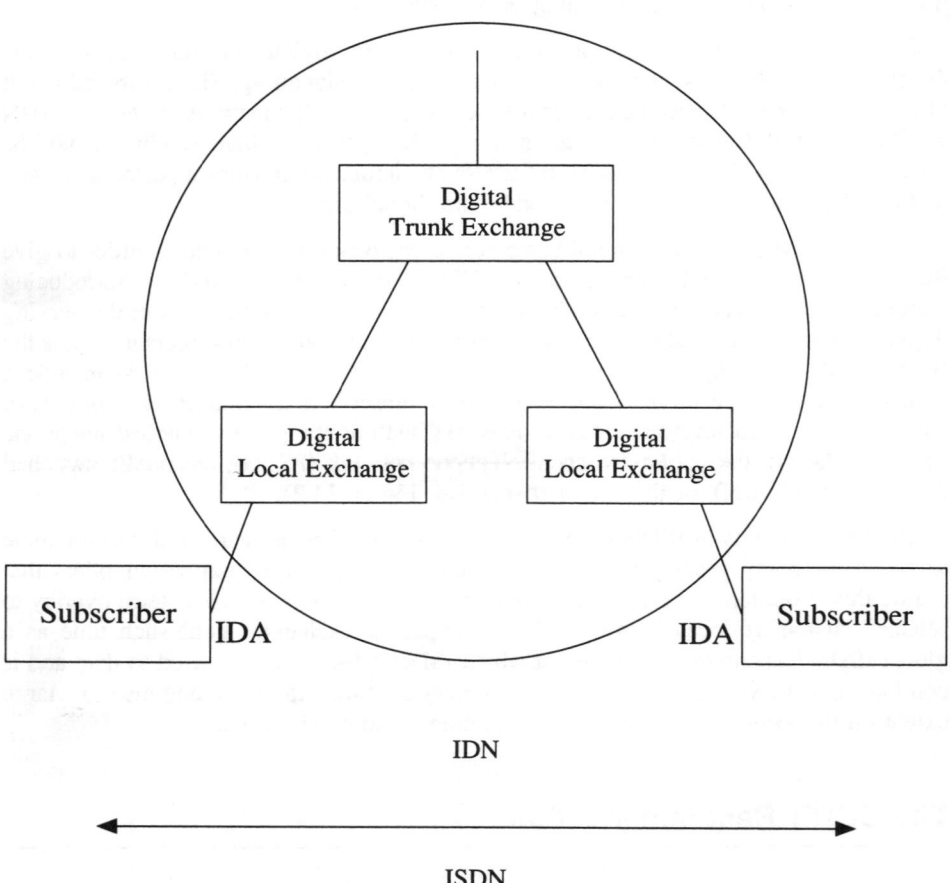

Figure 11.1 Fully digital telecommunication

Some of the existing recommendations for telephony and other dedicated service networks (such as packet-switched networks) may also be directly applicable to ISDN, although they may need to be developed to cover ISDN applications.

The Development of ISDN

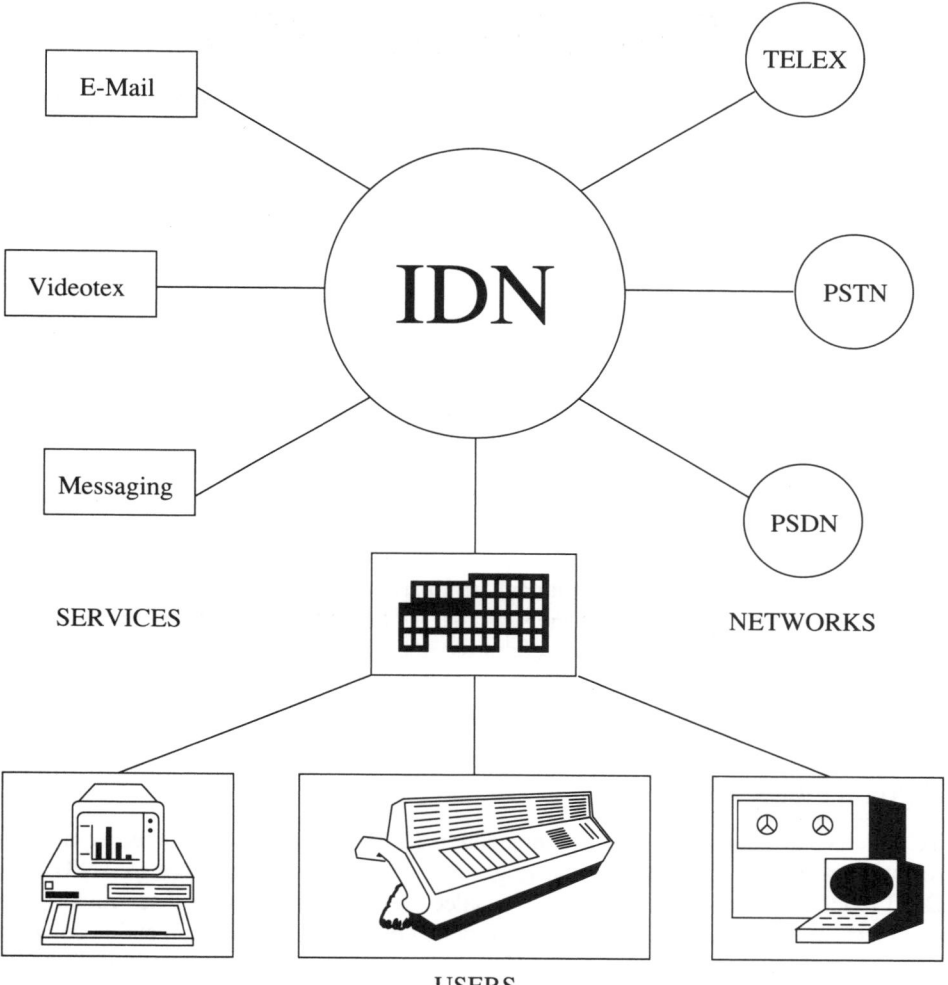

Figure 11.2 Applications which may be used with digital telecommunication

ISDN Services and Applications

There are three generic groups of services provided by an ISDN: *bearer services,*
supplementary or additional services and *teleservices or standard telecommunication*
services.

Bearer Services

Recommendation I.230 defines a bear service as being a type of telecommunication service which provides the capability for the transmission of signals between user network interfaces. The ISDN connector used to support bearer services may be identical to the connector used for current telecommunications services.

Bearer services provide the high speed digital connections and fast call set up on switched circuits and also provide the support of existing packet or circuit mode telecommunication services. Together with the high speed of ISDN, bearer services make it possible to develop new applications (or teleservices).

The CCITT lists the following bearer services:

— circuit mode speech;

— circuit mode 3.1 kHz audio;

— circuit mode alternate speech/64 kbits unrestricted information transfer;

— circuit mode 64, 2 x 64, 384, 1536 and 1920 kbits unrestricted information transfer;

— packet mode virtual call and permanent virtual circuit;

— packet mode connectionless; and

— packet mode user signalling.

Teleservices

Teleservices are described in Recommendation I.240 and can be considered equivalent to the applications supported by existing value added network services (VANS). The CCITT defines a teleservice as a type of telecommunication service that provides the complete capability, including terminal equipment functions, for communication between users according to protocols established by agreement between administrations. Because of the nature of teleservices, they can be thought of as the applications of *narrowband* ISDN (the more powerful applications of *broadband* ISDN will be described later).

CCITT Recommendation I.241 specifies the following as teleservices:

— *telephony service,* provides speech transmission at an audio bandwidth of 3.1 kHz. The communication which takes place does so in both directions, continuously and simultaneously;

— *teletex,* is an international service which can be viewed as an enhanced telex service, enabling subscribers to exchange office correspondence, via the ISDN, in the form of documents containing teletex coded information on an automatic memory-to-memory basis;

— *telefax 4*, is an international service which enables users to exchange correspondence, via the ISDN, in the form of documents containing facsimile information. The terminals used for telefax 4 are known as Group 4 facsimile machines, categorised into sending and receiving machines (Class I), sending machines (Class II) and generating, sending and receiving machines (Class III);

— *mixed mode*, is a service which allows combined text and facsimile communication for end-to-end transfer of documents containing mixed text and fixed image information. The attributes of this service are based on the CCITT Recommendation for teletex and telefax 4;

— *videotex*, is an ISDN videotex service which is an enhancement of the existing teletex service with retrieval and mailbox functions for text (alphanumeric) and graphic information; and

— *telex*, is a service that provides interactive text communication following the agreed recommendations for telex.

The distinction between bearer services and teleservices can be clarified by considering the use of a Group 4 facsimile machine on a circuit mode of 64 kbits line. In this case, the fax application is the teleservice and the circuit mode 64 kbits line is the bearer service.

Many other applications of ISDN can be found using the set of teleservices. The only real limits to what can be achieved by using an ISDN are the maximum transfer rate of information, and the imagination of the user.

Supplementary Services

Supplementary services fall into one of seven categories which are described in Recommendation I.250, and can be likened to the optional extras which can be found acting within and between some contemporary PABXs. Supplementary services are defined by the CCITT as being services which are offered by a PTT to its customers in order to satisfy a specific telecommunication requirement. These services will be made available across the ISDN either as a permanent feature of the network, or invoked by the user during an ISDN call.

Number Identification Supplementary Services

Number identification supplementary services are described in Recommendation I.251 and include:

- *direct dialling in (DDI)*, enables an ISDN user to call another user directly on a PBX or other private system without intervention by a supervisor;

- *multiple subscriber number (MSN)*, provides a user with the possibility of assigning multiple ISDN numbers to a single interface;

- *calling line identification presentation (CLIP)*, provides the called party with the calling party's ISDN number and sub-address (extension) information by displaying it on a small screen on an ISDN telephone before the call is answered;

- *calling line identification restriction (CLIR)*, enables the calling party to prevent their ISDN number from being presented to the called party before the call is answered;

- *connected line identification presentation (COLP)*, provides the called party with the connected party's ISDN number and sub-address information during the progress of a call;

- *connected line identification restriction (COLR)*, enables the calling party to prevent their ISDN number from being presented to the called party during the progress of a call.

There are also two other number identification supplementary services mentioned in Recommendation I.251: *malicious call identification* and *sub-addressing*. Both of these supplementary services have no formal descriptions or definitions in Recommendation I.251 in the Blue Book version, which requires further study.

Call Offering Supplementary Services

Call offering supplementary services are described in Recommendation I.252 and include:

- *call transfer* (CT), enables the called party to transfer an established incoming or outgoing call to a third party;

- *call forwarding busy (CFB)*, permits a user with an already established call to have all subsequent incoming calls transferred to another number;

- *call forwarding no reply (CFNR)*, permits a user to have all incoming calls which remain unanswered for up to one minute transferred to another number;

- *call forwarding unconditional (CFU)*, allows a user to have all incoming calls transferred to a third party, regardless of the state of the third party's line; and

- *line hunting (LH)*, enables incoming calls to a specific ISDN number to be distributed over a group of ISDN interfaces.

Call deflection is also mentioned in the Blue Book version of Recommendation I.252 but this has no formal description or definition and therefore needs further study before the 1992 CCITT assembly.

Call Completion Supplementary Service

Call completion supplementary services are described in Recommendation I.253 and include:

- *call waiting*, permits a busy user to be notified of an incoming call which can then be accepted, rejected or ignored; and

- *call hold*, allows a user to interrupt communications on an existing connection and then subsequently re-establish the connection if required.

This recommendation also lists completion of calls to busy subscribers as a call completion supplementary service, but again no full description is given in the Blue Book and further study is required.

Multiparty Supplementary Services

Multiparty supplementary services are described in Recommendation I.254 and include:

- *conference calling*, allows a user to communicate simultaneously with multiple parties, which may also communicate among themselves; and

- *three party service*, enables a called user to hold a call, make another call to a third party, switch from one call to another, and then release either of the calls.

Community of Interest Supplementary Services

The only fully documented community of interest supplementary service in I.255 is the closed user group (CUG). This service enables users to form groups, to and from which access is restricted. A specific user may be a member of one or more CUGs. It is also possible to prevent specific members of a specific CUG calling other members of that CUG, or receiving calls from other members of that CUG. Also included in I.255 is a private numbering plan, but as this requires further study, it has no full definition in the Blue Book.

Charging Supplementary Services

The only fully described charging supplementary service within I.256 is advice of charge. This service allows the user paying for the call to be informed of the usage based charging information. Charging information can be provided at the end of a call, during a call, or at call set up time. This Recommendation also mentions credit card calling and reverse charging as being set aside for further study.

Additional Information Transfer

This supplementary service category, found in Recommendation I.257 describes user-to-user signalling. This service allows an ISDN user to send and receive a limited amount of user generated information to and from another ISDN signalling channel in association with a call to the other ISDN user. The user information must be passed transparently by the network.

Access to the ISDN

Access to narrowband ISDN is available in the form of two structures. The CCITT have called these basic rate access and primary rate access (see Figure 11.3).

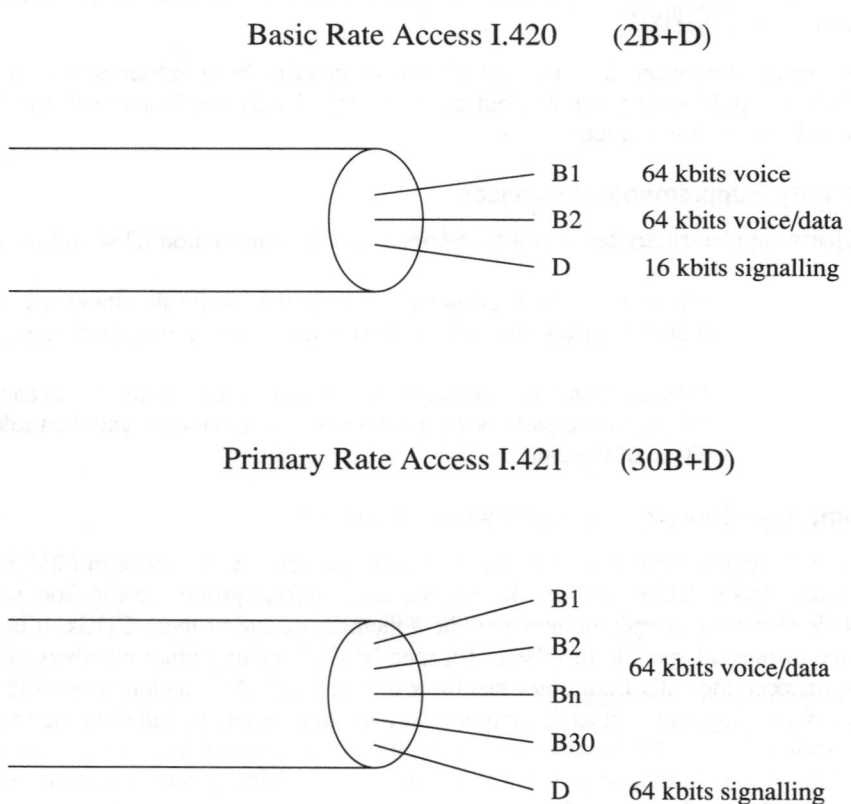

Basic Rate Access I.420   (2B+D)

B1   64 kbits voice
B2   64 kbits voice/data
D   16 kbits signalling

Primary Rate Access I.421   (30B+D)

B1
B2   64 kbits voice/data
Bn
B30
D   64 kbits signalling

Figure 11.3   Accessing the ISDN

Basic Rate Access

Basic rate access (also referred to as BRA, 2B+D or I.420) provides an ISDN user with access to two simultaneous channels using standard twisted pair copper cables. Recommendation I.420 states that this single circuit link should operate at 192 kbits, providing a user data rate of 144 kbits and 48 kbits of overhead not available to the user. The user data rate is split to give two 64 kbits voice/data channels (the B-channels) and one 16 kbits channel (the D-channel) for signalling control based on high-level data link control (HDLC) protocols. It is also feasible to utilise the spare capacity of the D-channel to provide a slower 9.6 kbits user data channel. This channel can be routed to the same termination as either of the B-channels, or to a completely different termination on the network.

Subscribers connect to the BRA service via network terminating equipment (NTEs). The NTE is effectively an interface between the IDN and the subscriber (see Figure 11.4) and is able to provide standard CCITT interfaces such as X.21 for the connection of terminal equipment (for example, a digital telephone, as well as ISDN terminal equipment interfaces).

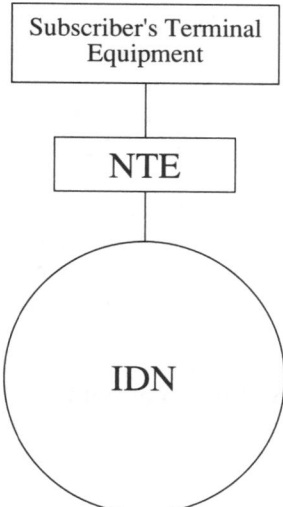

Figure 11.4   Network terminating equipment

Primary Rate Access

The second form of ISDN access is called primary rate access (PRA) and is also known as *30B+D*. It is described in Recommendation I.421. Primary rate access is primarily intended for use by larger organisations making voice and data calls via a private automatic branch exchange (PABX). A terminal of this type is provided by a new generation of PABXs known as integrated services PABXs (ISPBX). An ISPBX can provide voice, data or integrated services to any of its extensions and connects to the

ISDN via a 2.048 Mbits digital link such as the British Telecom (BT) Megastream link. Each PRA link has the capacity to carry 30 independent voice/data channels at bit rates of 64 kbits, in addition to one 64 kbits channel used for signalling (the D-channel) and another 64 kbits channel for synchronisation and monitoring functions.

In North America and Japan, the primary rate transmission bearer is based upon systems operating at 1.544 Mbits (23B+D) and is known as *T1*. This type of multi-line access is also covered by Recommendation I.421. Primary rate access can also use higher speed channels known as H_{10}, H_{11} and H_{12} channels operating at speeds of 384, 1536 and 1920 kbits respectively. Interface structures using H_0 channels can be specified with or without the 64 kbits D-channel and so primary rate access at 1.544 Mbits can have a $3H_0+D$ structure, a $3H_0+6B$ structure or a $4H_0$ structure. When the D-channel is not provided (as in $4H_0$), signalling for the H_0–channels must be provided by the D-channel in another interface structure (ie in another T1 link in this example). At 2.048 Mbits, the H-channel structure is $5H_0+D$-channels in another interface which is not using a D-channel.

The CCITT Reference Model for ISDN

As an aid to configuration, the CCITT introduced an ISDN reference model. This is a conceptual model used to identify various possible physical user access arrangements to an ISDN. To make this possible, various reference points (often referred to as interfaces), and the functions which can be performed between those reference points are defined. The reference configuration is shown in Figure 11.5 and comprises the following:

— *functional groups, which* are sets of functions (such as line termination or switching) which may be needed in ISDN user access arrangements. It must be noted that specific functions within the functional groups may be performed by more than one piece of equipment.

— *reference points,* which are the conceptual points dividing the functional groups. A reference point may be found either between two pieces of equipment, or within a single piece of equipment. The CCITT does not recognise physical interfaces that do not correspond to a CCITT reference point (for example, transmission line interfaces). The reference points S and T are covered by the I-series Recommendations whereas the R reference point is covered by the non I-series Recommendations such as the X-series.

The first network termination or NTE1, includes functions broadly equivalent to layer 1 (physical) of the OSI model (refer to Chapter 9), and in this respect the NTE1 acts as a physical termination only.

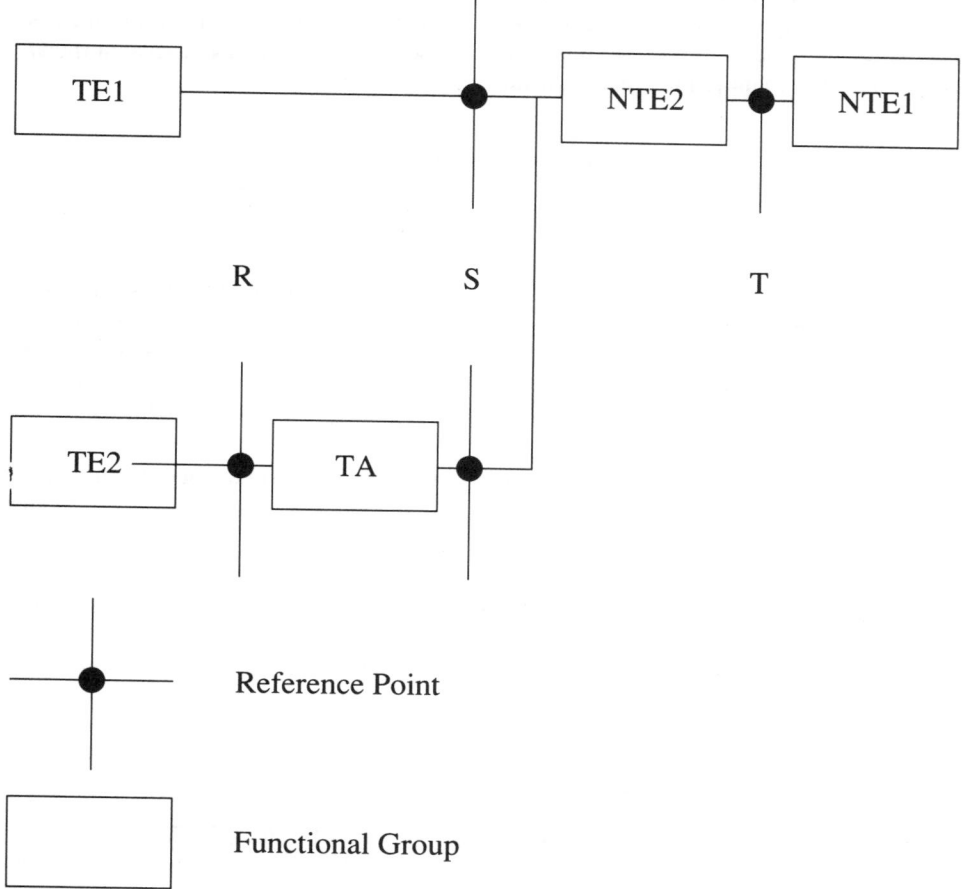

Figure 11.5 CCITT model for ISDN access

Its functions include:

—   transmission line termination;

—   physical diagnostics and performance monitoring;

—   timing;

—   power transfer to the terminals;

—   multiplexing at purely the physical (bit) layer;

—   interface termination.

NTE2, or network termination 2, performs functions related not only to layer 1 but also layers 2 and 3 of the OSI model. Thus, it would be responsible for such tasks as, say, terminal selection. PABXs and LAN controllers are both examples of NTE2 equipment, and provide functions such as:

- layers 2 and 3 protocol handling and multiplexing;
- switching and concentration;
- higher level maintenance and interface termination.

To illustrate these functions, a PABX can provide NTE2 functions at layers 1, 2 and 3, whilst at the other end of the NTE2 spectrum, a simple time-division multiplexer provides NTE2 functions at layer 1 only. It is also useful to note that in some arrangements the NTE2 will just be a series of physical connections and therefore will effectively not exist (such as when a single device terminates a single channel).

Terminal equipment such as digital phones, data terminals or integrated workstations, provide functions in layers 1, 2 and 3 of the OSI model, and whatever else above this that they may need for their operation.

The terminal equipment functions include:

- protocol handling;
- maintenance;
- interfacing;
- connection to other equipment.

Two types of terminal equipment have been specified by the CCITT. The TE1, or terminal equipment type 1, is a TE supporting the ISDN interface requirements specified by the I-series. TE2 is also a TE, but supports interface Recommendations other than the ISDN interface Recommendations. These other interface Recommendations include V-series and X-series Recommendations. The terminal adaptor (TA) allows a TE2 terminal to be served by an ISDN user network interface. Adaptors between physical interfaces at reference points R and S or R and T are examples of equipment providing TA functions.

ISDN Signalling

The CCITT have defined signalling as the exchange of information concerned with the establishment and control of connections, and management, in a telecommunications network.

User-Network Call Control

Call control between the subscriber and the exchange is done using the network layer of the Digital Subscriber Signalling System No.1 (DSS1) as described by CCITT

Recommendation I.451 (equivalent to Recommendation Q.931). Call control is reliant upon the signalling built into the ISDN. It is the signalling which sets up calls, connects calls, sends acknowledgements of calls and so on. Recommendation I.451 (Q.931) therefore lays out network connection methods and also the transfer of packetised data (for example, X.25) through the D-channel.

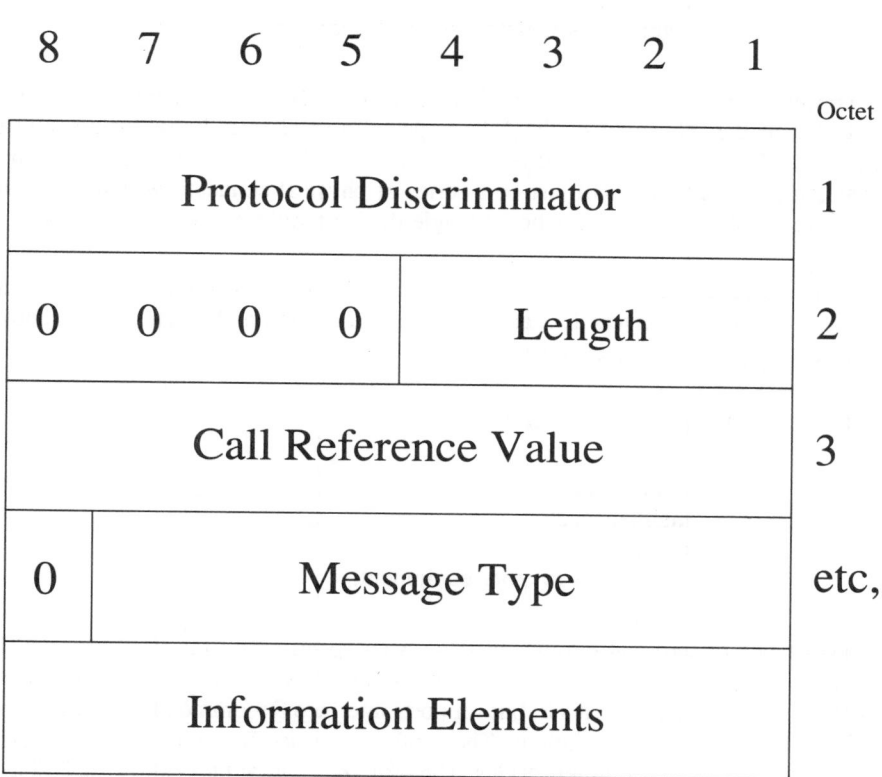

Figure 11.6 ISDN call control messages

Signalling or call control messages are carried across the D-channel of the user-network interface by layer 2 frames. The call control messages are based on a series of octets (groups of 8 bits) whose structure is shown in Figure 11.6. The first of the octets is called the protocol discriminator. This octet contains the bits 0000 1000 to distinguish the message as an ISDN signalling message. After the protocol discriminator is the call reference value which uniquely and independently identifies the call to which the control message is associated. This is prefixed by an octet which specifies length of the call reference value. The message type code after the call reference value describes the intention of the message. This code falls into one of four categories: call establishment, call information, call clearing or miscellaneous. The next octets which appear contain information on, for example, transfer mode (eg packet- or circuit-switched) and transfer (or bit) rate. To ensure smooth signalling between the subscriber and the exchange the call control messages are sent across the network in a specific order and are denoted by the uppercase letters in Figure 11.7.

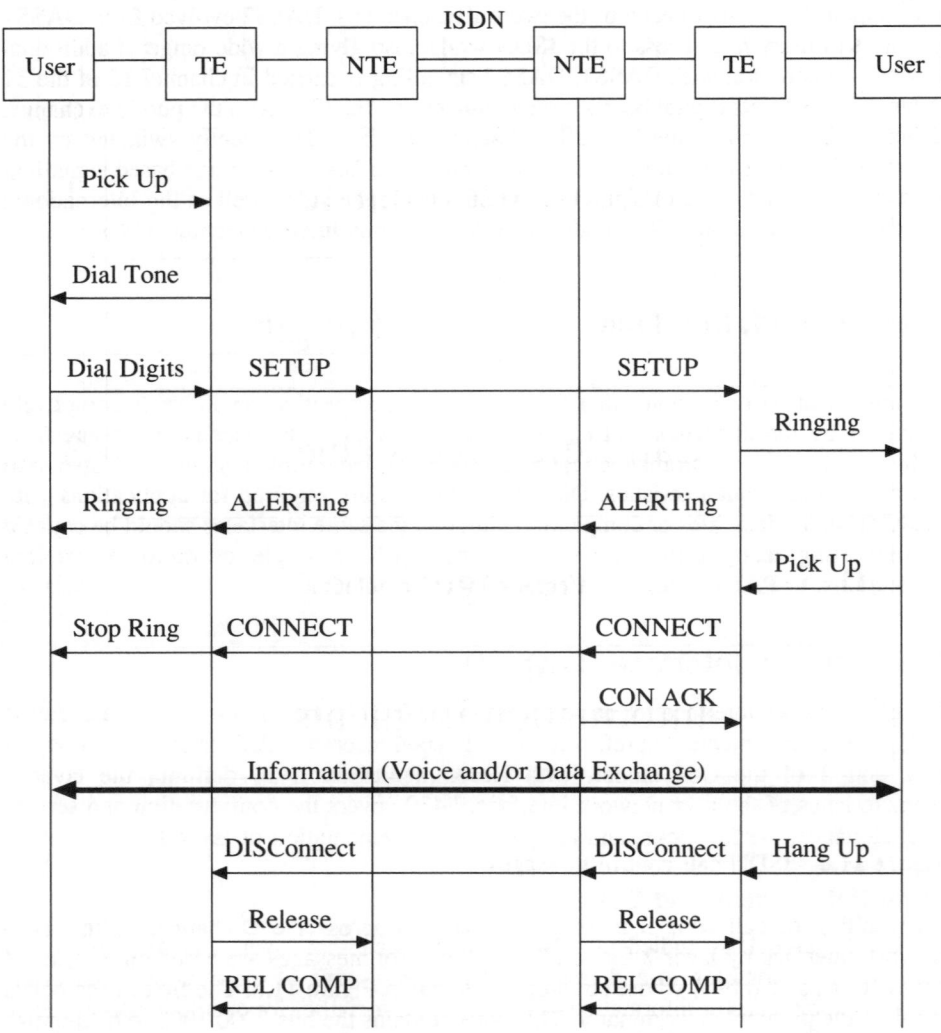

Figure 11.7   Circuit-switched message sequence

Digital Access Signalling System 2

In the UK, BT worked ahead of CCITT by developing their own digital signalling system known as digital access signalling system number 2 (DASS2). British Telecom's original digital access signalling system was used for BRA signalling between ISDN terminals and digital local SPC exchanges. DASS1, as it was known, transferred variable length signal frames over the D-channel and in this way it achieved its goal of enabling independent switching of each of the two traffic channels. DASS2 evolved from DASS1 to enable primary rate access to the ISDN whilst also giving a wide range of additional services. In common with DASS1, DASS2 signalling is carried in channel 16 of the 31 channel 2.408 Mbits digital bearer circuit, connecting the ISPBX to the public exchange. Using DASS2, each of the 30 traffic channels can be independently switched by the common signalling D-channel. Also incorporated into DASS2 message based signalling were the principles of other signalling system developments as well as the International Standards Organisation (ISO) model for Open Systems Interconnection (OSI).

User Network Interfaces

Recommendation I.411 states that from the user's perspective, an ISDN is completely described by the attributes that can be observed at an ISDN user network interface, including physical, electromagnetic, protocol, service, capability, maintenance, operation and performance characteristics. Different interfaces are required for applications with vastly different data rates or complexity, although the same interfaces should be capable of supporting many different configurations (such as single or multiple terminal connections, to PABXs etc) or different national regulations.

The Physical Interface (Layer 1)

The user network interface for basic access to the ISDN is covered by Recommendation I.420. This Recommendation refers to more detailed recommendations concerning layers 1, 2 and 3 of the OSI model, with Recommendation I.430 defining the layer 1 characteristics of the user network interface. I.430 covers the configuration and service characteristics, the functions, and the physical implementation of layer 1.

Basic Rate Access and OSI

If the OSI model is studied closely it can be seen that layer 1 is responsible for transferring information to and from layer 2 on each side of the interface. Basic rate access systems transfer information via twisted pair metallic (for example, copper) local loop cable. The intrinsic electrical characteristics of this cable may induce errors into the information (it is the responsibility of layer 2 to error detect and correct these errors), but regardless of this fact, layer 1's prime responsibility is still to transport information at the physical level with acceptable quality.

In an ISDN configuration, layer 1 has the following service characteristics:

— the ability to transport information through the S/T interface using two symmetrical twisted pair (ie 4-wire) cables that can support data rates of up to 192 kbits; and the ability to provide to layer 2 the transmission capability for the B-channels and D-channels as well as: related timing and synchronisation; capability to be activated and de-activated; a procedure for allowing access to the D-channel and the ability to communicate with other entities.

The examples of user network configurations (see Figure 11.8) show the NTE1 acting as the termination of the ISDN network. In the UK, the socket in the NTE1 (a PTT supplied device) marks the boundary between BT's network and the liberalised area. Into this socket is plugged a 4-wire bus, often referred to as the *S/T bus* because it is at the S/T reference points. This bus can operate in one of two modes. In point-to-point mode, one TE is connected to the end of the bus which can be up to 1 km in length; in point-to-multipoint (passive bus mode), up to eight TEs can be connected in parallel anywhere along a reduced length (about 200m) cable (see Example 2 in Figure 11.8).

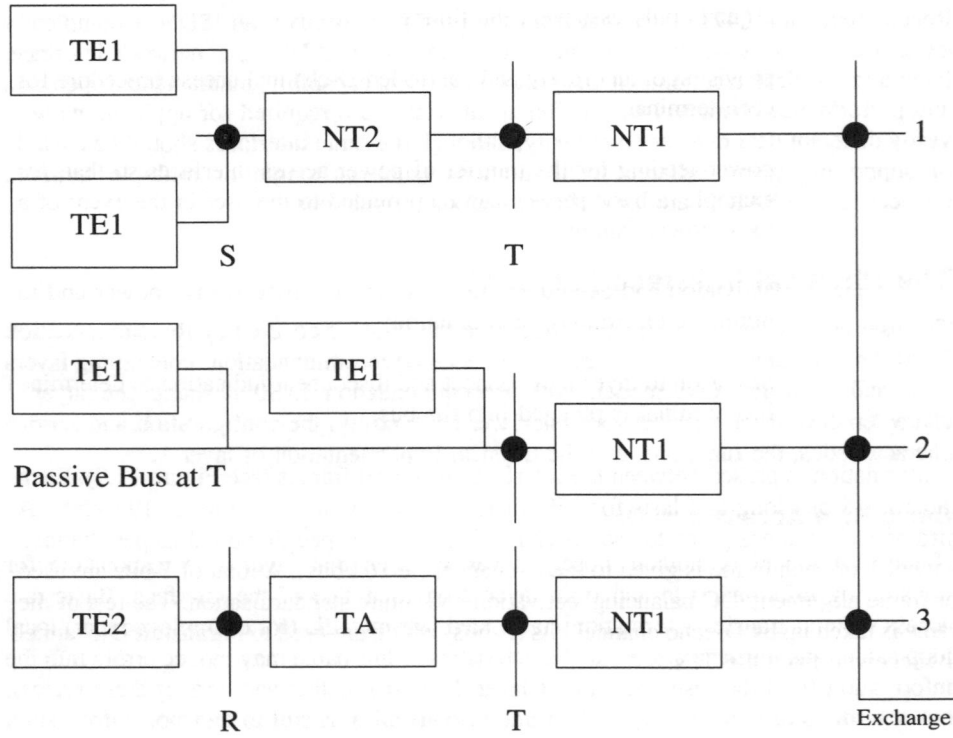

Figure 11.8   Example of ISDN access

Over the bus passes the data from the two B-channels and the D-channel. The B-channel path is established by signalling over the D-channel. In a multi-terminal situation, all terminals have access to the D-channel by using an access procedure. Hence, the passive bus is a special kind of LAN. However, each B-channel is allocated to a specific terminal on the bus at call set up time and is not capable of being shared by other terminals. This means that a maximum of two terminals can use the B-channels at any one time.

To provide the services required of layer 1, the following functions must be performed at the S/T reference point:

—   provision (for each direction of transmission) of two independent 64 kbits B-channels with the associated bit timing (to enable the NTEs and TEs to recover data from the bit-stream), octet timing (since the B-channels are byte oriented) and frame alignment (to aid the proper working of TDM);

—   provision (for each direction of transmission) of a 16 kbits D-channel for signalling information. Since this information is packetised, it requires only associated bit timing;

—   provision of an orderly and controlled D-channel access procedure for each terminal;

—   power feeding for the transfer of power across interfaces so that, for example, a basic service can be provided to the user in the event of a local power failure;

—   activation and de-activation of the line in order to save power and to minimise electromagnetic radiation;

—   provision to layer 2 of connect and disconnect indication to determine if a terminal is plugged into the bus.

Information is passed between the interfaces using bit frames (see Figure 11.9). Each frame is 48 bits long and lasts for 250 microseconds, giving a bit rate of 192 kbits. A third of each frame is given to each B-channel, giving the specified 64 kbits per channel. A twelfth of each frame is given to the D-channel (ie 16 kbits). A total of 7 bits are used for frame alignment, DC balancing, activation and future standardisation. The rest of the frame is taken up by D-echo channel bits (E-bits) which are used to retransmit D-channel bits received from the TEs.

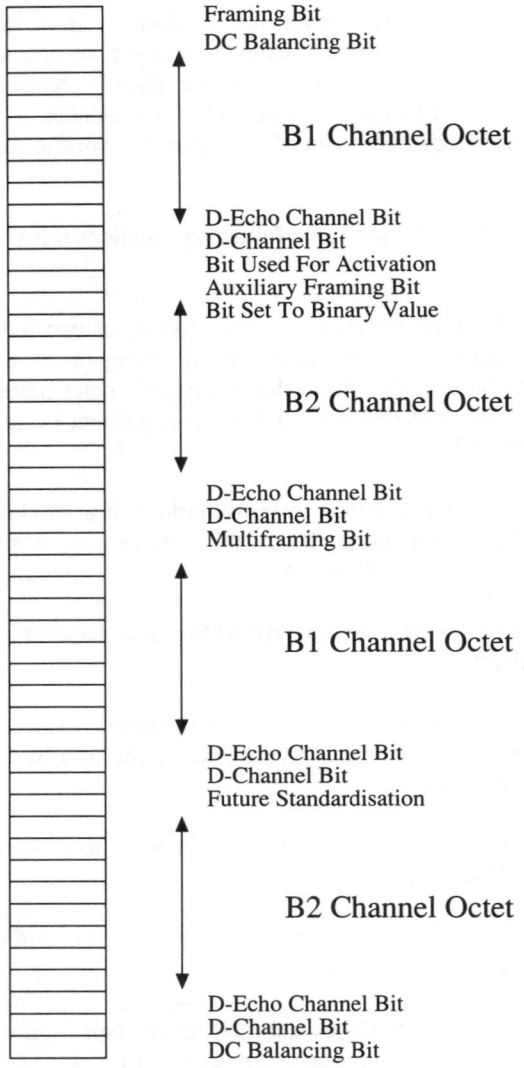

Framing Bit
DC Balancing Bit

B1 Channel Octet

D-Echo Channel Bit
D-Channel Bit
Bit Used For Activation
Auxiliary Framing Bit
Bit Set To Binary Value

B2 Channel Octet

D-Echo Channel Bit
D-Channel Bit
Multiframing Bit

B1 Channel Octet

D-Echo Channel Bit
D-Channel Bit
Future Standardisation

B2 Channel Octet

D-Echo Channel Bit
D-Channel Bit
DC Balancing Bit

Figure 11.9   Physical layer frame structure

Primary Rate Access and OSI

Primary rate access at layer 1 is described in Recommendation I.431. Description of the electrical, format and channel usage characteristics at the S and T interfaces are given in Recommendation G.703 (a specification of the physical and electrical characteristics of a hierarchical digital interface).

Although Recommendation I.431 describes both I.544 Mbits (23B+D) and 2.048 Mbits (30B+D) rates, the CCITT decided to keep the differences between the two to a minimum. Primary rate access supports only the point-to-point configuration, that is, there is only one transmitter and receiver on each side of the interface, the distance between the ends of the link being governed by the electrical characteristics (defined in G.703) of the cable.

The functions of the primary rate interface (ie the channel types and timing) are the same as those for the basic rate interface given in Recommendation I.421. The only major difference between the two types of access is the frame structure. The I.544 Mbits rate frame structure is based on G.704 and comprises 193 bits per frame. The frame is composed of an F-bit for frame alignment followed by 24 time-slots of 8 bits each, with the last time-slot being used for D-channel signalling when this channel is present at the interface. The 2.048 Mbits frame consists of 32 time-slots of 8 bits each with time-slot 16 being assigned to the D-channel. Both the 1.544 and the 2.48 Mbits rates use the first time-slot for synchronisation and have a repetition rate of 800 frames/s.

The U Reference Point

The U reference point is the connection that links the customer's premises and the digital local exchange. This interface exists on the exchange side of the NTE1 functional group. Primary rate access at the U interface is via a standard 2.048 Mbits link (described in the G-series Recommendations). Basic rate access at the U interface is via a single cable pair which has to provide a full duplex digital transmission link using one of two methods.

Historically, the first method is known as burst mode (or *ping pong*) transmission, in which each end transmits alternately in short bursts, allowing a short time for line turn-round. The bursts from each end are transmitted at such a speed (in excess of 300 kbits) that to the user the link appears to be full duplex at 144 kbits. However, if the bursts are slowed down, it can be seen that this method is really only half duplex since at any one time only one end is transmitting.

The latest, and more generally used method, is known as echo cancellation in which both ends transmit simultaneously in both directions, and use echo cancelling devices to ensure separation of the two data streams.

The Data Link Layer (Layer 2)

The CCITT Recommendation for the data link layer of an ISDN user network interface describes the necessary procedures for accessing a data link. The link access procedure for a D-channel (LAPD) provides a data link (D-channel) between two end points physically connected by a medium described by the layer 1 Recommendation. This link must be independent of the rate at which information is sent over it, transparent to the information sent over it and fully error protected. It is the responsibility of layer 2 frames (which are guaranteed error free by layer 2) to transport the layer 3 call control information.

The definition of LAPD given in Recommendation I.440 (a reference to Q.920 and Q.921) uses the principles and terminology of a number of other Recommendations and standards including:

— X.200: OSI reference model;

— X.400: Message handling service;

— X.25: LAPB user network interface for packet mode terminals;

— ISO 3309 and ISO 4335: High-level data link control (HDLC) standards for frame structure and procedures.

Layer 2 functions provide the means for transferring information between point-to-point or point-to-multipoint systems where information frames are directed to one end point or many end points respectively. (See Chapter 9 for a detailed explanation of layer 2 operation).

These functions include:

— the provision of one or more D-channel connections, with each separate connection being identified by a data link connection identifier (DLCI) contained in each layer 2 frame;

— frame delimiting, alignment and transparency to ensure that a sequence of bits transmitted over the D-channel are recognisable as a frame;

— sequence control to ensure that frames are kept in sequence while being transmitted across the data link;

— detection of errors in transmission, format and operation;

— recovery from transmission, format and operation errors whilst notifying the link managing function of any unrecoverable errors;

— flow control.

LAPD was primarily developed for ISDN (since LAPB is not able to handle multiple channels), but it can also be considered as an independent data link layer and applied to other applications. One potentially major application of LAPD is in broadband ISDN where it may form the basis of layer 2 signalling systems.

The Network Layer (Layer 3)

The general characteristics of layer 3 functions and protocols used by the D-channel are described by CCITT Recommendation I.450, and in more detail in Recommendation I.451 (a reference to Recommendation Q.931). The role of layer 3 is to establish, maintain and terminate network connections across an ISDN. It also provides to an ISDN user the functions which are associated with the operation of a network connection, without the user having to be aware of any functions relating to lower layers (ie the physical and data link).

To carry out its role, layer 3 utilises the following functions and services provided to itself by the data link layer:

- establishment of data link connections;

- error protected transmission of data;

- notification of unrecoverable data link errors;

- release of data link connections;

- notification of data link layer failure;

- recovery from certain error conditions;

- indication of data link layer status.

The following general functions are performed by the network layer:

- *routing and relaying* for determining an appropriate route between layer 3 addresses (users) and for performing relaying of information between sub-networks;

- *network connection mechanisms* for providing connections by utilising the data links set up by the data link layer;

- *conveying user-user and user-network information*, with or without establishment of a circuit-switched connection;

- *network connection multiplexing* of call control information for multiple calls onto a single data link;

- *segmenting and blocking* of layer 3 information to facilitate its transfer across the network;

- *error detection*: these functions use error notification from the data link layer to check for procedural discrepancies in the layer 3 protocol;

- *recovery from detected errors;*

- *sequencing* to ensure an ordered delivery of layer 3 information over a given network connection when requested by a user, and in the sequence submitted by the user;

— *flow control* for user-user signalling messages (as described in Recommendation I.451 and Q.931);

— *reset.*

ISDN and Existing Interfaces

As ISDN is still evolving, it has to be able to support existing interfaces and terminals. The CCITT have recognised this and have produced a number of Recommendations pertaining to existing interfaces and the functions performed by a terminal adaptor (TA).

Rate Adaption and B-channel Sub-rate Multiplexing

Rate adaption is the process by which a transmitting device (such as a DTE) operating at data rates lower than those required for transfer across a network can be interfaced to that network. In an ISDN, this enables a data stream with a bit rate of less than 64 kbits to be carried in the B-channel structure, and therefore enables an ISDN to support existing terminals and interfaces. In the CCITT Blue Book, four types of terminals are mentioned: X-series, packet mode X-series, V-series and statistical multiplexing V-series DTEs.

Detailed procedures for supporting these DTEs are given in Recommendations X.30 (as stated by Recommendation I.461), X.31 (I.462), V.110 (I.463) and V.120 (I.465) respectively.

B-channel sub-rate multiplexing is the process by which several data streams with bit rates of less than 64 kbits can be combined and transmitted across a single 64 kbits B-channel (see Figure 11.10). The bit rates of less than 64 kbits can be in one of two forms. First, there are binary rates of 8, 16 and 31 kbits and, secondly, there are other rates which include those pertaining to DTEs conforming to the X- and V-series Recommendations.

Multiplexing a number of 8, 16 and 32 kbits streams up to a total of 64 kbits into a single 64 kbits channel is done by interleaving groups of bits from the sub-rate (8, 16 or 32 kbits) streams within each B-channel octet (8-bits). This can be done either by fixed format or flexible format multiplexing (I.460). Fixed format multiplexing can often result in some bit redundancy in the B-channel. For example, multiplexing an 8 kbits stream with a 32 kbits stream into a 64 kbits channel will result in 3-bits of each octet being redundant. This redundancy can be eliminated by using flexible format multiplexing which always allows sub-rate streams to be multiplexed up to the 64 kbits limit of the B-channel.

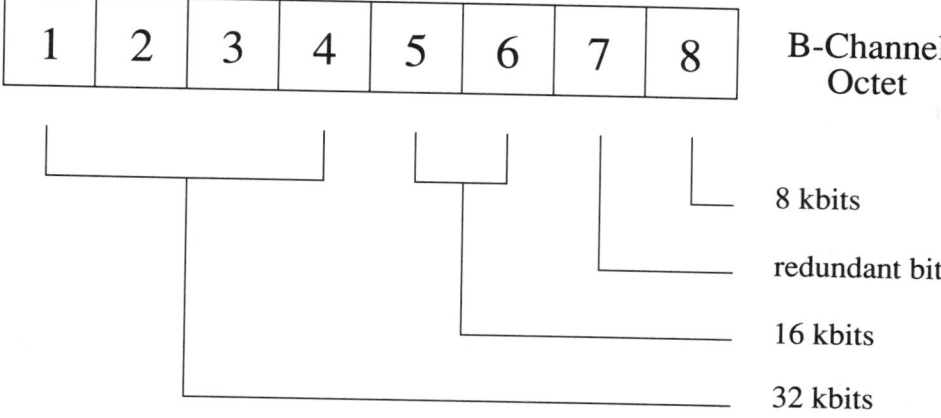

Figure 11.10   Channel sub-rate multiplexing

Supporting X.21 and X.21 *bis* DTEs

There are two configurations which involve the support of X.21 type DTEs (see Figure 11.11). The first configuration is the *minimum integration scenario* which allows two X.21 terminals to intercommunicate over an ISDN using transparent X.21 call handling. In this case, the data path is a semi-permanent connection established by the network provider and so the terminal adaptor performs only the necessary physical channel rate adaption between the user rate (at reference point R) and the 64 kbits rate. The second configuration is known as the *maximum integration scenario* which provides processed X.21 call handling. In this scenario, the terminal adaptor performs all the necessary bit rate adaption, and the signalling conversion to and from X.21 signalling to I.451 (Q.931) D-channel signalling.

Both integration scenarios use the same method of rate adaption (see Figure 11.12). The general approach for adapting user rates to the ISDN 64 kbits within the TA is given in recommendation X.30. It involves two functional blocks. The first block (RA1) adapts the user rate to the next highest rate (either 8 kbits, 16 kbits or 32 kbits) by inserting additional bits (bit stuffing) with the second block performing adaption to 64 kbits by repetition of the 8, 16 or 31 kbits frames.

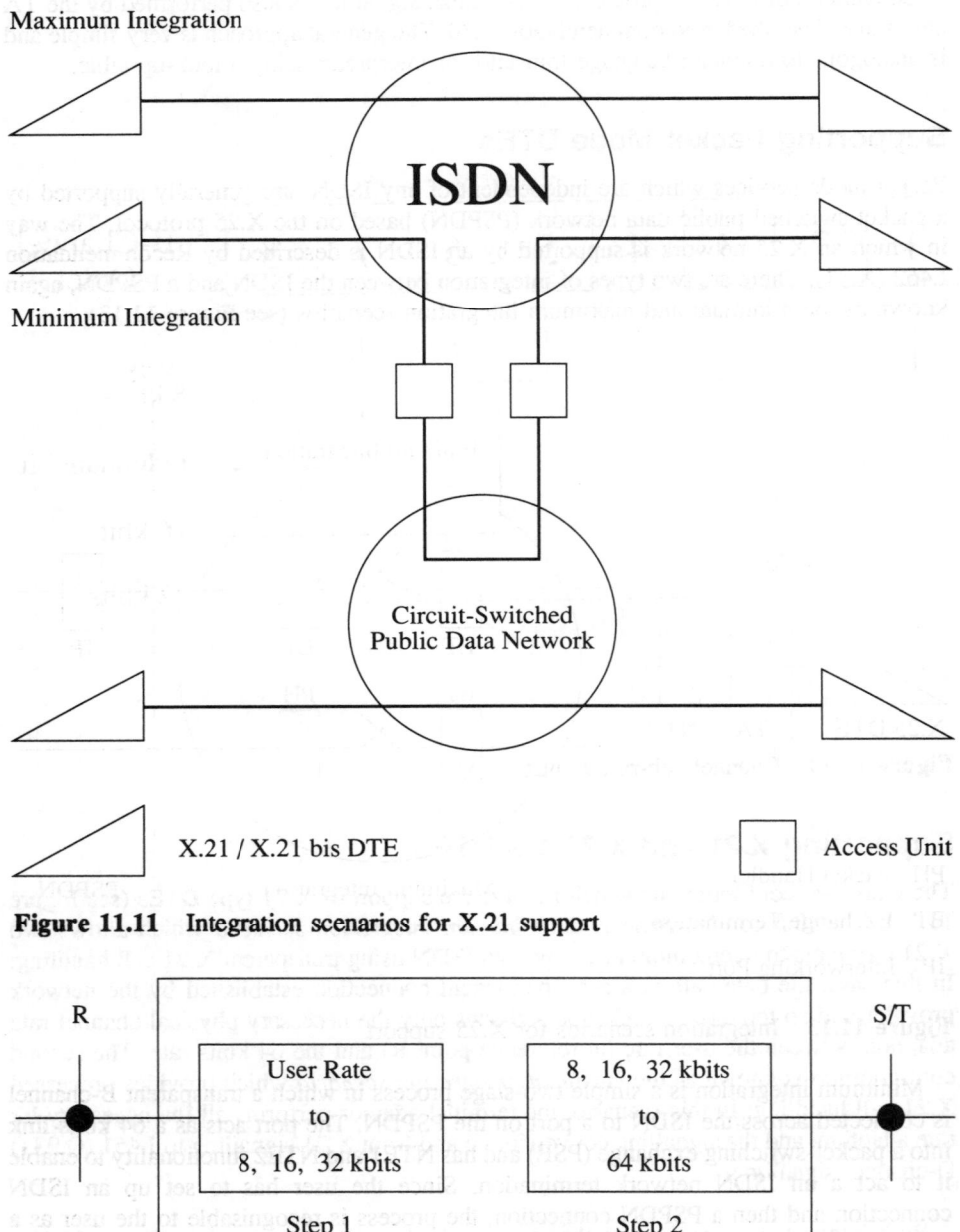

Maximum Integration

ISDN

Minimum Integration

Circuit-Switched
Public Data Network

X.21 / X.21 bis DTE   Access Unit

Figure 11.11 Integration scenarios for X.21 support

R			S/T
	User Rate	8, 16, 32 kbits	
	to	to	
	8, 16, 32 kbits	64 kbits	
	Step 1	Step 2	

X.21 DTE   TA   ISDN

Figure 11.12 Rate adaption for existing interfaces

Conversion of the X.21 protocol to D-channel signalling is also performed by the TA and is also described in Recommendation X.30. The general approach is very simple and is analogous to a human language translator or interpreter using a look-up table.

Supporting Packet Mode DTEs

Packet mode services which are independent of any ISDN, are generally supported by a packet-switched public data network (PSPDN) based on the X.25 protocol. The way in which an X.25 network is supported by an ISDN is described by Recommendation I.462 (X.31). There are two types of integration between the ISDN and a PSPDN, again known as the minimum and maximum integration scenarios (see Figure 11.13).

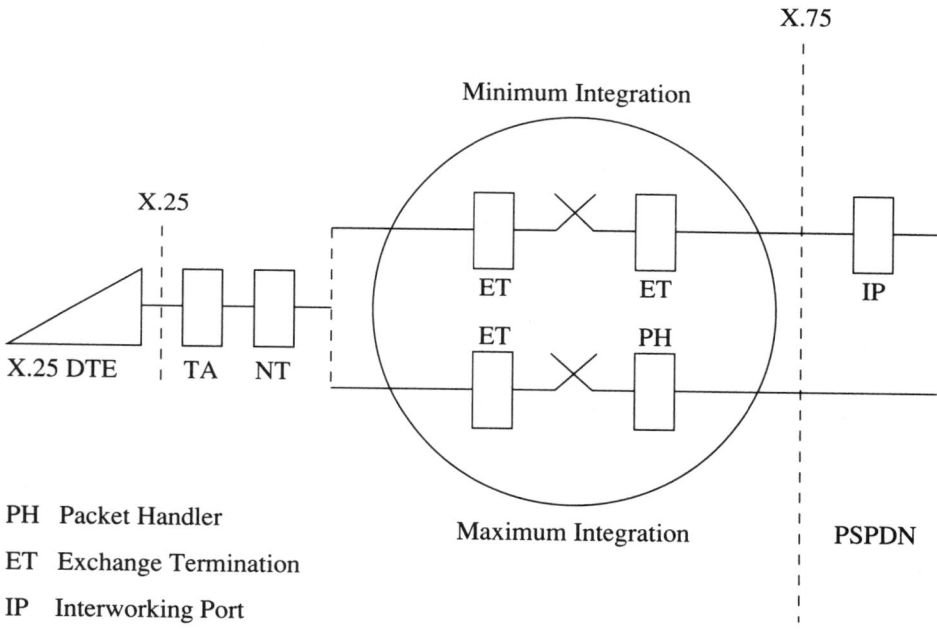

PH   Packet Handler

ET   Exchange Termination

IP   Interworking Port

Figure 11.13   Integration scenarios for X.25 support

Minimum integration is a simple two-stage process in which a transparent B-channel is connected across the ISDN to a port on the PSPDN. The port acts as a 64 kbits link into a packet-switching exchange (PSE) and has NTE1 and NTE2 functionality to enable it to act a an ISDN network termination. Since the user has to set up an ISDN connection and then a PSPDN connection, the process is recognisable to the user as a two-stage process. The ISDN terminal in this situation has just one PSPDN address (described in Recommendation X.121).

Maximum integration is a single stage process which provides a higher degree of integration and more flexibility than minimum integration. In this set up, the packet handler is located within the ISDN and because of this, the packet mode terminal on the

ISDN can signal directly to it over the D-channel. The transfer of information can then be done over the D-channel or over the B-channel (which is effectively minimum integration since there is a recognisable two-stage process). In this case the ISDN terminal has a single ISDN number (as described in Recommendation E.164) and possibly a sub-address or extension number.

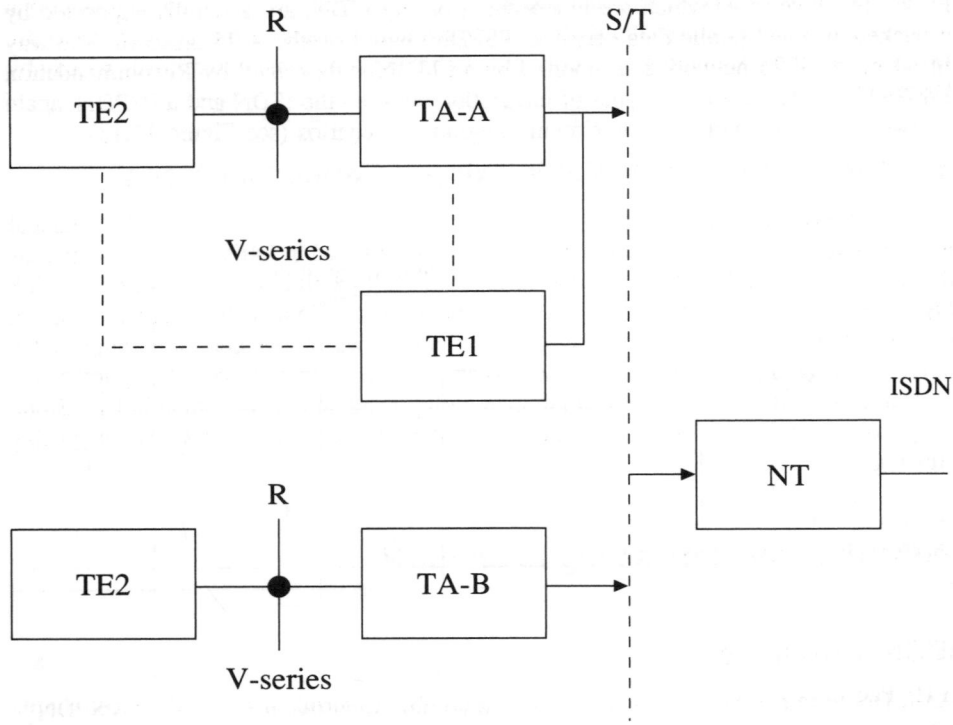

TE1   ISDN Terminal Function

TE2   DTE with V-series Interface

TA-A   Alternative Voice/Data TA

TA-B   Automatic Dial/Answer TA

NT   Network Terminating Function

Figure 11.14   Configurations for V-series support

Supporting the V-Series Interfaces

The V-series interfaces are supported by an ISDN in much the same way as the X-series interfaces are. Both use the same two-step rate adaption and protocol conversion approaches. The only real difference between supporting the two types of interface is in the functionality of the TA (see Figure 11.14). When supporting V-series DTEs, the terminal adaptor A (TA-A) provides manual call control functions associated with alternate voice/data, circuit-switched services. These functions include conversion of the electrical, mechanical, functional and procedural characteristics of the V-series interface to those required by an ISDN at the S/T reference point. In addition, or as an alternative to the functions of TA-A, terminal adaptor B (TA-B) provides the mapping functions necessary to convert automatic dialling and/or automatic answering procedures to the ISDN's D-channel signalling protocols, although these functions require further study by the CCITT.

Supporting Statistical Multiplexing V-Series Interfaces

The methods for supporting V-series interfaces which are capable of statistical multiplexing are given in Recommendation V.120 (as stated in I.465). This recommends the use of an HDLC based technique to achieve multiplexing at layer 2 (the data link layer). Each LAPD frame contains a field known as the logical link identification (a number between 1 and 254 inclusive). This field allows proper demultiplexing of the data. Once again, rate adaption as a two-stage process is involved. The first stage assembles the HDLC frames which are then multiplexed onto a B-channel in the second stage. Any spare time on the channel is filled with flags (a process known as flag stuffing).

Apparatus for Connecting to an ISDN

ISDN Terminals

An ISDN terminal can be categorised into a number of terminal types, the most simple one being the digital telephone. A typical digital telephone looks very much like an analogue telephone with a built-in liquid crystal display (LCD) screen. It is attached to the ISDN at the S/T reference point and transmits high quality digitally encoded voice data through one of the 64 kbits b-channels. Digitisation of the voice data is performed within the telephone using a technique known as adaptive differential pulse code modulation (ADCPM) as described in Chapter 4. This type of encoding produces a very clear natural voice communication service. Some digital telephones also have the ability to interconnect data terminals to computers or public data networks by providing, for example, V.24 or X.21 interfaces for use by the associated device.

By using both of the B-channels provided by basic rate access, a digital videophone can be used as an ISDN terminal. This terminal is also attached to the ISDN at the S/T interface and allows enhanced voice and video communications. Built into the videophone is a miniature screen, such as an LCD screen or a cathode ray tube (CRT), and a small video camera, based on charge coupled device (CCD) technology, as found in modern home video cameras. The images received from the camera are sent to the called party's screen through one B-channel, while voice communications take place using the other B-channel. Videophones can be used in applications such as teleconferencing and remote video monitoring. In the domestic ISDN environment, the videophone is expected to emerge as the major ISDN terminal.

An ordinary personal computer can be transformed into an ISDN workstation by purchasing an ISDN PC card which can be installed into one of the expansion slots generally provided with PCs. Once again, the card is attached to the basic rate ISDN at the S/T interface and allows inter-computer communications in the form of facsimile or telex by using specialised software. Voice communications can also take place via an ISDN card by plugging a standard analogue telephone into the card itself. Because there are two B-channels, voice and data communications can occur simultaneously.

ISDN workstations are currently being used in most countries with ISDN capabilities. The workstations take the form of PC compatible computers with integral ISDN adaptor cards, light pens and digital telephones. The workstations provide the ability to use applications such as remote interactive CAD/CAM, remote terminal control and desktop conferencing. The emergence of the PC as the *de facto* office terminal means that this device is expected to become the *de facto* ISDN terminal in the business environment.

It is important to note that any ISDN terminal which is to be connected to an ISDN in a country which belongs to Conference of European Posts and Telecommunications (CEPT), must be approved as conforming to the Normes Européenes de Télécommunications (NET3) requirement.

ISDN Terminal Adaptors

As mentioned earlier, terminal adaptors (TA) are used as an interface between existing terminals and the ISDN network terminating equipment, and take the form of a small desktop box similar in size and shape to a desktop modem. Using TA software on a data terminal, the TA is able to configure either or both of the basic rate B-channels to suit the user's needs. Non-ISDN terminal equipment is plugged into one of the interface sockets on the TA, which can then be used to set up calls either from buttons on the front of the TA or by using terminal software.

ISDN Cords

The basic rate ISDN cord which connects terminal equipment to the network terminating equipment across the S/T interface is described in Recommendation I.430 and is made up of four symmetrical twisted cable pairs. Two of these pairs are used for transmitting and receiving digital data. Power may be transferred from the network terminating equipment (NTE) to the terminal equipment (TE) by using the digital signals as a power source (this is known as phantom power transfer). In this case the two remaining cable pairs (used solely for power transfer) are redundant unless additional power is needed by the terminal equipment. Recommendation I.430 also states that the cord should be terminated at both ends in plugs with individual conductors being connected to the same contact in the plug at each end.

ISDN Connectors

The type of connectors used for basic rate access via symmetrical twisted pair cables, are specified in the international standard ISO 8877, and are known as RJ-45 connectors. This same type of connector can also be used for primary rate access via symmetrical twisted pair cable, using a different key shape (a small protrusion on the plug) to prevent it from being confused with a basic rate plug. Primary rate ISDN can also be accessed via 75 ohm coaxial cable pairs, terminated at both ends with 75 ohm BNC type connectors.

ISDN Numbering and Addressing

Since ISDN is evolving from the telephone network, it can also be expected that the ISDN numbering plan should evolve from the current telephone numbering plan described in I.331 (and E.164). This is recognised by the CCITT in Recommendation I.330 which makes the following points:

— the international numbering plan should provide for substantial capacity to accommodate future network requirements;

— it should use the numeric 0-9 character set throughout and should unambiguously identify a particular country and particular networks or ISDN within that country.

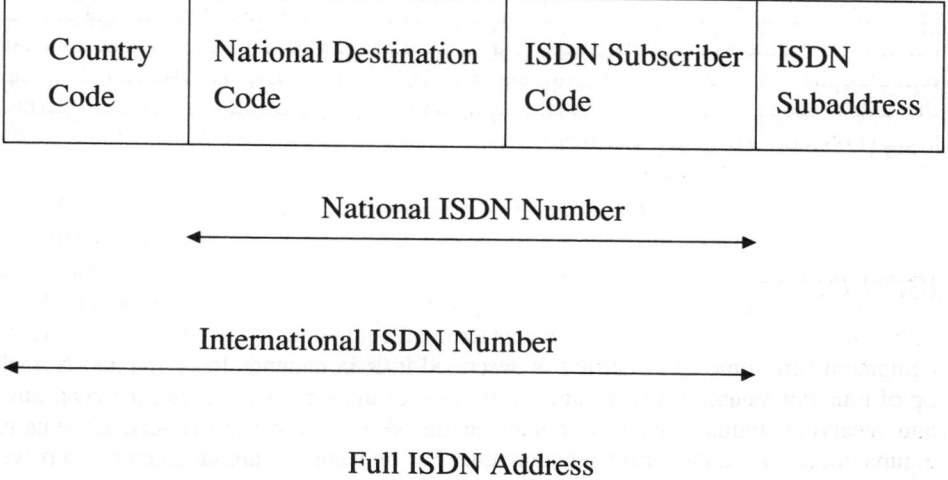

Figure 11.15   The ISDN address

The international ISDN number (see Figure 11.15) can be a maximum of 15 decimal digits, although national and international requirements can determine the exact number of digits up to this maximum. An international ISDN number is a variable length decimal number split into specific fields: the *country code* and the *national ISDN number*. The country code selects the country to which the call is being made. The national ISDN number selects the destination, subscriber's address and therefore comprises a national destination code and the subscriber's ISDN number. This type of numbering scheme is analogous to that of the PSTN. As a suffix to the subscriber's ISDN number is the ISDN sub-address which can be a maximum of 32 decimal digits and must be conveyed transparently across all ISDNs. Although the maximum length of the ISDN number is 15 digits, a network provider (such as British Telecom or France Télécom) can increase this limit to 17 if the extra capacity is required.

Private ISDN

Private ISDN is the term applied to ISDN services which are available solely within one organisation (for example, a corporate network). Private ISDN has the ability to offer all the services and handle all the applications of a full ISDN. At the heart of private ISDN is an integrated services PBX (ISPBX) whose prime function is to terminate one or more PRA links and convert them into BRA links to individual extensions within the organisation. Since it acts both as a concentrator and a switching device, the ISPBX is equivalent to an NTE2 functional block which also contains NTE1 line termination functions (see Figure 11.16).

Digital Private Network Signalling System (DPNSS)

Private ISDN is dependent on the ability of remote ISPBXs to communicate with each other. To meet these requirements, a UK consortium of digital PABX manufacturers (headed by British Telecom) began the development of DPNSS ahead of any CCITT Recommendations regarding inter-ISPBX signalling. Despite this, DPNSS does provide facilities which are compatible with current ISDN Recommendations, as well conforming to the ISO reference model for OSI.

DPNSS is the latest addition to the DASS family of signalling systems and is similar to DASS2 with separate mechanisms for private inter-ISPBX communication. DPNSS is a common channel signalling system which enables ISPBXs to intercommunicate across private leased 2.048 Mbits point-to-point digital networks (see Figure 11.17).

Signalling information is carried between ISPBXs in channel 16 of the 32 channel digital link, thus enabling a wide range of supplementary services and features to be used transparently, as though the network was a single ISPBX. The other 30 channels are used to carry voice and data traffic between ISPBXs with the remaining channel used for timing and synchronisation purposes.

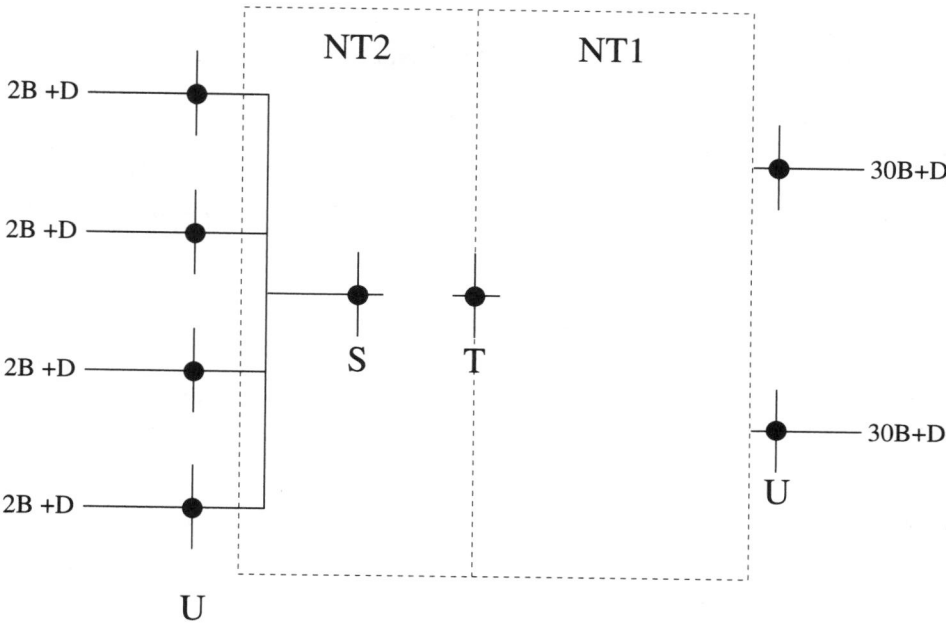

Figure 11.16 The ISPBX functional block

DPNSS signalling supersedes the previous inter-PABX signalling system in which voice and signalling information was combined to give in-band signalling. The commonality between DASS2 and DPNSS allows DPNSS signalling to be transmitted over public networks by DASS2. This, in turn, enables remote private networks to be interlinked across the public network.

In the international arena, DPNSS is becoming a *de facto* standard for inter-PABX signalling. Many of the major PABX manufacturers have incorporated DPNSS signalling into their systems, and the success of this signalling system means there is little point in migrating to a new system. The private network signalling equivalents of DPNSS, being developed by a CCITT study group is not a fully compatible specification. Using DPNSS signalling between ISPBXs offers an organisation many benefits which existing intra-organisation communication networks are not able to offer. These benefits include dramatically reduced internal and external call connection times, increased quality of transmission (particularly when teleconferencing), increased network reliability, the ability to handle voice/data simultaneously and other features such as call divert and camp-on.

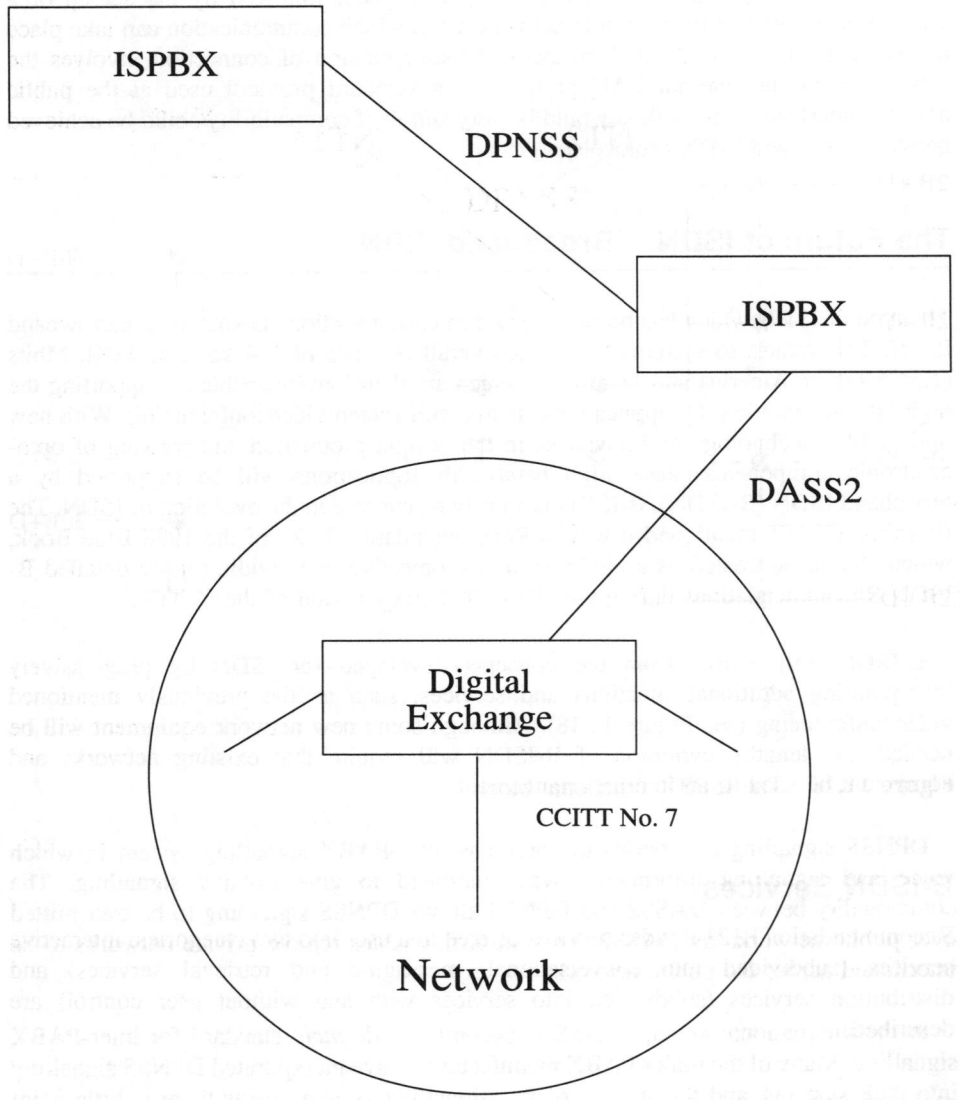

Figure 11.17 Networking ISPBXs within the UK

ISDN and Local Area Networks

If it is to be viable in the business world, private ISDN will have to be able to cope with the connection of LANs via ISPBXs. A LAN connected by the ISDN is very similar to one which is not connected by the ISDN except that it must be able to handle both circuit and packet mode connections, as well as offering the distributed control and high bandwidth required to maintain the essential functionality of a LAN. To connect an

ISDN LAN to a public network a gateway is needed, followed by the set up of a transparent B-channel in the public network down which communication can take place according to the internal LAN protocol. Another method of connection involves the conversion of the internal LAN protocol to a standard protocol used at the public network interface. This method would be very simple if compatibility could be achieved between LAN and ISDN protocols.

The Future of ISDN — Broadband ISDN

The type of ISDN which has been discussed in earlier sections is known as narrowband ISDN. This relates to systems limited to overall bit rates of 144 kbits or 2.048 Mbits (1.54 Mbits in America and Japan) and which are therefore incapable of supporting the high bit rates required by applications such as full screen videoconferencing. With new optical fibre technology and advances in the design, production and packing of opto-electronic components, these high bandwidth applications will be supported by a broadband ISDN (B-ISDN). B-ISDN is a fairly recent step in the evolution of ISDN. The first time CCITT mentioned it was in Recommendation I.121 of the 1988 Blue Book, which should be treated as a guideline to the objective of providing more detailed B-ISDN Recommendations during the 1989-1992 study period of the CCITT.

B-ISDN will evolve from the concepts developed for ISDN by progressively incorporating additional functions and services, such as the previously mentioned videoconferencing (see Figure 11.18). Although some new network equipment will be needed, the lengthy evolution of B-ISDN will require that existing networks and equipment be used as an interim arrangement.

B-ISDN Services

Recommendation I.121 divides services offered to a user into two categories: interactive services (subdivided into conversational, messaging and retrieval services) and distribution services (subdivided into services with and without user control) are described.

Conversational Services

Conversational services provide the means for bi-directional, real-time, end-to-end communication from user-to-user (voice) or from user-to-user (data). There are four types of information included in the conversational services:

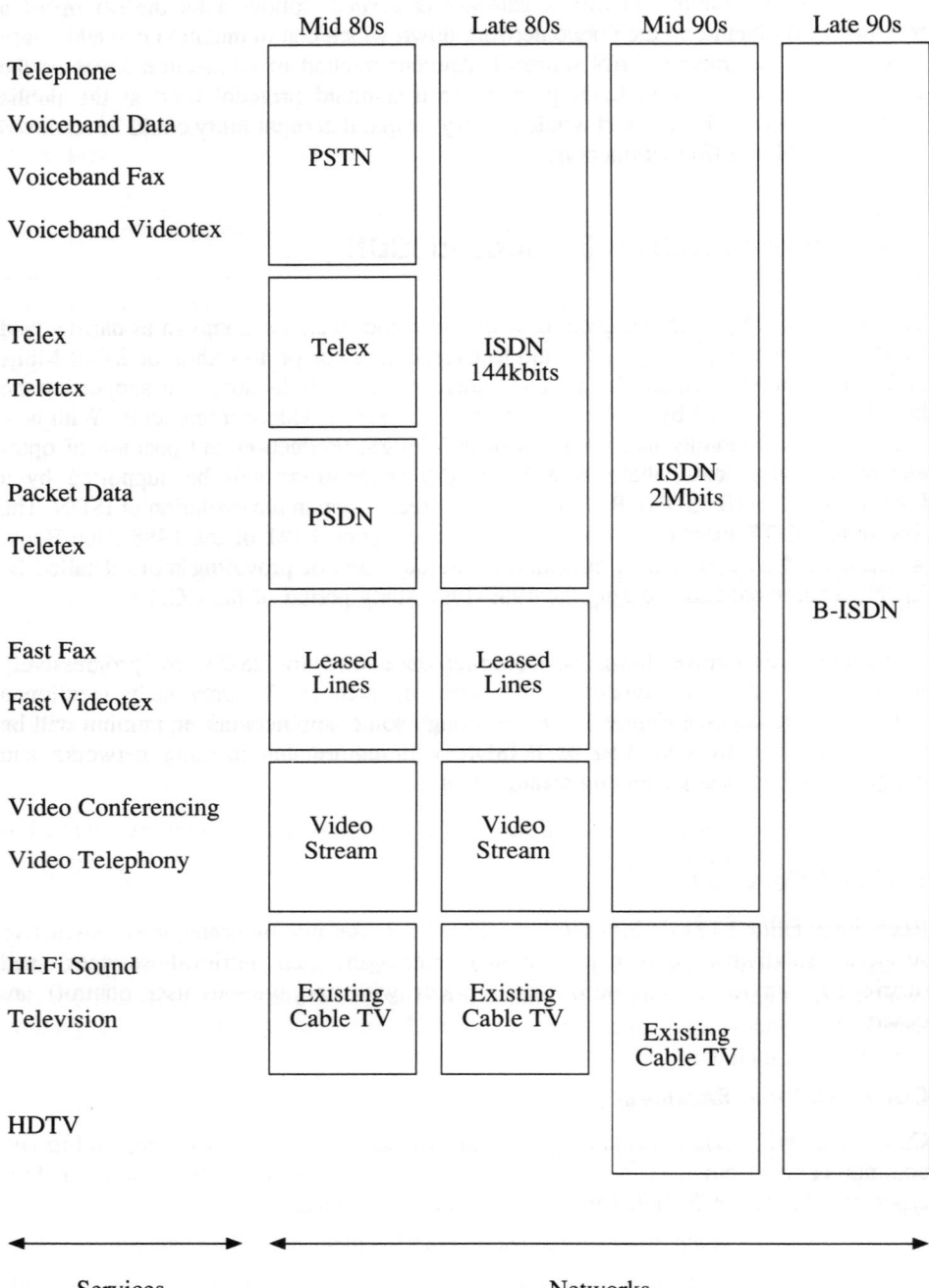

Figure 11.18   The evolution of broadband ISDN

- *moving pictures (video) and sound*: applications of this type of information include transfer of sound, moving pictures and video scanned still images and documents between two locations (either point-to-point or point-to-multipoint); tele-education, video surveillance for use in building security and traffic monitoring; and video/audio information transmission for use in TV signal transfer and audio dialogue;

- *sound:* this information can be used in multiple sound programme signals to allow multilingual commentaries;

- *data:* high speed transmission of digital data has applications in LAN interconnection, transfer of video and other information types, still image transfer, multi-site interactive CAD/CAM, and real-time control of alarm and metering systems;

- *document:* information of this type is used for user-user transfer of text, images, drawings and mixed documents.

Messaging Services

Messaging services offer user-to-user communication between individual users via storage units with store-and-forward, mailbox, or message handling functions.

Two types of information are included in messaging services:

- *moving pictures (video) and sound:* applications of this information include video mail services for the transfer of moving pictures and accompanying sound;

- *document:* this information is used in electronic mailbox applications for mixed documents.

Retrieval Services

Retrieval services enable the user to retrieve information stored in information centres which are, in general, provided for public use. The type of information handled by these services includes text, data, graphics, sound, still images and moving pictures. Applications include:

- *broadband videotex* for use in remote education and training;

- *tele-shopping and tele-advertising;*

- *news retrieval;*

- *video retrieval* for use in entertainment and tele-education;

- *high resolution image retrieval* for entertainment purposes, tele-education and image communication;

- *document retrieval* from information centres.

Distribution Services without User Control

Applications of this service are distribution of existing quality, high definition and pay-per-view television (video information); electronic newspapers and electronic publishing (text, graphics and still image information); high speed digital data transmission; and distribution of video/audio signals (moving picture and sound information).

Distribution Services with User Control

One example of this service is full channel broadcast videography in which a user can select and retrieve video information from a central store. This has applications in remote education and training, tele-advertising, news retrieval and tele-software.

B-ISDN Channel Rates

In addition to the narrowband ISDN access structures using B, H_o and H_1 channels, B-ISDN will support broadband channels H_2 AND H_4. These channel rates will be known as H_{21} (32 Mbits), H_{22} (43 to 45 Mbits) and H_4 (132 to 138 Mbits). The rates do, however, have to conform to the following stipulations: H_{21} must be less than or equal to a quarter of H_4 and H_{22} must be less than or equal to a third of H_4.

Technology will probably dictate that the high speed streams will be separated from the low speed streams at all ISDN exchanges and will be separately switched. This will, in turn, allow systems designed for the narrowband ISDN to be attached at specific points on the local loop. Independent switching of the various streams will also assist in the interfacing of existing digital systems to the new B-ISDN at the ISDN exchanges.

Asynchronous Transfer Mode

A transport mechanism known as asynchronous transfer mode (ATM) has been developed for transferring information over B-ISDN. The ATM mechanism is based on a simple frame or cell structure, in which each cell has a header and an information field. The header contains only the information required to transport user information via ATM through the B-ISDN. The user information field comprises between 32 and 120 octets of information, the exact requirement being determined by the quality of service required, and the efficiency of the transmission link. The cell can be used in both circuit and packet modes, and can be processed and switched at extremely high speeds by the ISDN exchange, giving broadband ISDN using ATM methods, a very high throughput of information and makes it very flexible.

Development of B-ISDN

B-ISDN is being developed by three major groups, all of which aim to develop integrated broadband networks. In 1983, it was announced that the development of an information network system in Japan would be controlled by the Japanese PTT, Nippon

Telegraph and Telephone Corporation (NTT). Two years later in 1985, AT&T announced that they would be developing universal information services for the USA. It was in 1984 that the European Economic Community decided to go ahead with a community-wide development of an integrated broadband communications (IBC) network. To carry out the development, 1985 saw the European Council of Ministers authorising a programme called Research in Advanced Communications in Europe (RACE). This programme was very different from the two other development programmes because it comprised PTTs, the telecommunications industry and IT companies, as opposed to just a PTT in Japan or an industrial organisation in America. Its objective is the '*introduction of integrated broadband communication taking into account the evolving ISDN and national introduction strategies, progressing to community-wide services by 1995.*'

Conclusion

Despite the problems associated with ISDN, it is still the communications network of the future. The European Commission is committed to a European-wide ISDN by 1992, with at least 80 percent geographical coverage across its member countries. The commitment has been taken on by 22 of the 26 European PTTs in the form of a Memorandum of Understanding.

There are, however, some major stumbling blocks for PTTS trying to introduce ISDNs; the slow introduction of ISDN standards; the apparent lack of industry commitment and the indecisiveness of other PTTs with regard to their tariffing structures. These problems should be resolved during the early part of the 1990s as the pan-European ISDN evolves. On the other hand, the communication equipment suppliers have been reluctant to supply money for research and development of ISDN equipment and applications until the PTTs have decided on the exact format of their ISDN implementation and until the ISDN standards have become ratified. There has therefore developed a catch-22 situation, with suppliers waiting for PTTs to introduce ISDNs and with the PTTs waiting for the suppliers to produce ISDN terminals and new applications. Once again, this problem should diminish in the mid-1990s since by this time it is anticipated that the majority of European PTTs will have introduced CCITT conformant ISDNs.

There is limited availability, in the UK, for two ISDN services from BT & Mercury:

-   ISDN 30 (BT) 2100 Premier (Mercury);

-   ISDN 2 (BT).

ISDN 30 and 2100 Premier are primary rate services suitable for termination on ISPBXs.

ISDN 2 is a basic rate service terminating on an NTE6 (BT equipment). Terminal adapting equipment (for example, standalone devices, cards for PCs) are available so DTEs can connect. User interest is in the availability of other services from ISDN. Packet-switching, for example, would be of great interest.

Appendix 11.1: The I-Series Recommendations

I.100 series — General structure:

- framework of I-series Recommendations and terminology;
- description of an ISDN;
- general modelling methods;
- telecommunication network and service attributes.

I.200 series — Service capabilities:

- general aspects of services in an ISDN;
- common aspects of services in an ISDN;
- bearer services supported by an ISDN;
- teleservices supported by an ISDN;
- supplementary services in an ISDN.

I.300 series — Overall network aspects and functions:

- network functional principles;
- reference models;
- numbering, addressing and routing;
- connection types;
- performance objectives.

I.400 series — ISDN user network interfaces:

- ISDN user network interfaces;

- application of I-series Recommendations to ISDN user network interfaces;

- user network interfaces — layers 1, 2 and 3 Recommendations;

- multiplexing, rate adaption and support of existing interfaces.

I.500 series — Inter-network interfaces.

I.600 series — Maintenance principles.

12

Local Area Networks (LANs)

Since its beginnings in the late 1970s, the Local Area Network (LAN) market has grown rapidly, projected by the increasing demand for distributed applications and bolstered by the proliferation of PCs. Today LANs have become common and are a basic resource within most organisations.

A LAN is a generic expression for a shared data communications facility controlled within a small geographical radius, usually a few kilometres. Its main purpose is to interconnect heterogeneous computers, terminals and office machines together. This obviously leads to it having one principal objective, the sharing of all the resources available to the LAN, ie the sharing of data, information, hardware and software. They provide resource sharing cost effectively and have proven to be very reliable.

The main characteristics of a LAN can be summarised as follows:

— economical transmission media;

— inexpensive devices (for example, modems, repeaters and transceivers) to interface to the media;

— high data transmission rates; a LAN is not subject to the speed limitation of common carriers;

— expensive peripherals like laser printers or hard disk storage may be shared among all the network users, thereby reducing the effective unit cost;

— single networked versions of software may be held centrally and located into each user's station when the application needs to be run. The most popular LAN applications to date are: word processing, spreadsheets, file transfer, electronic mail and database sharing;

— limited geographical scope; usually confined to a business office building, although a maximum distance limitation, depending upon the technology, of 80 kilometres can be achieved;

— privately owned high speed data communications system with efficient data and resource sharing;

— may serve a department, an entire building or a cluster of buildings.

365

With all these characteristics, it can be seen that a LAN has evolved into a versatile private communications medium. The effects of a LAN on the working practices of a business environment are phenomenal. Not only does it unify the activities of disparate and often unco-ordinated business functions (in terms of hardware and software utilisation), but a LAN can also increase the effectiveness and efficiency of the enterprise by facilitating a co-operative environment in the sharing of information and facilities.

A Brief History of LANs

The history of LANs since the 1970s has been one of evolution, from university based investigation into digital transmission of information to the development of business co-operative software and hardware. Initially, most computer networks were designed around the concept of conventional telecommunications, ie the transport of analogue speech signals. Digital information was translated into an analogue stream of tones which represented the bit pattern being transmitted. The various analogue signals were re-translated by the recipient back into the corresponding digital information. This method of transmission has been used not only for connecting terminals to a local central processor, but is utilised currently in the public telephone network to connect to a remote device. This type of connection uses the modem to translate the digital signals to analogue signals and vice versa, so that the carrier can transmit the signal.

The advent of packet-switching transmission was one of the determining factors in the design of local area networks, as this method *shares* a single transmission system across multiple users. Packet-switching has been covered separately in Chapter 10. The development of packet-switching brought about the first local area network (LAN), although this implementation adheres more readily to a wide area network (WAN) implementation in its geographical distribution.

The university of Hawaii adapted this method to connect remote users to its computer centre. The Hawaiian Islands at the time had an unreliable telephone system, distributed across a very difficult geographical archipelago. The university developed a radio broadcasting system called ALOHA, where every device broadcasted a packet of information whenever it had one to transmit. All the devices continuously listened to the radio channel and read into storage any packets addressed to them. They then acknowledged receipt by sending back another packet. If any packets collided then this would be detected, as no acknowledgement was received.

ALOHA was an important development as a medium was shared in which there was no direct control over who transmitted packets. Due to the high number of collisions a lot of time was wasted in coping with detecting these. This problem was remedied by introducing time-slots for each transmission which later became known as slotted-ALOHA.

This packet transmission methodology has since been adopted by designers of Ethernet (and also developed by Digital Equipment Corporation, Intel Corporation, and Xerox

Corporation) and other local area networks. Other major developments that have influenced the evolution of the local area networks were technological progress and the effective cost reduction in implementing these tools.

The introduction of the IBM personal computer in 1981 was the catalyst for many LAN developments. Among the major LAN software implementations were Corvus's Omninet (which was used largely as a means for sharing a hard disk among several micros) and Novell's NetWare. By 1983 the IBM PC had saturated the market and dozens of IBM PC networks evolved, including 3COM's Ether-Series, IBM's PC Network (1984) and IBM's token ring LANs (1985).

Underlying Technologies

PC LANs have three components:

— workstations;

— servers;

— connection.

Both hardware and software are associated with each of these components.

Workstations (usually a personal computer) are used to communicate with the network. This element of hardware comprises three parts, the single-user PC side, the hardware network interface side (which is usually a card) and network software (which makes the connection between the applications and the network services).

Servers are computers that provide services to the workstations on the network. These services could be accessed to disk storage, printers, public networks, etc. A server may be a dedicated piece of hardware or it could be shared with user-oriented applications. Dedicated servers usually provide faster response than a general purpose server, as they are designed to serve the network only. The sharing functions usually provided by a server include:

— dividing resources among the workstations, for example, hard disks;

— controlling sequential access to a piece of hardware, for example, a printer;

— keeping information secure so that only authorised users can read or modify it after supplying the correct password;

— managing the stored information reference to the network users, for example, who has special privileges;

— providing shared applications service, for example, databases, spreadsheets, etc.

Connections are the backbone of the LAN, as it connects the workstations and the servers together. This system is a combination of cables (for example, wire or fibre) and special purpose hardware (a network interface card or transceiver). For further discussion of this topic the reader is referred to the sections on signalling and medium access.

These three components together make up a local area network, but for it to function effectively and efficiently it requires three further properties.

A Specific Topology

The way a network cable is run between workstations and servers, that is, the conceptual layout, is called its topology. Topologies are not to be confused with physical wiring layouts; the topology of a network can be confirmed by removing the network user nodes (for example, workstations) from any physical point and investigating the conceptual structure of the servers and their connections.

The four most common topologies are the bus, ring, star and tree.

Bus and Tree

These two topologies are discussed together because of the great similarities between them. A tree could be referred to as a logical extension to the bus or the bus could be simply referred to as a special case of tree, in which there is only one trunk, with no branches. The main feature of a bus is that it consists of a *backbone highway* to which the nodes are linked via a tap. (See Figure 12.1).

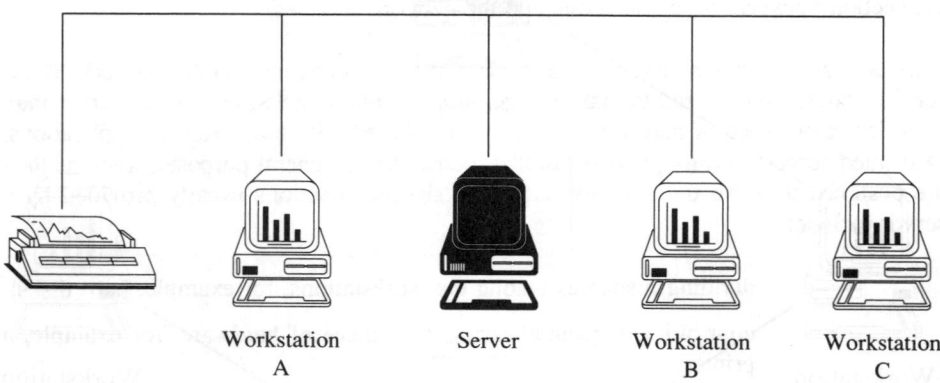

| Workstation | Server | Workstation | Workstation |
| A | | B | C |

Figure 12.1 Bus network topology

Since all nodes share a common communications medium and are configured as a multipoint network, (ie there are more than two devices connected to the medium and capable of transmitting on the medium), only one pair of devices on a bus or tree can communicate at a time. A distributed medium access protocol (refer to medium access control section for further information on protocols) is used to determine which node may transmit next, employing a packet containing source and destination address fields to do so. Each node monitors the medium and copies packets addressed to itself. One of the most common communications media utilised on a bus or tree is twisted pair wiring, although most recently fibre optic cabling has been used.

The bus and tree topologies have been the most common ones used to implement LANs, with Ethernet (using a baseband bus architecture) being the earliest and best known local network. The multipoint character of the bus and tree topology raises two particular questions in determining the efficiency of these types of LANs. First, there is the question of determining which node may transmit at any point in time. With a node-to-node link it is simple: if the line is full duplex, both nodes may transmit simultaneously; if the line is half duplex then they take turns on the line. A variety of strategies, referred to as medium access control protocols have been designed for the bus/tree topology, and this is discussed in detail later in this chapter.

Secondly, is the signal strong enough to reach its destination? This is known as signal balancing. The signal must be strong enough (but not too strong to overload the channel) so that after attenuation across the medium it meets the receiver's minimum signal strength requirements. It must also be strong enough to maintain an adequate signal to noise ratio. This is easily done for node-to-node links, but for a multipoint line it becomes very difficult as the signal balancing must be performed for all permutations of nodes taken two at a time. Hence, the bus/tree topology is usually utilised in a small geographical area, for a small number of nodes.

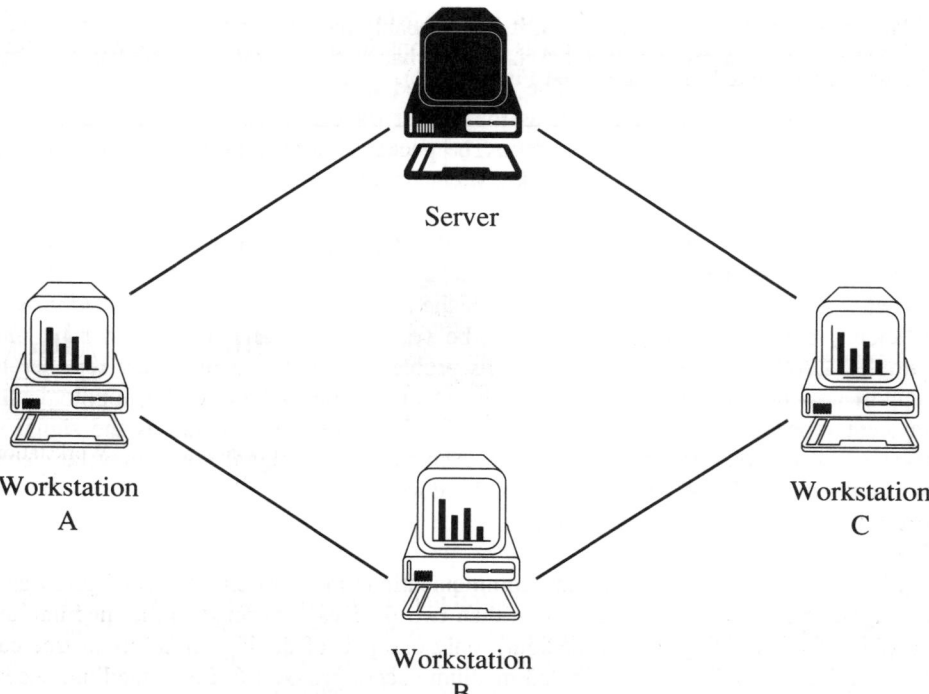

Figure 12.2 Ring network topology

Ring

The ring topology is the main alternative to the bus/tree topology (see Figure 12.20). The ring topology consists of a closed loop, with each node attached to its adjacent partner via a repeater, but note a ring is not a bus with its ends tied together. The signal propagates around the ring either in a clockwise or anti-clockwise direction from node-to-node between uni-directional repeaters. The repeaters have four major tasks to perform:

- *listening* where each bit received by the repeater is retransmitted;

- *transmitting* where it basically outputs everything it receives from the node to the ring;

- *bypass relay* which has two major functions, first it is a method of propagating the signal with a minimum delaying factor (ie medium propagation) and secondly, if one of the repeaters on the ring fails, then this facility will also come into play;

- *device attachment point.*

Note: The LANs network software will decide which direction the transmission will take.

Hence, if a node wants to transmit, it will probably have to wait for its next turn even though the ring is a multi-access LAN. Once it has access it transmits its data out onto the ring in the form of a packet. The wait period will be based on the ring's medium access control protocol. Therefore, as the packet travels around the ring it passes its destination node which copies the data. The packet continues around the ring until it returns to the source node with an acknowledgement.

You can identify various problems with this topology. A break in any link or the failure of a repeater normally disables the entire LAN or requires very expensive software (ie this cost is not only the cost of the software but also the associated cost of efficiency and time as the software will be very complex and difficult to run), and complex LAN software to cope with this problem. Installation of a new repeater to support new devices or nodes requires the identification of two nearby, topologically adjacent repeaters. Timing jitter must also be dealt with; this occurs as the signal is regenerated in every repeater, and where even though the signal levels may be *clean*, marginal errors will be introduced to the bit duration.

Star

This topology is based on the familiar central mainframe computer system configuration. In other words, a central switching system (or central hub) is used to link all the nodes in the network via a *private* dedicated line. The centre of the star is in charge of two major functions, first, processing and secondly, the switching of messages from one incoming line to another. Note this method is also used in a telephone exchange system, in which the central hub (the PABX) is a switch which interconnects the different users on the network. (See Figure 12.3).

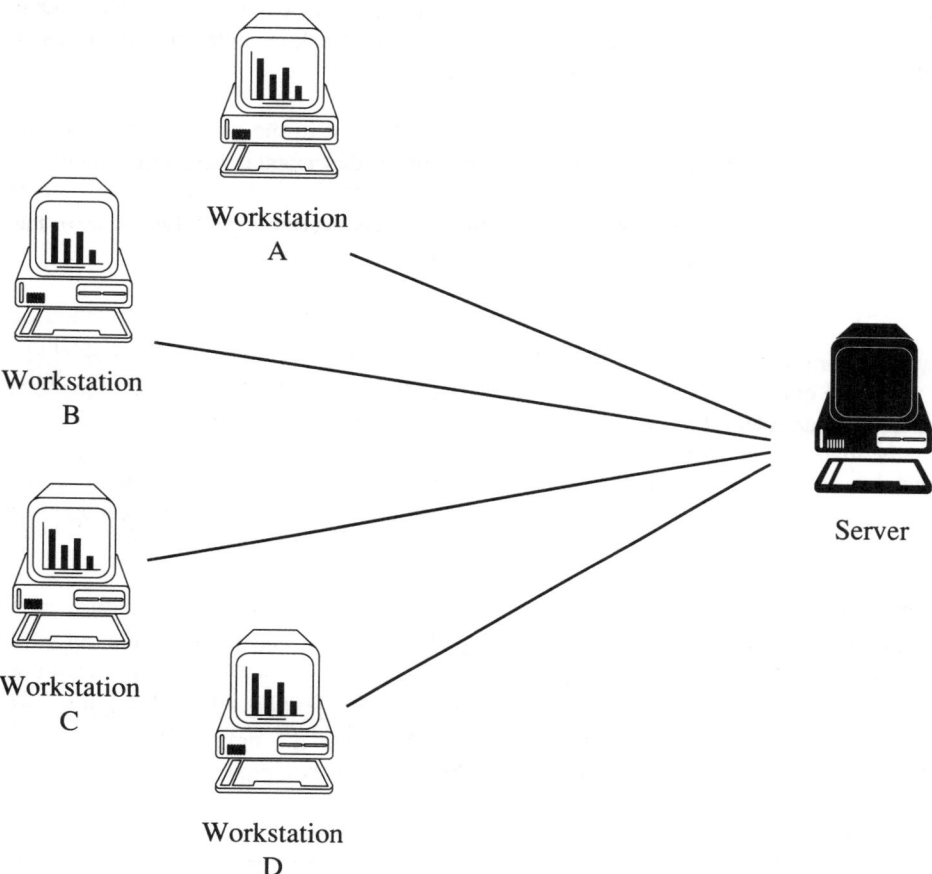

Figure 12.3 Star network topology

In theory, the attachment of a new device (another user or peripheral) is easy in a star as each device only needs to be tapped into an existing multidropped line, or a new line put in between the device and the hub. In practice, this is not the case. The installation of each new cable requires that the whole route between the two points be accessible so that the wire can be placed in the appropriate ducts. Furthermore, the communications handling device at the hub also needs to have a suitable port available and the necessary communications software to handle the particular device being put on.

A star network generally operates in a polled mode with each outlying device being asked, in turn, if it has any information to transmit.

Type of Cable

The cable that is utilised will seriously influence which topology will be used, whether the LAN can be easily installed, how it performs and its rate of efficiency. In many local area networks this medium can be one of several options, or even combinations of media, although most LANs stick with one type.

There are basically three types of cable that LANs use in the market today: *twisted pair, coaxial* and *fibre optic*. These cable types will be described in terms of: bandwidth (the width of the frequency band that can be handled efficiently by the system — this is important as it affects the speed of transmission), connectivity, geographic coverage possible, suitable topology, interference immunity, security and cost.

Twisted Pair

Twisted pair cabling is the standard telephone or telex wire, in which one or more pairs of wires are contained within a single outer case. Each pair of wires is twisted together in a helix in an attempt to provide fairly constant electrical characteristics; the wires are usually made of copper and are currently the least expensive medium owing to its extensive use by the telephone companies. This cable is suited to both analogue and digital transmission. It is a fairly attenuative medium and displays a high capacitance which makes it distortive too. This limits its usefulness to shorter distances.

Coaxial Cable

The coaxial cable provides large bandwidth and capability supporting high data rates. Its make up provides protection against electrical interference and has a low error rate. It has been used by mainframe networks to connect a wide audience. It is also utilised by the telephone companies to multiplex their traffic onto a single cable and is also employed in the transmission of TV signals by cable TV companies.

This cable consists of an inner carrier wire enveloped by fine copper wire mesh and an aluminium sleeve. Due to its capacity, low error rates and configuration flexibility, it is extensively used in LANs today. It can be a good replacement for the twisted pair cable, especially point-to-point links in the star or ring topologies.

There are two main types of coaxial cables used in LANs, broadband and baseband. This terminology is rather misleading, as the terms broadband or baseband are signalling techniques which are independent of the physical medium. In baseband LANs the capacity of the coaxial cable is utilised essentially to transmit a single baseband signal at very high speeds, for example, from 10 to 12 Mbits. In a broadband LAN, however, the capacity is utilised to establish numerous sub-channels from the one physical channel.

Fibre Optic Cable

Optical fibres are made of plastic or glass and can serve as a very high performance transmission medium. This type of cable transmits light instead of electrical signals which are transposed into light pulses by a modulator, transmitted over the fibre optic cable by a light source, and detected and converted back into electrical signals by photoelectric diodes.

A single optic fibre is basically a one way transmission medium, with a light source at one end and a detector at the other. For two way transmission a pair of cables is required. Since optical fibre filaments are very small compared with traditional copper wires, more optic filaments than copper cables can be put in ducts and hence there is a space and weight saving.

There are a variety of optical fibre available, from single mode to multimode, in varying thickness and densities, with bandwidths of up to 3.3 billion Hz and 1 Gbits. The quality of cable and the type of light source used will influence the cost and rate of transmission. A benefit of optical fibre is their immunity to electrical interference. Furthermore, the data they carry cannot be easily tapped into.

Signalling Techniques

Messages are transmitted in either digital or analogue form. Digital signals are *independent*. They are made up of two mutually exclusive choices, either ON or OFF (ie either there is an electrical signal or not). This is the basic method of communication between computers transmitting binary digits or bits (1s and 0s). Analogue signals consist of an unbroken wave like signal. Each complete wave is called a cycle. The signal's frequency in Hz is defined as the number of complete cycles per second.

Analogue and digital are the transmission modes used to transmit data successfully from one node to another; the data must be coded in a way appropriate for the medium transmission mode and data type involved. In the LAN arena two types of signalling techniques are predominantly used: *baseband* and *broadband*.

Bandwidth is the terminology used to describe capacity or the deviation between the highest and lowest frequencies of a signal. The human voice is transmitted over the telephone network in the range of about 300 to 3400 Hz, and hence the bandwidth for voice is 3100 Hz (3400 − 300 = 31000). With digital transmission, bandwidth is defined as the amount of data that can be transmitted over the channel in bits per second (bit/s).

Baseband

Baseband signalling is the most common and uncomplicated method of transmission in LANs. As this is the only signal present on the cable at any one time, it effectively uses all of the available spectrum of the transmission channel. This is the only transmission method used with twisted pair cables.

To describe this type of transmission, we could compare digital transmission with a garden hosepipe. Here, the signalling would be accomplished by turning the water on and off, the water is considered to be the binary 1s and 0s of the data. Since the hose can only handle one gush of water, only one tap can modulate the signal. If we have a hose with various taps, only one message could be passed at one time. If two taps tried to send messages at the same time, the signal would be garbled.

This analogy may describe the nature of the signal but it is not simply a matter of squirting the water (voltages) through the hose (medium). There are several problems with digital signals. First, they are more likely to suffer from attenuation than analogue signals, which obviously can lead to distortion. Secondly, you need to ensure that the receiving end can synchronise itself to the pulses. These problems can be solved with the appropriate encoding scheme. There are various ways of encoding the signals onto the medium, with *Manchester encoding* being the standard method for baseband systems. Manchester encoding is one of the easiest to implement and has an in-built clocking scheme which enables every node on the network to remain in synchronisation.

In Manchester encoding the time period is split up into identical cells and each is utilised to represent a single bit, each cell is also divided into equal portions. During the initial half of the cell, the signal transmitted is the complement of the bit value being sent in that cell. In the second half of the cell, the uncomplemented value is sent. In this way there will always be a signal change during a cell, at the half way point, which guarantees that nodes are kept in synchronisation without the need for separate synchronisation signals.

Broadband

Broadband communication uses a medium which can support a large bandwidth or frequency spectrum. Basically, it is a way of frequency multiplexing many users onto one cable. In other words, the cable is sub-divided into channels; usually a broadband system has a bandwidth of about 300 Mhz.

There are several significant advantages to broadband systems:

— capability of mixing totally different networks, for example, the telephone network with a LAN, where the broadband cable allows them to be assigned a separate channel — this type of transmission keeps the one cable concept of a LAN;

— attenuation of the cable has less effect on the analogue signal and so a larger area can be covered;

— it is the only feasible approach to a tree topology signal splitting/joining being fairly straightforward.

Nevertheless, broadband is not bi-directional, hence for every transmission there has to be a defined direction signal flow.

But how does a broadband system function? One or two cables are used to link all the systems on the network. If two cables are used one cable is assigned the task of transmitting while the other is assigned the task of receiving. Each node is linked to both cables through a radio frequency transmitter/receiver which both listens and transmits and is the interface between the node and the cables. At the *headends* of the cables a transmitter/receiver is attached. It is the headend's responsibility to listen to the transmissions on the transmission cable and redirect them out on the receive cable.

If one cable is used, then the accessible bandwidth is split into two distinct transmission bands with a frequency guardband in between them (for example, a frequency gap between the bands). One band is responsible for the data being transmitted, while the other is for data being received. Consequently, in broadband two distinct frequency bands are utilised for transmission, one for the transmitted data and the other for the retransmission of the data. Note: all the nodes on the network listen to the receiving frequency and take the data assigned to them.

Other Elements of a LAN

There are various devices which permit LANs to communicate with non-standard devices or the outside world.

Repeaters

The majority of LANs are limited to a physical size in terms of the maximum allowable length of cable, the number of nodes and the distance between each node to be utilised. Nevertheless LANs can be extended beyond their configuration limit by the use of repeaters. (Note: repeaters are only possible on identical LANs as they are transparent to the links and have no intelligence to it apart from being a form of relay). A repeater is simply a black box which amplifies and boosts the signals so that they can reach their destination.

Most LANs allow this type of extension, for example, Ethernet. In Ethernet its main trunk cable has a maximum configurable size of 500 metres, the limit of signal propagation. This system can be extended up to 2500 metres, split up into five segments of 500 metres each, with the links to each segment being connected via a repeater.

Gateways

A gateway connects two physically distinct LAN architectures or connects a LAN to some other type of communications network.

Gateways facilitate different types of conversions, it may just convert between the different protocols of two distinct LANs or convert speeds from one LAN to another type of network, as in X.25 gateway. An example of this conversion can be seen if we look at a LAN gateway into an IBM mainframe environment. A 3270 gateway would be required so that the LAN could connect to the IBM mainframe. This gateway would provide protocol and link conversions to allow IBM screens and functions to appear on the PCs. The PC operator would continue to operate through the LAN, but their terminal would have the screens and appearance of a 3278 terminal through to the host system.

Hence, the gateway carries out a conversion between the communications protocols used in the LAN and the protocols used in the IBM 3270 network.

Bridges

A bridge would be used to connect two physically distinct homogeneous LANs. A bridge just captures frames on one network and relays them to the other. Bridges are transparent to users. Within a bridged environment all nodes are regarded as belonging to one logical network and with no address translation taking place it means that the medium access control (MAC) addresses must all be unique on the whole logical LAN. As the bridge connects two networks at the MAC sub-layer it means that the two networks connected via the bridge must have identical MAC and logical link control (LLC) levels, although a different physical specification could be used.

A bridge could connect either local LANs or remote LANs. A local bridge is just a single unit physically linking the two networks whilst a remote bridge consists of two units (half bridges) connecting the two LANs either over a point-to-point link or via a PSDN. Other characteristics of bridges include:

—   learning capability, ie to ignore locally addressed packets;

—   working at layer 2 so that different protocols are handled, for example, TCP/IP and LAT;

—   limitations of connections, ie two hour connection of the same two networks is not allowed, but 'spanning tree algorithms' can be used to form loops for resilience.

Router

A router provides a connection at layer 3 of the OSI model, and a layer 3 device is utilised to connect heterogeneous sub networks with different protocols at layer 2. The router makes a layer 3 decision on whether or not to pass on the network address. The LANs can operate as both physically and logically separate networks. A router could be used in the connection of a LAN to a WAN (at layer 3) and, therefore, could mean that they need to be multi-protocol routers.

FDDI

Fibre distributed data interface (FDDI) was defined by the American National Standards Institute (ANSI) as a high speed fibre optic networking standard which conforms to OSI.

The standard describes a 100 Mbits network working over fibre optic cabling. FDDI utilises a dual counter-rotating ring topology which means that it can continue working even when there has been a break in the cable or a node has failed. It is also capable of using most of the available bandwidth via an append token protocol technique. Furthermore, by using multimode fibre optic cabling a total fibre length of 200km can be reached which creates a 10km dual ring. FDDI can support up to 1000 physical connections with a maximum inter-node distance of 2km.

FDDI will become more and more important as more optical fibre backbones are installed. Presently, Ethernet and token ring systems represent the bulk of the main company backbone network topologies. Standalone or clusters of Ethernet and token ring networks can be linked together on an FDDI backbone network via bridges and routers.

Security

Security is one of the most important issues in the use of any IT equipment, not least LANs. Without a security function there is a potential for:

—   loss of information through negligence;

—   access to unauthorised files on the servers;

—   access to mainframe computers and their files;

—   hacking and eavesdropping.

The simplest network security function would be to control the physical access to the PCs by either locking them up in a secure environment and/or providing a key on every terminal. Without the ability to unlock the terminal a user could not access the network.

In most networks the security function is provided through software or hardware and the use of *passwords*. Passwords can be multi-level which allow the user access to various functions based on the level of their password. These passwords must always be encrypted using a *one way* technique, ie they should be irreversibly scrambled when stored. The network should also keep a list of the attempted and failed log-ons and notify the administrator of suspected attempts to break into the network.

If everything fails and there is a breach in security, you must be prepared for this with an adequate contingency plan.

Medium Access Control

As mentioned earlier the one objective of LANs is to share resources, from a database to a piece of hardware. This also includes the sharing of a common transmission medium. Hence, when sharing any part of the network there must be something in charge of managing its usage. Medium access techniques are the management tool used to control the way in which nodes literally gain access to the network channel. These techniques are employed to mediate between nodes that contend for the use of a channel and are commonly known as either *polling* or *contention* techniques.

Polling Techniques

Polling is the type of protocol commonly used for any type of network containing multiple communication points. The polling protocol is designed to stop two nodes from transmitting at the same time by telling each node when to send or receive data, in other words each node waits its turn. In this configuration there is a master node that sends out a message to the other nodes. The node that received this message will then proceed to transmit if required. If not, it will respond with a message saying something like 'nothing to transmit'. The master node then continues to the next node inquiring whether it wants to transmit or not, and so on. This type of polling technique is rather too restrictive in a distributed adaptive network like the LAN. However, some systems still use this type of protocol in the centralised networks particularly where a high degree of control and security are required.

Managing the access to a network can also be achieved by *distributed polling*. The two most frequently discussed methods of distributed polling used in LANs are *token passing* and *slotted rings*.

Token Passing

Token passing is a unique example of a protocol that uses polling techniques. In this type of network, a packet is used to poll all of the nodes; the packet is called a *token*. In this network each node, in turn and in a predetermined order, receives and passes the right to use the channel.

Tokens are unique packets that travel around the ring from node to node. Current ownership of the token gives a node the sole right to transmit its message around the network. As soon as the node gets control of the token, it transmits its message, but it only has an allotted time duration to do so, after which it must surrender the token and pass it on to the next recipient. If it completes its transmission it obviously relinquishes the token sooner. In order to complete the transmission the receiver must copy the data and regenerate the frame for the sending station to detect and transmit the free token. This method obviously avoids the problem of conflicting simultaneous transmissions, as they do not occur.

There are two implementations of token passing systems, the token bus and the token ring. The topologies require different methods of execution. In a ring topology there is a physical ring configuration allowing all the nodes to have a turn in sequence. In the bus, added sophistication is required as it needs to establish a *logical* ring giving each node three items of information: addresses of the previous node, next node and its own identity.

In the token ring topology each system is independent without any one node being the master. The token ring network allows computers to move data at high speeds between them. Usually token rings use either coaxial or fibre optic media, although recent developments have seen token ring implementations on twisted pair, for example, IBM token ring running at 4 Mbps.

Slotted Rings

The slotted rings protocol uses a modified token passing access control method. Normally a number of slots, or packets, of fixed size travel around the ring. Each packet will normally contain data on the source and destination addresses, control and parity information, and data. If the node needs to transmit, it waits for a free packet, inserts data into the appropriate field, sets a bit to indicate that the packet is full, and indicates the source and destination address. As in token passing, nodes along the way check to see if the packet is addressed to them. The destination node copies the data into its buffers and marks the packet to indicate that it is empty and can be used by other nodes.

There are certain problems with slotted rings. First, the number and size of packets on a ring is limited to the ring's capacity (its size and propagation time). The need to break down the data into smaller units means that the packets must contain a vast amount of control information in proportion to the actual data carried. Secondly, the lack of support by the big network suppliers, who usually prefer CSMA/CD or token passing, has meant that there exists little knowledge or direct support for this type of network and hence little prospect of future enhancements.

Contention Techniques

Polling systems have certain problems associated with their throughput and operation:

- polling systems can be somewhat wasteful in terms of the actual data that is transmitted versus the overhead for the protocol;

- central polling systems have to direct all messages through the master node which means double transmissions for nodes on the same channel.

To overcome the problem of polling overhead, a method of *contention* could be utilised. Contention techniques anticipate conflicts or collisions and actually use them as part of the design to allocate the common channel. The most common contention technique is carrier sense multiple access detection (CSMA) with CSMA/CA being an adaption of this method but with 'collision avoidance'.

CSMA/CD has three basic principles: if we split up the mnemonic into three parts, CS, MA and CD we can analyse how this protocol actually works.

- *Multiple access (MA)* allows any node to send a message immediately upon sensing that the channel is free of any traffic. In this way, a considerable amount of the wait-time characteristic of non-contention techniques is cancelled.

— *Carrier sense (CS)* is defined as the listen before speaking *polite* habit. Each node examines the medium first to make sure that no one else is using it at that time. If the medium is in use it waits a random amount of time before it tries again. This does not always work because of the time if takes for a signal to travel across the network (its propagation delay). Two nodes could detect that the channel is free at exactly the same time, as each node will not yet have detected the signal of the other; in this situation a collision will occur.

Hence, once a node starts transmission it cannot forget about what it has sent, it must continue to listen to the medium to see if any collisions have taken place. If it does the *collision detect* stage comes into effect.

— *Collision detect (CD)* is the facility a node has for sensing any collision of its message. If this occurs, it should back off and send a jamming signal to warn others that a collision has taken place. Then it must stop transmitting, and wait a random amount of time to attempt any retransmission.

It becomes apparent that the fewer collisions that take place, the more efficient the network as a whole will be. The efficiency of this protocol depends on the available time it has for transmitting a packet, its *slot time*. When the slot time has elapsed during the transmission of a packet, the node is said to have taken the channel, therefore there is no longer the likelihood of a collision, as all the nodes have had the opportunity of detecting the traffic on the channel.

The most efficient use of CSMA/CD therefore is made when using large packets, as the more big packets are transmitted, the fewer collisions there will be relative to the total amount of time the channel is in use.

As mentioned earlier, instead of collision detection you could try to avoid them. CSMA/CA tries to minimise collisions with a procedure of *collision avoidance*. Here each node waits a random period after any transmission on the network (rather than just after collision).

The advantage of the CSMA/CD or CA protocols is that they allow distributed control of the transmission link. Each node decides when to transmit and can send directly to the addressed node. CSMA/CD and token ring techniques presently hold the major share of the LAN market.

CSMA/CD versus Token Ring

The algorithm required to run CSMA/CD is much simpler than the one required by token passing, but its throughput seems to decline as the number of nodes increase. In terms of accessibility of the network with CSMA/CD everyone gets an 'equal bite at the cherry', while with token passing this is regulated according to the size, direction, time slot and type of packet.

Access to the channel is determined in two different ways. CSMA/CD is non-deterministic while with token passing each node knows the amount of time it needs to wait between the opportunities it gets to transmit. These access methods affect the overheads of both systems. In CSMA/CD a light load suffers from little overhead but as the load increases, the retransmissions caused by collisions will constitute some overhead. With token passing the overhead is mainly dependent on getting the token around the network. This occurs even with no loading at all, as a node needing to transmit still has to wait until the token gets to it.

With regard to the network signal tolerance, token passing is by far the superior method of transmission, as it is not vulnerable to signal fluctuation levels, due to node-to-node data transfer. With CSMA/CD, however, its collision detection function poses some problems which affect the continuance of a moderately even signal level throughput the network. Furthermore, this capability dictates the frame sizes and some wastage can occur through packets needing to be packed out to meet the minimum length requirement.

In terms of popularity and practicality, CSMA/CD is a more accepted protocol due to the marketing of Ethernet. This popularity obviously brings with it the benefit of experience and market support.

LAN Representation within the OSI Model

In the early days of the LAN, standardisation seemed to be conflicting with the principle it was promoting, the desire to provide a communications technique which was free from the constraints typical of normal communications. But it soon became apparent that, due to the proliferation of intelligent equipment and computers from many different manufacturers, LANs could not be left out of the standardisation process.

Communications between these systems and networks can only be possible if they abide by some common set of rules. In other words, networks require two levels of standards for communications:

— *standard protocols and interfaces* to facilitate a common method for communication among systems;

— *a standard method of network design,* that is a network architecture that defines the relationships and interactions between network devices and functions via common interfaces and protocols.

The International Standards Organisation (ISO) recognised the importance of, and the need for, universality in exchanging information between and within networks and across geographical boundaries.

In 1978, ISO issued a recommendation which stimulated greater conformity in the design of communications networks and the control of distributed processing. The recommendation, which has gained wide acceptance, is in the form of a seven-layer model for network architecture, known as the ISO model for open systems interconnection (OSI).

OSI is a philosophy which embraces international standards for the exchange of information. It was developed to facilitate open communications among a broad conglomeration of mixed computers, free from the idiosyncrasies of their suppliers and so freeing the user from having to tie themselves to a single vendor and his proprietary offerings.

OSI comprises a reference model, a set of service definitions and a set of protocol specifications. The reference model is described in the ISO document 7498 parts 1 to 4. For further information please refer to Chapter 9 of this book. With this model in mind we can proceed to investigate what we mean by an OSI LAN.

OSI LANs will use acknowledged international protocol standards in communications either between LANs and WANs, or within LANs. The majority of the major LAN manufacturers have announced migration plans to introduce international standards into their networks. Hence, in the future OSI LANs should be able to support communications between multi-vendor computers within a multi-vendor network environment.

The OSI reference model relates to LANs in the following way, taking each layer in turn:

Physical Layer

This provides the electrical encoding and decoding of information for transmission over the physical medium. For example, a method of putting data onto the network could be CSMA/CD.

Data Link Layer

This administers the access to the physical links between two nodes. Within the LAN environment, the data link layer is divided into two sub-layers: logical link control (LLC), and media access control (MAC). The LLC provides error and point-to-point flow control, and the MAC takes care of node addressing, frame formatting and error detection.

Network Layer

This layer is in charge of routing and relaying between communicating end nodes. This could be on the same sub-network or on sub-networks connected through intermediate systems.

Transport Layer

This layer in conjunction with the lower three layers is the whole transport service. It is responsible for facilitating an error free end-to-end data path to the upper three layers.

Session Layer

This layer comprises a set of tools used by the upper layers, or session users. The primary aim of the session layer is to provide for an orderly transfer of data and control information between the peer open systems.

Presentation Layer

This layer is in charge of how data is represented within the open systems.For example, a number can be represented in many ways, from binary form to two's complement.

Application Layer

This provides a number of services, for example, virtual terminal (VT) which allows remote access to different resources across the OSI environment, file transfer access (FTAM) describes a very advanced and powerful method for manipulation of data across a network, and message handling system (MHS) defines a store-and-forward message system to allow users to exchange information electronically.

When considering LANs it is advisable to concentrate on levels 1-4, since most standards adhere to these levels.

LAN Standards

If we are going to develop the use of LANs within our organisations then we must adhere to the standards that have been set in this area. Standards should be viewed as the best method to interwork with other equipment.

The setting of LAN standards for the electrical and basic link transmission is the responsibility of the Institute of Electrical and Electronic Engineers (IEEE). Traditionally this was not the case with the local post and telephone companies setting local standards, but LANs fell outside their jurisdiction as they are designed not to go beyond a company's boundary and are not classed to be a common carrier network. Consequently, the IEEE had to try and set some standards of electrical characteristics for the development of LANs.

IEEE 802 committee developed the IEEE 802 standard. This is a collection of standards for several different types of LANs and the link control procedures for communications on the networks. Each network or procedure is defined with a number IEEE 802.*n* (where *n* denotes the individual standard for the network or procedure).

IEEE categorises LANs into five areas:

—   privately owned;

—   geographical coverage of less than 20km;

—   data rates 1 to 100 Mbits;

—   error rates of 1 in 100 or 1000 million;

—   transmission delays of 10s of milliseconds.

IEEE then categorises LANs into a three communications layers mapping to the lower OSI/RM layers.

—   *Logical link control (LLC).* This layer is in charge of maintaining communications between one of its service access points to another remote point. As data is passed through this layer it attaches a header to it. This is then passed on to the MAC below.

—   *Medium access control (MAC).* This section manages and controls protocol compliance onto the network. Once it receives the message with the right protocol header from LLC, it envelopes it with its own protocol information and produces a frame. This frame is then passed onto the network.

—   *Physical layer (PHY).* This specifies the route in which the messages are to be carried to the chosen medium.

With this broad IEEE definition of LANs, its communications proprieties and size, the IEEE went on to formulate the 802 standards. It defined the following standards.

802.1 Architecture, Management and Inter-networking

This standard describes how the other 802.*n* standards relate to each other. It also describes the correlation between the various parts of the 802 standard with the OSI/ISO reference model.

802.2 Logical Link Control

This is the logical link control (LLC) standard which stipulates how messages are framed and passed onto the links. 802.2 is universally applied to all forms of LANs and effectively defines the procedure for moving data. This layer conforms to the data link layer in the OSI model.

The following mechanisms are included within the standard:

- *frame numbering* allows the reconstruction of larger entities from the transmitted blocks, and permits recognition of lost data conditions;

- *flow control* allows the receiver to regulate his incoming traffic so that he does not get flooded;

- *error control* allows measures to be taken when a transmission error is detected;

- *addressing* identifies peer bodies.

As defined earlier the LLC can be seen as the interface between the MAC layer and its service access points. This interface gets requests or report messages which contain different information depending upon their respective uses.

Through these messages it can then provide three levels of services:

- *An unacknowledged connectionless service.* The LLC simply sends a block of data and no acknowledgement, nor failure indication, nor error control is given.

- *A connection mode service.* Here the communication between two devices occurs in three stages: connection, data transfer and disconnection. Both ends monitor the transmission and the LLC certifies that the transmission is ordered indicating where there has been any failure.

- *An acknowledged connectionless service.* Data is transmitted a block at a time, following each transmission, an acknowledgement is requested, note this can only operate on a one-to-one basis.

Note: An LLC implementation does not have to provide all the above services but must at least offer an unacknowledged connectionless facility, which is normally the case.

802.3 CSMA/CD Networks

This standard defines the physical layer for a CSMA/CD system based on a coaxial cable, although unshielded twisted pair and cable television (CATV) cable can also be used. Originally this network was supposed to be identical to Ethernet, but there was a conflict in the cable grounding scheme. Ethernet later changed their specification to Ethernet II which is compatible with 802.3.

This standard defines a bus topology with CSMA/CD medium access control. Its physical specification defines an attachment to the medium split into four elements:

- attachment unit interface;

- medium attachment unit;

—　medium dependent interface;

—　physical medium attachment.

At present there are four media defined in the standard:

—　10BASE5 (Ethernet 50 ohm coaxial cable);

—　10BASE2 (Cheapernet, thin grade 50 ohm coaxial cable);

—　10BROAD36 (Broadband Ethernet/Mitrenet 75 ohm CATV cable);

—　1BASE5 (StarLAN unshielded twisted pair cable).

Note: 10BASE5 is: 10 Mbits, Baseband, 500 metres.

In each of the above media the four component parts of the physical specification differ according to the needs of the technology.

802.4 Token Bus Networks

This describes the physical layer for a token passing bus standard that is used in a broadcast (rather than ring type) mode. The transmission medium in this standard is a 75 ohm coaxial cable.

There are three physical medium specifications:

—　*phase continuous carrierband with frequency shift keying* runs at 1 Mbits with an unbranched semi-rigid central trunk cable. The drop-cables have to be less than 3cm with the bi-directional taps;

—　*coherent carrierband with frequency shift keying* runs at 5 or 10 Mbits with drop-cables as long as 30;

—　*broadband* is a mid-split single cable which is recommended, although a dual cable could be used. It runs at 1, 5 or 10 Mbits; pre-coding is required to overcome phase ambiguities in detection.

There are two aspects to the management procedures of a token bus. First, it must ensure that the token is not lost. It does this by making every node responsible for checking that the token reaches its destination. Secondly, it must cater for the priority system specified in 802.4, where there are four priorities: 6, 4, 2, 0 — with 6 the highest.

802.5 Token Ring Networks

This specification defines the physical layer for a token ring network. The transmission medium in this standard is 150 ohm balanced twisted pair cable using Manchester encoding and operating at speeds of 1 to 4 Mbits.

The management procedures of this network are rather complex, with one station adopting the status of *active monitor* and the others *standby monitor*. While in an active status it checks for erroneous situations, like lost token, persistent data frame, priority failure (the priority scheme is based on eight levels with the higher numbers indicating greater priority) and duplicate monitor. (See earlier part of this chapter for further details of token ring networks).

802.6 Metropolitan Area Networks (MANs)

This is the specification for a metropolitan area network. This committee is investigating integrated voice and data network facilities which can stretch through a whole city with a radius in excess of 50 kms.

Other specifications are:

- 802.7 Broadband Technical Advisory Group;

- 802.8 Fibre Optic Technical Advisory Group;

- 802.9 Integrated Data and Voice Networks.

ISO Standards for Fibre Distributed Data Interface (FDDI)

The FDDI standard is still in its early stages, but the demand for greater bandwidth will undoubtedly accelerate the standardisation process. This forthcoming standard contains four significant sections:

- physical medium dependent (PMD);

- physical layer protocol (PHY);

- media access control (MAC);

- station management (SMT).

The first three are already full ISO status. The fourth, and final, section is the most complicated and important section. It deals with network configuration and control and is only at draft ISO status.

Although many manufacturers claim to offer FDDI compliant products, true compliance is not possible until the completion of SMT, which is expected to be published sometime during 1993.

Application Areas

PCs have brought us many marketable applications, from word processing to the spreadsheet. These applications have made the PC a very important part of the business environment. The introduction of LANs saw many of these PC standalone applications being used in a co-operative manner, but applications that take full advantage of networking capabilities are few.

Network versions of applications originally developed as a standalone PC application are the most successful form of network application. The main reason for this popularity has been the need for more memory and the ability to share information and applications across a work group. Exploiting these network capabilities in this manner has its drawbacks. If applications are centralised on an unintelligent server which merely exports these programs to each workstation on request so that they may be run there the process is both time and memory consuming. Also, the workstation is tied up for the duration of that program's run time.

These drawbacks can be overcome by establishing a client-server architecture. This concept uses intelligent workstations and intelligent servers. The way it operates can be defined as clients (the end users) sending requests to the server to perform basic functions, such as communicating with other LANs, and the server processes the requests and sends the client the results. This sets the workstation's memory free from having to run memory or time consuming programs itself and also allows management and security to be centralised on the network around the server machines.

At the heart of this client-server methodology is the process of providing the users the reliability and high performance transaction throughput offered by minis and mainframes with the flexibility and all round user-friendliness traditionally associated with PCs.

These requirements allied with the ability to access whatever, from wherever, is a major undertaking and probably explains in part why application software for networks has not been forthcoming, for example, real-time database access. However, future LANs developments, particularly in the *client-server* field, indicate that this situation is being addressed.

13

Networking considerations

In the earlier chapters we have been largely concerned with the basic principles, the underlying transmission technology and the fundamental concepts of data communications systems. We have started to consider how all the components might be assembled together to provide a system which meets the applications requirements for which it was designed. We now turn our attention to the subject of network design and discuss some of the more important aspects which have to be considered.

Once a properly designed system is installed and operational, it is of paramount importance to ensure that it consistently supplies the level of service for which it was designed. Communications-based systems are decidedly more vulnerable to performance shortcomings than were batch processing systems. This arose partly because of the rise in expectation when batch applications were transferred to a real-time environment; lead times to provide processed information were to be reduced to seconds and a whole range of company activities and procedures modified to take advantage of this vastly improved response time and processing power.

It is therefore not sufficient to have a well designed system, it must also be effectively managed, not only in terms of management skills, but also in regard to the management support resources and facilities provided within, and external to the system. A comprehensive discussion of these subjects is beyond the scope of this book, and we shall therefore confine ourselves to a review of the more important issues.

Network Design

The most difficult part of the network design process is defining the scope of the problems to be solved. The basic demands on the network will depend on the computer systems design. With the advent of distributed processing techniques, the network is an increasingly vital component in the system's functionality. It is therefore vital that the designs of both computer systems and data networks should be coordinated closely. While it will be possible to quantify the load that an individual transaction places on the network (this is a function of computer system design), it will be very difficult to forecast the numbers of transactions which will need to be supported per day or how that load will be distributed between the network sites. The crucial demand of any network design is that it should meet the peak loads placed on it.

It seems the very best forecasts of network loading will inevitably prove to be inaccurate in practice. So it is vital that the system and network, once installed, should

389

access and record as much information as practicable about actual network use. This will allow the network design to be reviewed on a regular basis with a view to maintaining the service to the users in the most cost-effective manner possible.

Acceptance by the user is a vital ingredient in the success of data communications systems. The level of success achieved will depend upon the extent to which the total system meets the user's perceived needs. However, the network has its part to play in delivering data to the user in a manner which is timely, available and reliable. We will now go on to look at how the choice of network components can provide a cost-effective solution to network requirements in:

- performance;

- reliability and availability;

- cost;

- expandability.

Performance

The data transmitted through a network will be of two basic types: either essentially one way reporting or interactive transaction. Where bulk data is being passed between nodes without direct human involvement, other than issuing the command to start transfer, short term delays in delivering the data may not be important and can be tolerated; for example, where data is needed at the central site in time for the beginning of overnight processing at 6 pm. An interactive transaction involves on line applications running on a question-answer basis between a user terminal and another network location. In such applications, response time is a major concern here. Although there are several definitions, *response time* is commonly defined as the elapsed time from depressing the send key on the terminal, to receiving the first character of the response. In general, response time consists of four elements.

- time taken and time waiting to propagate the user's request through the network to the computer;

- time taken and time waiting to execute program code on the central processor;

- time taken and time waiting in accessing the data required from file (or database) storage devices;

- time taken and time waiting to propagate the computer system's response through the network to the terminal.

Network design, as opposed to distributed system design, can only affect two of these elements of response time: the propagation times to and from a remote site. For example, where a dedicated data link connects a computer system to a terminal, the propagation time is the time taken to transmit the question and the answer. However, where the link is used to serve multiple devices at either or both ends, contention protocols are necessary and users have to queue for service. As already discussed, the characteristics of the link protocol used have an important bearing on transmission efficiency.

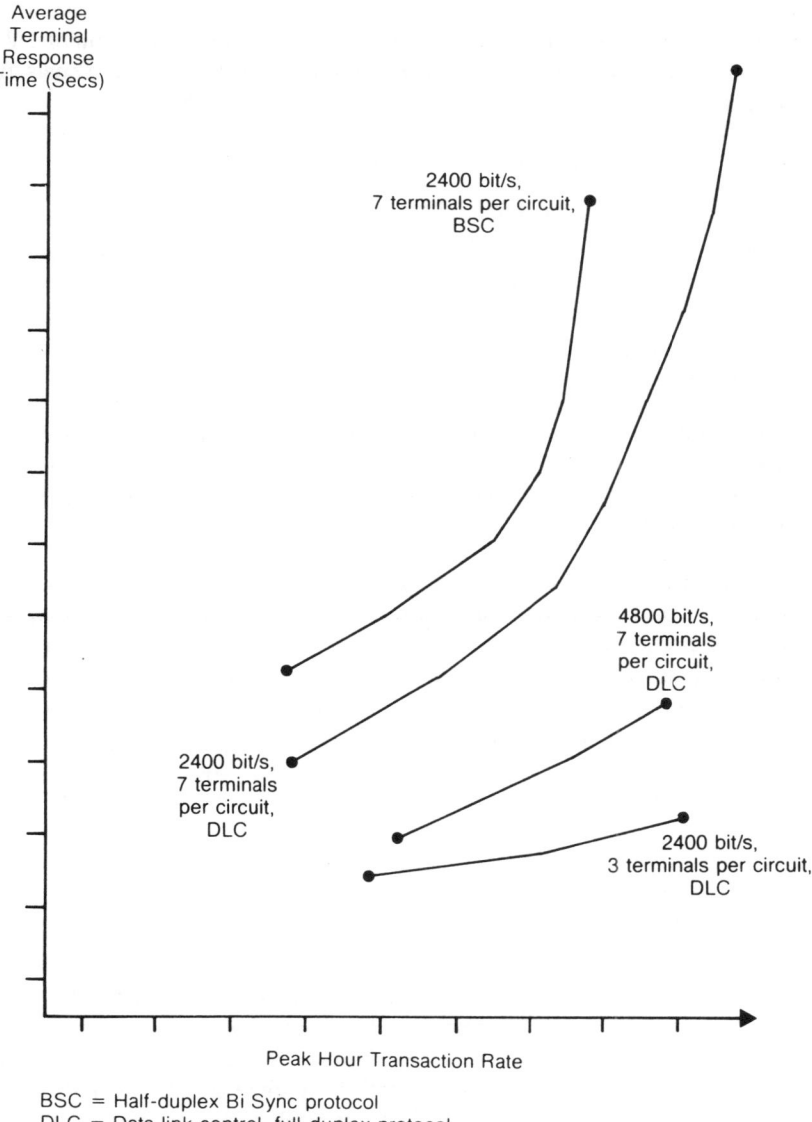

BSC = Half-duplex Bi Sync protocol
DLC = Data link control, full duplex protocol

Figure 13.1 Graph of performance versus peak hour transaction rate

Queues of messages waiting for service can build up on various points in the system, waiting for circuits to be available, a disk access arm to be free, a CPU program to be released, and so on. Queue formation is inevitable and must be allowed for in the design of a system. A whole body of *queuing theory* exists which can be utilised in predicting queue sizes and their effect on response time under various assumed traffic levels and system capacities.

The average communications component of response time is found to have an exponential relationship to the level of utilisation of the link. Figure 13.1 shows how the average response time varies with the transaction rate of a data link. If we are to maintain reasonable response times for terminal users the peak transaction rate must be restricted to the part of the graph before it starts steepening and average response time rises sharply.

On the steep portion, not only are response times likely to be extended beyond acceptable limits, but they would also be sensitive to marginal variations in the transaction rate. The standard speeds, up to 9600 bit/s, are typical speeds of access from terminals onto a wide area network (WAN). Access from a mainframe's FEP or minicomputer at 64 kbits onto, say, a packet-switched network is also typical. It would also be quite common for the links between the packet-switched nodes to run at 64 kbits. Local area networks will run at speeds of up to 10 Mbits. These higher speed networks will have more terminals, PCs, etc connected to them and higher traffic volumes to handle, but essentially they will react in the same way as lower speed networks when demand for service is high.

Deterioration in response time will generally be caused by a component in the system being too slow to service the traffic demands. In a packet-switched network an example would be a link from the network to a mainframe where the link carries traffic from a sizeable terminal population. This problem can usually be remedied by adding extra links. In a LAN, the remedy may necessitate splitting the LAN into two parts with a bridge between them. This may not be the answer if traffic routing is not thoroughly analysed. It may be necessary to attach extra printers or servers to provide a solution to the problem.

Reliability and Availability

Reliability

An important aspect of network design is estimating the likely levels of reliability, both in terms of the quantity of the data transferred between nodes and the non-availability of systems as a result of network component failure. We have discussed the nature of errors in transmission systems, and their detection, at some length in Chapter 8. In this chapter we noted that, so far as is practicable, transmission errors should be detected and corrected low down in the system as remote as possible from the application itself, and that one important function of a *link protocol* is to convert an inherently unreliable link into a reliable one. Nevertheless, there will be errors which can only be resolved, or only manifest themselves within the application itself. The design of procedures for recovering from such errors is perhaps one of the most difficult and complex parts of data communications system design.

By the time the error is detected, it might already have had a widespread impact. For example, individual transactions might have got out of sequence, files may have been updated in the wrong sequence, and so on. In the meantime, the terminal user may well be wondering what has gone wrong, but is almost certainly expecting to recommence at the point where he left off. This is where the difficulties really start, because it may be necessary to backtrack to a point preceding the error.

In designing recovery procedures very careful attention is paid to ensuring that this is achieved with the absolute minimum of disruption, that the terminal user is kept fully informed, and that the recovery process occurs as gracefully as possible. Error recovery is a major function of all network architectures, ensuring that the data is delivered correctly and in the order it was sent. In the OSI model, error detection and recovery may be present at a number of layers. In a wide area network it will normally function at the data link layer. It will feature at the transport layer, if class 4 services are used and it has particular importance here as this is the lowest layer that interfaces between end user systems.

Availability

The *usefulness* of a system is a function of the time it is available for use relative to the time when it is required for use. The measurement of usefulness, termed *availability*, can be calculated for individual components and complete systems, from the expression:

$$\text{Availability (A)} = \frac{\text{MTBF}}{\text{MTBF} + \text{MTTR}}$$

where: mean time between failures (MTBF) = mean time to repair (MTTR).

A typical data transmission link is made up of several components, modems, terminals and circuits. The availability for the system can be arrived at by considering each discrete component as an element in a series electric circuit (see Figure 13.2). The availability of each of these components will be a factor in the overall link availability as they are all mutually dependent on the functioning of the link.

Example 1 — availability of a component (ac)

$$ac_1 = \frac{\text{MTBF}}{\text{MTBF} + \text{MTTR}}$$

where: MTBF = 200 days
        MTTR = 2 days

then: $ac_1 = \dfrac{200}{202} = 0.99009$

A component will have an availability of 0.99 or be usable for 99 percent of the time.

Example 2 — availability of a link

If this availability is applied to all components (1-5 that form the complete transmission path (see Figure 13.2) then the availability for the link will be:

$$
\begin{aligned}
\text{Availability} \quad (A) &= ac_1 \times ac_2 \times ac_3 \times ac_4 \times ac_5 \\
A &= 0.99 \times 0.99 \times 0.99 \times 0.99 \times 0.99 \\
A &= 0.9510 \text{ or } 95 \text{ percent}
\end{aligned}
$$

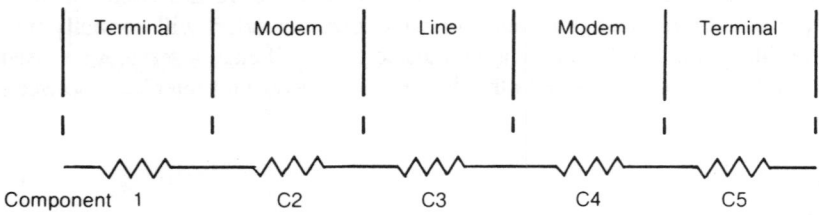

Figure 13.2   Schematic (component) representation of a data transmission link

In a year of 365 days there is a probability that the circuit will be unavailable for 5 percent of the time, ie 18.5 days during the year.

Example 3 — availability of two components in parallel

If we chose to duplicate a component we would improve the overall availability, because if one component fails its duplicate (or standby) will be there to replace it.

$$
A = 1 - \left[\ (1 - ac_1) \times (1 - ac_2) \ \right]
$$

where: A = availability of the similar 2 components in parallel.

  c_1 = primary component

  c_2 = spare or standby component

The composite availability of two components, each with an individual availability of 0.99 or 99 percent will be:

$$
\begin{aligned}
A &= 1 - \left[\ (1 - ac_1) \times (1 - ac_2) \ \right] \\
A &= 1 - \left[\ (1 - 0.99) \times (1 - 0.99) \ \right] \\
A &= 1 - (0.01 \times 0.01) \\
A &= 1 - 0.0001 \\
A &= 0.999 \text{ or } 99.99 \text{ percent}
\end{aligned}
$$

The same methodology can be applied to the duplication of a whole circuit, where each circuit has an availability of 95 percent as shown in Figure 13.3.

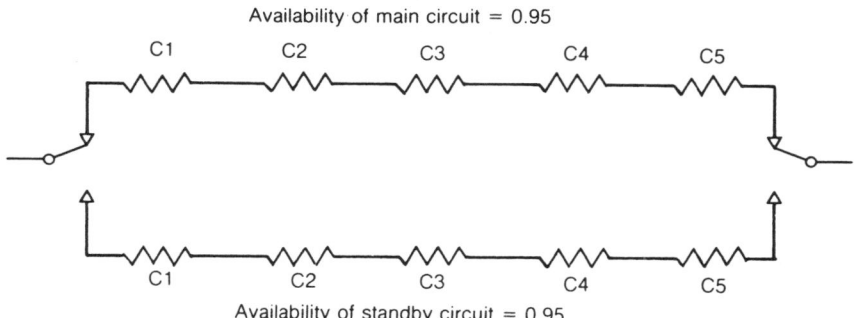

Figure 13.3 Availability of parallel circuits

Example 4 — availability of parallel circuits

$$A \quad = \quad 1 - [\quad (1{-}a \text{ (main circuit)} \quad x \quad (1{-}a \text{ (standby circuit)} \quad]$$

Based on the availability of each circuit (main and standby) of 9.95 or 95 percent then:

$$A \quad = \quad 1 - [\quad (0.05) \quad x \quad (0.05) \quad]$$

$$A \quad = \quad 1 - 0.0025$$

$$A \quad = \quad 0.997 \text{ or } 99.75 \text{ percent}$$

A further improvement in the overall availability can be achieved by duplicating each component and providing full flexibility between them as shown in Figure 13.4.

Example 5 — availability with full flexibility.

In example 3 we calculated the availability of a component when that component is provided with a standby. This was shown to be 0.9999. Working on the same figure and with full flexibility we have:

$$A \quad = \quad ac_1 \quad x \quad ac_2 \quad x \quad ac_3 \quad x \quad ac_4 \quad x \quad ac_5$$

where: ac_1 to ac_5 includes a spare component

$$A \quad = \quad 0.9999 \times 0.9999 \times 0.9999 \times 0.9999$$

$$A \quad = \quad 0.9995 \text{ or } 99.95 \text{ percent}$$

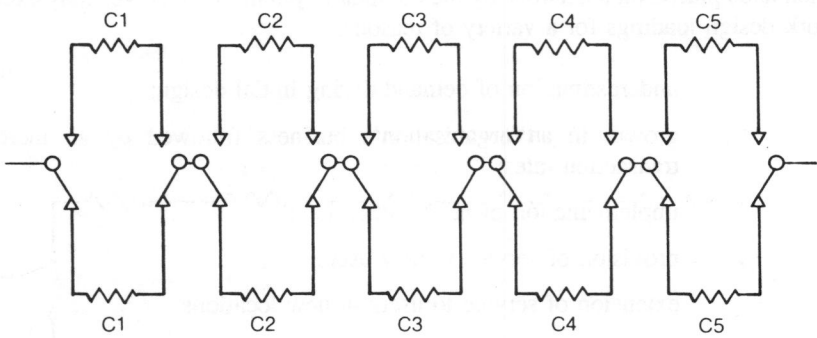

Figure 13.4 Availability when full flexibility between all components is provided

It is common in both simple and complex networks to provide the availability users require by configuring redundant components selectively. Redundancy may be brought into operation automatically or manually, depending on the elegance of the system employed and the network management requirement. Loss of availability, as seen by a user, can be reduced to a negligible amount if required. The expression for user availability can be changed to:

$$A = \frac{MTBF}{MTBF + MTTD + MTTIR + MTTRE}$$

where: MTBF is mean time between failures;

MTTD is mean time between failures to diagnose;

MTTIR is mean time between failures to invoke redundancy;

MTTRE is mean time between failures to restore the user back on to the repaired component.

The way to provide cost-effective redundancy is not always by duplicating leased lines and network switches. Dial-up standby can provide an effective and cheap alternative to the duplicate private circuit, but the availability for the switched telephone network is far more difficult to assess as the average fault rate per annum for each telephone is low. Many faults, particularly dialling failures, are not reported to the UK PTT.

This approach depends upon the suppliers offering MTBF and MTTR figures and their confidence in those figures (for example, a well known modem supplier quotes MTBF as $2\frac{1}{2}$ years of 24 hours/day operation and states that in practice it could be exceeded by 25 percent). In the absence of suppliers' MTBF and MTTR figures, users may sometimes obtain valuable experience of the equipment from other users.

Expandability

The demands placed on a network by the computer system that it serves may exceed the network design loadings for a variety of reasons.

- underestimation of demand during initial design;

- growth in an organisation's business followed by an increase in transaction rates;

- implementation of new applications;

- provision of service to new users;

- extension of service to users in new locations.

It is therefore important to ensure that the system has sufficient spare capacity to cope with anticipated expansion of workload, and in both the physical and functional sense, has the potential for easy expansion of capacity.

Anticipating future business needs, even over a five year period, is by no means easy, particularly in a world which is expecting such rapid rates of technological and socio-economic change. But, at the very least, the design of the system should take into account the corporate long term plans, so far as these are known or can be prudently revealed.

Shortcomings in the performance of a system and its components may be more difficult to rectify. When evaluating the network requirements it is important to consider the following:

- The risk of loss of service if relatively small amounts of network growth are necessary. Service loss, especially in the early days of a network's life, can lead to a massive loss of user confidence which could prove to be more of a problem than a reasonable amount of over-configuration and disruption.

- The change and cost involved in the network have to be upgraded by an order of magnitude, say, doubling the throughput or number of end users. This scenario can be used as a means to compare different supplier solutions.

Network Management

Communications networks are of fundamental importance to businesses, so successful management is critical. It is important that the network is designed to meet users' service level requirements in terms of:

- availability;

- response time;

- reliability;

- throughput.

It is also important that the level of service can be measured, otherwise network management is not possible. Network management is so important that redundant systems are often configured.

Availability

Actual availability is measured and compared against the expected availability calculated using the mean time figures, discussed earlier. This requires:

— a fault management system which records the occurrence of faults including when faults were reported by users, when they were reported to maintainers, when service was restored;

— a performance analysis system to analyse the fault data and provide figures and graphs illustrating required versus actual availability. It is more useful to have such information over a given period so that the development of trends can be seen.

As well as providing information on the level of service to users, the system can calculate the service level of suppliers' systems and equipment. Again, this can be checked against the level agreed with the supplier. The systems to provide these functions do not have to be complicated. Some very good systems have been developed on PCs using industry standard databases and spreadsheets.

Diagnosis of faults is an important factor in availability. The time taken to diagnose can be kept low by:

— skilled operators;

— good systems and tools;

— good procedures.

Good systems and tools are necessary even for skilled operators. They need to monitor the interfaces (for example, V.24, data link, network level) in the network and test components and links. Monitoring can require bringing control signals and data from remote interfaces back to a central site. Testing can mean invoking a central site from link or loop. A test message is then sent around the loop and back to the test point where the number of bits in error are counted, and the result is sent to the central site. These functions can be provided in the network equipment, for example in packet-switches, multiplexers, modems and LANs and controlled from a central management system. Where this is not adequate third party management equipment can be supplied. Third party equipment adds to design problems. It has to be configured in the interfaces it is monitoring and connected to an operator's terminal, over the network or via dial-up links. The problems should be outweighed by the benefits a diagnostic system brings.

It is also important to have good procedures. These are not to describe how an operator fault finds, but to define what his responsibilities are and the extent of his authority to say, change configurations.

Response Time

Response time can be measured by an intelligent terminal starting a timer when a message is ready for sending and stopping it when the response is received. The problem of gathering this data from such terminals must also be overcome. If they cannot provide this function a protocol monitor can be used, but of course, it can only be configured into one interface at a time. Network management is the transit delay of messages across the network and also knowing how long a message has to queue to enter the network.

Reliability

This information can be collected by terminal/host equipment and by network equipment, for example, the packet-switches. It will show such data as the number of retransmissions and the number of messages undelivered.

Throughput

The network components have certain capacity thresholds above which response times will deteriorate rapidly. Throughput of each application should be measured against expected throughput. This can be done on the application mainframes and terminals. The actual utilisation of the network components should be measured and compared with the capacity thresholds. Some network suppliers' management systems will do this in real-time and provide alarms as thresholds are approached. Others will provide off line statistics from volume and transaction counts. Ideally both types of measurement are required.

Support

Support of network equipment and systems is usually done by the supplier. The level of support service required is specified using the same parameters used to specify the service required by the communication system users. The level of support required in hours and days covered and time to respond to call out will depend on the level of service demanded by users and the configuration of redundant components.

Most of the equipment available today will also require software support (for example, LANs, packet-switches, multiplexers and modems). Essentially, it is no different from the software support required by mainframes and terminals. Network software is supplied subject to having new releases made which the customer may have to take, whether he requires the new feature or not, to ensure continued support. Customers should ensure, from the outset, that they are given:

- release information including how to migrate;

- regular updates on bugs on current releases;

- details of testing carried out on releases;

- lists of modules and revision numbers used in a customer's release. There should be a mechanism for ensuring that these are the latest, tested modules, and known deficiencies are documented.

This level of documentation will help diagnosis if a fault occurs.

OSI Network Management Standards

OSI network management standards are largely based on:

- Part 4 (OSI Reference Model) — OSI Management — ISO 7498-4;

- Common Management Information Service — ISO 9595;

- Common Management Information Protocol — ISO 9596;

- System Management Overview — ISO 10040.

A framework, under which the model can be broken down, is provided by defining a number of functional areas. These are: faults, configuration, accounting, performance and security management (FCAPS). There is another model which compliments FCAPS called carrier, node and product (CNP) domain model, which splits the network in a hierarchical way. The functions of the domains are:

- carrier which moves the information about the network;

- node which is the logical domain that can be reached by a host and extends to every system attached to the network;

- product which covers a set of services provided by the objects attached to the network.

The FCAPS functions are carried out in different ways in the different CNP domains. These include the use of testing tools and procedures; network functions can be segmented by combining the FCAPS and CNP models to provide systems that are well suited to users.

Other Network Management Architectures

The other prominent open architecture is simple network management protocol (SNMP), which has become the main *de facto* standard protocol for multi-vendor network management.

SNMP originated from the internet family of protocols. These came from the US Department of Defence who produced them for reasons of system interoperability and procurement. The best known of these protocols is transmission control protocol and internet protocol (TCP/IP), which provide functions similar to the transport and network layer protocols in the OSI model. As its name suggests, SNMP is designed for simple implementation to encourage its use. It is a diagram protocol which is designed to be independent of the network architecture it runs over. Unlike common management information protocol (CMIP), SNMP is designed with the idea that most of the management work will be done by a central manager, with remote devices, rather like managed modems functioning in existing proprietary systems. The criticisms levelled at SNMP are that there is no security management or efficient bulk data transfer. Hand in hand with SNMP is the management information base (MIB), which is a virtual data architecture populated with information about network devices. It can be thought of as a tree structure in which some of the sub-trees contain standard information such as on the operation of protocols like TCP/IP. Other sub-trees are available for vendors to add-in their own data trees. Network management protocols can then retrieve or change data items held in the MIB. Over 100 vendors have built up private MIB trees.

Open network management systems are initially being developed where a particular supplier makes a number of agreements with other suppliers to exchange network management information using OSI interfaces. The interface specifications are made public. Examples are:

- DEC's enterprise management architecture (EMA);

- AT&T's unified network management architecture on which their accumaster integrator product is based;

- British Telecom's Concert.

Large numbers of suppliers have agreed to exchange information in this way. Although this sounds very promising, users will have to take those precautions that must be taken whenever procuring products that are meant to interface to other systems. They must ensure that the products have the level of conformance required and the versions of the standards adhered to are the same.

Proprietary Network Management Architectures

These fall broadly into two camps, systems from host equipment manufacturers and systems from network communications manufacturers. In the first category would be manufacturers like IBM and DEC. Under IBM SNA there are a number of useful products such as network performance analysis (NPA), network performance monitor (NPM) and netview which can provide information on:

- interface errors;
- numbers of message retransmissions;
- volumes of data on links;
- response times.

The real domain of these products is the SNA devices; hosts, FEPs, cluster controllers and terminals. Devices in the network such as modems, multiplexers and packet-switches, would not be managed in this way. IBM have specified interfaces for connecting third party network management systems. Through these, alarms can be sent to netview and netview can interrogate the third party systems for more information. This is about as far as these arrangements go; configuration changes for instance, being outside their scope.

The second camp is that of the network system supplier. The major functions of these systems are to provide control of the communication suppliers manufacturers' equipment, and to provide non-interference monitoring of interfaces attached to the equipment. Non-interference means not causing any disruption to the user data passing through the network. A number of methods, for communicating between the central control system and network devices, are used to achieve this by different suppliers.

Modem manufacturers

Because of the limited bandwidth on analogue lines modem manufacturers have used secondary channels. These are generally on the lower frequencies from which can be derived 75 or 150 bits (see Figure 13.5).

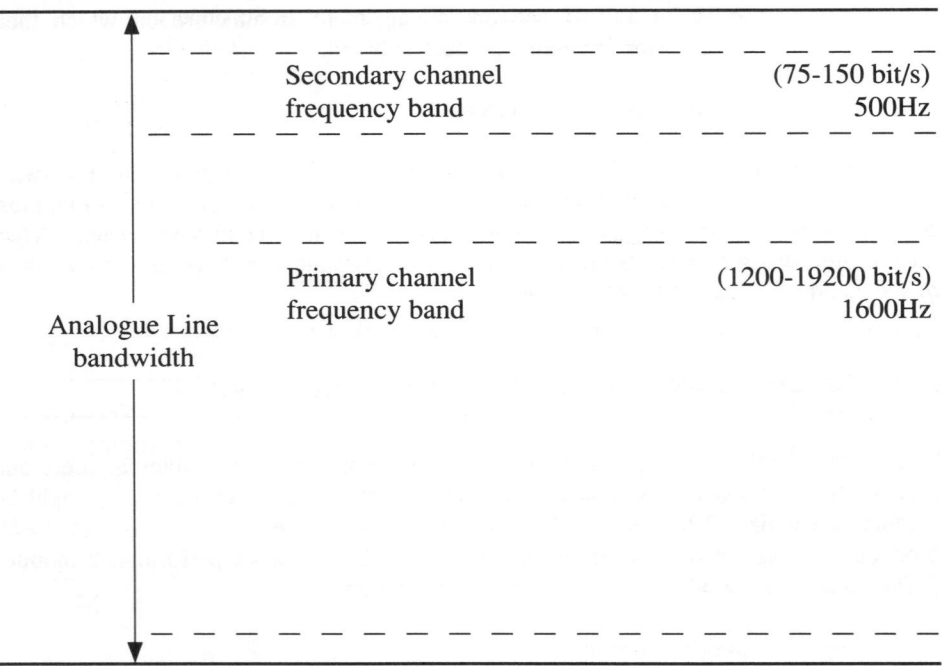

Figure 13.5 Primary and secondary channels

In the example networks (ie where there are remote FEPs or concentrator devices) the secondary channel has to bypass these devices (see Figure 13.6). The modems are polled from a central management system. The secondary channel on the modem is assessed by a separate V.24 interface from the main channel.

Another method is for the remote modem to dial-up the central network management system when it has an interface alarm and the central site to dial-up the remote modem when it wants to issue a command (see Figure 13.7).

An example of a modem manufacturer's secondary channel system would be Codex's DNCS.

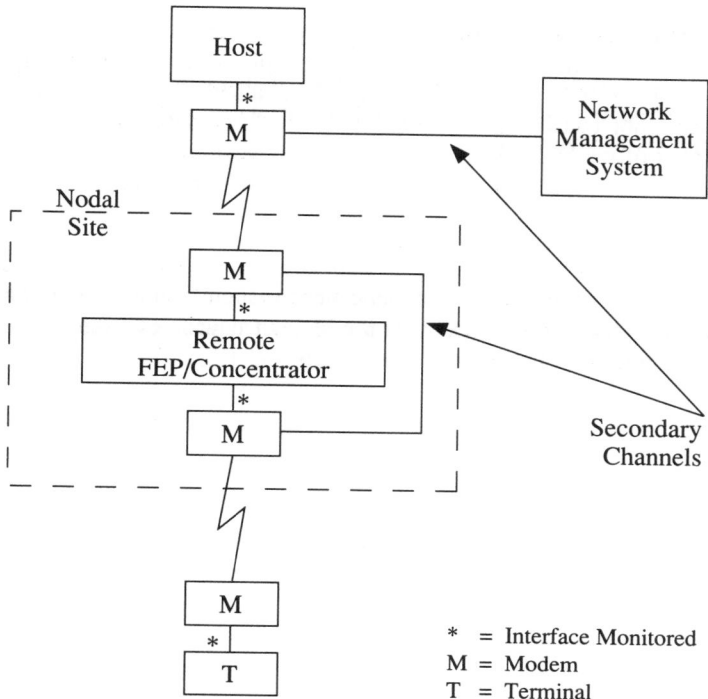

Figure 13.6 Using a secondary channel for network management

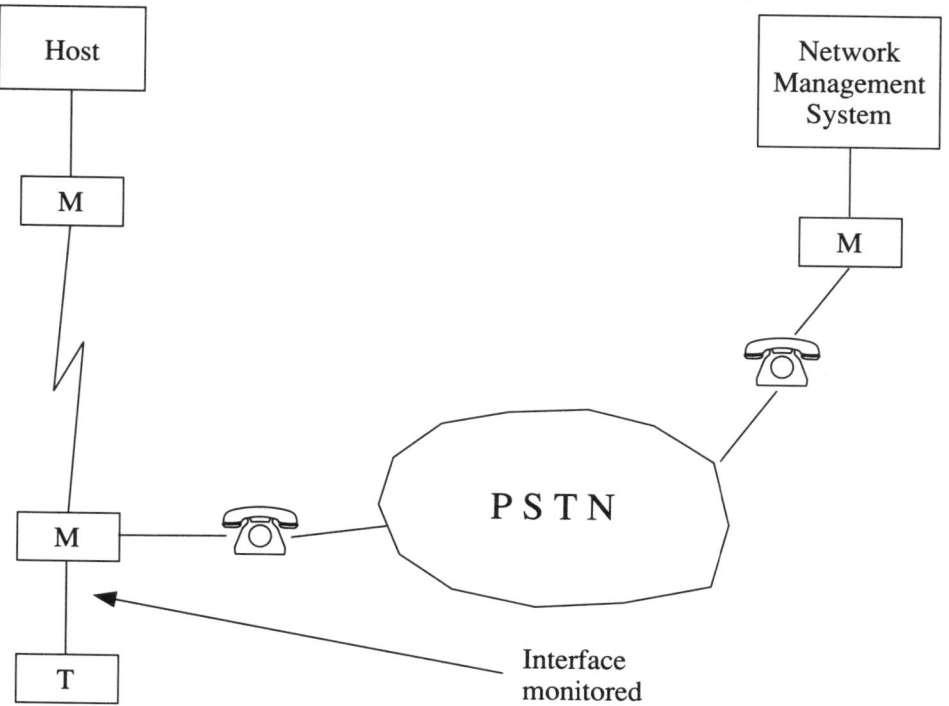

Figure 13.7 Using dial-up to make the network management connection

Multiplexer manufacturers

These systems have more freedom to insert network management data by utilising the multiplexing technique to share the link with the user data (see Figure 13.8). An example would be Timeplex's timeview (TV).

Packet-Switch Suppliers

These systems generate extra packets for the network management data. The packets are generated at switches with the network management system's address. Commands are generated by the network management system with switch addresses (see Figure 13.9), for example, Telematic's interactive network facilities (INF).

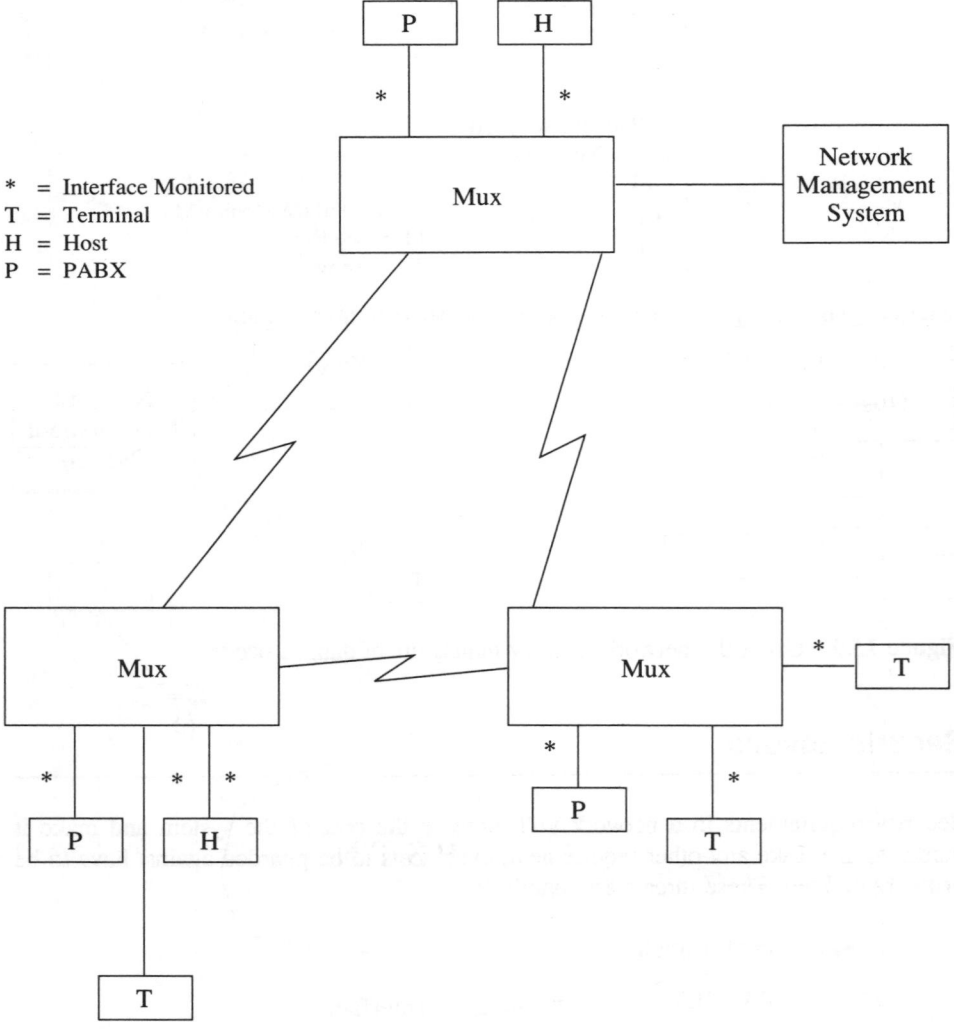

```
*  =  Interface Monitored
T  =  Terminal
H  =  Host
P  =  PABX
```

Figure 13.8   Using the multiplexer bandwidth for network management data

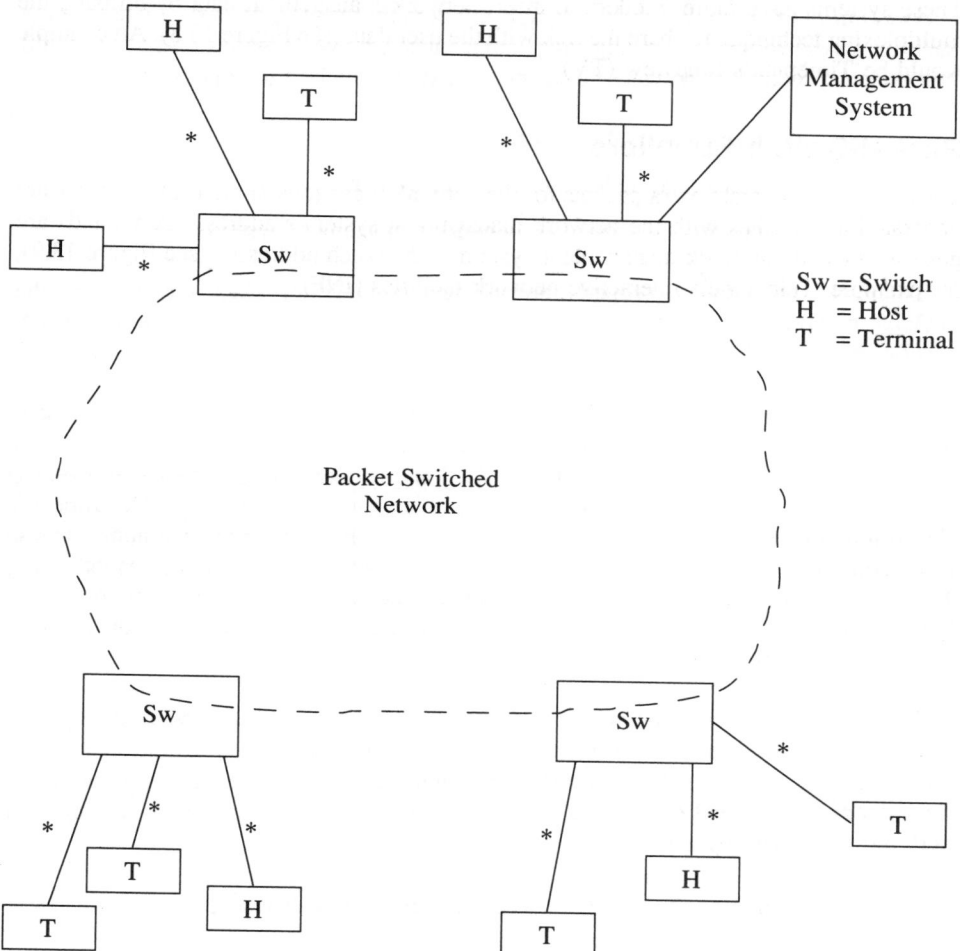

Figure 13.9 Using the network to carry management data in packets

Security Issues

Security requirements in a network will increase the cost of the system and make it harder to use. Like any other requirement, the threats to be guarded against have to be properly defined. These threats are mainly to:

— confidentiality;

— integrity;

— availability.

Here, dangers being guarded against include: fire, flood, sabotage, loss of confidentiality and fraud.

Availability has been discussed at length earlier in the chapter. After taking precautions such as access restriction, configuring uninterrupted power supplies and very reliable components, redundant network components may be required to cope with:

- loss of normal components and lines;

- connecting redundant end user systems;

- connecting redundant network management systems.

To invoke redundancy, properly documented contingency plans are required that include detailed written procedures describing actions to be taken and defining responsibilities and authority.

Integrity is largely concerned with making sure that what is received is what was sent. This threat is dealt with by the error detection and retransmission functions built into protocols operating at the link level (eg HDLC) and end-system to end-system level (eg ISO transport layer, class 4). Integrity goes hand in hand with authentication to ensure the validity of the source. In the OSI model security architecture *non-repudiation* is also discussed. This service is really the concern of end-systems to prevent the sender saying he did not send the data and the receiver denying he received data. The network has a part to play in authentication and non-repudiation by ensuring that access to the network is adequately controlled.

Confidentiality of information is at risk where it can be observed passing through the network or where the network routes it to the wrong destination. There is inherent confidentiality when the data is turned into complex transmission signals. In addition, there needs to be access restrictions to areas where data can be seen and to routing files and, ultimately encryption.

A key to minimising the threats is access control; this will restrict:

- access to network management functions;

- access to network equipment;

- user access;

- access to end-user systems;

- access to third party networks;

- locks and keys;

- passwords user IDs;

- closed user groups;

- access security systems;

- encryption.

Management of keys will be necessary for encryption and other mechanisms. Additionally, changes to the network must be properly authorised, documented and managed.

Security Framework for OSI Systems

This is described in the Open Systems Interconnection (OSI) security architecture document ISO 7498.

It is concerned about security at all layers of the seven layer model, not just those related to the network layer. Security services are categorised and mechanisms to provide them are listed (see Figure 13.10). The layers where the services should be provided are also listed (see Figure 13.11).

Service \ Mechanism	Encipherment	Digital Signature	Access Control	Data Integrity	Authentication Exchange	Traffic Padding	Routing Control	Notarisation
Peer Entity Authentication	Y	Y	*	*	Y	*	*	*
Data Origin Authentication	Y	Y	*	*	*	*	*	*
Access Control Service	*	*	Y	*	*	*	*	*
Connection Confidentiality	Y	*	*	*	*	*	Y	*
Connectionless Confidentiality	Y	*	*	*	*	*	Y	*
Selective Field Confidentiality	Y	*	*	*	*	*	*	*
Traffic Flow Confidentiality	Y	*	*	*	*	Y	Y	*
Connection Integrity with Recovery	Y	*	*	Y	*	*	*	*
Connection Integrity without Recovery	Y	*	*	Y	*	*	*	*
Selective Field Connection Integrity	Y	*	*	Y	*	*	*	*
Connectionless Integrity	Y	Y	*	Y	*	*	*	*
Selective Field Connectionless Integrity	Y	Y	*	Y	*	*	*	*
Non-repudiation, Origin	*	Y	*	Y	*	*	*	Y
Non-repudiation, Delivery	*	Y	*	Y	*	*	*	Y

Figure 13.10   Security services and mechanisms

Service	1	2	3	4	5	6	7°
Peer Entity Authentication	*	*	Y	Y	*	*	Y
Data Origin Authentication	*	*	Y	Y	*	*	Y
Access Control Service	*	*	Y	Y	*	*	Y
Connection Confidentiality	Y	Y	Y	Y	*	Y	Y
Connectionless Confidentiality	*	Y	Y	Y	*	Y	Y
Selective Field Confidentiality	*	*	*	*	*	Y	Y
Traffic Flow Confidentiality	Y	*	Y	*	*	*	Y
Connection Integrity with Recovery	*	*	*	Y	*	*	Y
Connection Integrity without Recovery	*	*	Y	Y	*	*	Y
Selective Field Connection Integrity	*	*	*	*	*	*	Y
Connectionless Integrity	*	*	Y	Y	*	*	Y
Selective Field Connectionless Integrity	*	*	*	*	*	*	Y
Non-repudiation, Origination	*	*	*	*	*	*	Y
Non-repudiation, Delivery	*	*	*	*	*	*	Y

Legend Y   Yes: service should be incorporated in the standards for the layer as a provider option.

o   It should be noted, with respect to layer 7, that the application process may, itself, provide security services.

*   Not provided.

Notes:

1 Entries are not of equal weight or importance; on the contrary there is a considerable gradation of scale within the entries.

2 Neither of the convergence roles of the network layer provide security services.

3 The position of the security services within the network layer does significantly affect the nature and scope of the protection services that will be provided.

4 The presentation layer contains a number of security *facilities* which support the provision of security services by the application layer.

Figure 13.11 OSI levels' security services

Conformance and Interoperability Testing

Making sure that new and upgraded systems provide the required conformance can be time-consuming and costly. Using standards can simplify the job of procurement and implementation. A number of organisations have produced specifications to help in this task. How this can be undertaken is illustrated by the development of the best known of these specifications, GOSIP.

The UK government OSI Profile is a procurement specification covering OSI based data communications for UK government civil administration. Its stated aims are:

- to facilitate procurement and acceptance testing of communications based products;

- to ensure that different and separately procured departmental systems can interwork to an assured level of functionality;

- to provide a clear specification to manufacturers on which to base strategic product development.

GOSIP was guided in its preparation by the Central Computer and Telecommunications Agency (CCTA), the IT advisory body to the government. At a simple level GOSIP has defined a subset of the available standards to suit the user environment, that is the UK government IT services. What has also been taken into account is the present and likely availability of OSI compliant products, so that it will be of value to users in the short term. The UK government is not alone in producing this type of OSI profile; many governments and other large IT users have done the same. In addition, there is also European Community legislation, directive 87/95, which requires OSI conformance in all public sector IT tenders over 100,00 ECUs. Some of the other standards and profiles with which GOSIP will attempt to be compatible, assuming UK government requirements are met, are:

- other government profiles, for example US GOSIP;

- British Telecom's open network architecture (ONA);

- General Motors' manufacturing automation protocol (MAP);

- Technical and office protocol (TOP) from Boeing;

- US National Bureau of Standards (NBS) implementation agreements;

- US Corporation for Open Systems which are intended for the development of testing services.

The GOSIP profile defines two levels of testing:

- conformance testing to determine that a product will meet the requirement in the standard to which it is claimed to conform;

— interoperability testing to determine that the product will interwork with other systems as intended.

For most sub-profiles (ie standards) GOSIP tabulates the selections and constraints required for conformance. In the actual procurement situation, additional user requirements, beyond the minimal conformance requirements, can be specified. This should fulfil the objectives of:

— providing a basis for contractual acceptance;

— providing a test specification;

— identifying the level of interworking that should be achieved with other systems.

GOSIP assumes that conformance testing is part of a product's normal development cycle rather than part of the procurement process. As this type of testing should be undertaken by a developer anyway, the additional requirement should be simply for properly documented testing details and results. In addition, there is a requirement to submit product details to a testing service, assuming a suitable one is available. The availability of suitable testing services will be monitored by CCTA and users and suppliers kept informed about them. CCTA will also encourage the development of internationally recognised *conformance test suites*. The advantages to the user, in terms of time and cost, are clear. There are similar advantages to suppliers as they should not not have to undertake conformance tests each time the product is supplied to a different user.

Interoperability testing is different in that covering all combinations of product interconnections is impractical. This aspect of testing will still have to be undertaken during acceptance testing.

14

Fibre optic communications

With the ever-increasing need to transmit large amounts of data in short periods of time, modern technology has become stretched to its limit due to the increase in transmission speeds needed to service this requirement. Communications networks embracing both voice and data had previously depended upon copper wire to act as the transmission medium. However, due to various factors, such as attenuation, noise and interference copper wire is not able to cope with very high speeds.

Fibre optics have been in use since the 1970s. They enable reliable information transfer at much higher rates than conventional copper cabling, thus allowing more traffic to be carried along one fibre than is possible using a copper cable. Communications companies have been quick to realise the benefits of such a medium; indeed, optical fibres are now installed between all of the major British Telecom trunk exchanges in the United Kingdom. The use of fibre optics in computer communications networks has also become apparent. Several vendors are currently implementing fibre-based networks, with improved versions imminent.

A Brief History of Fibre Optics

By the mid-1960s, coaxial cable was still used extensively to carry high volume data transfers. However, as the data rate increased so did the attenuation of the cable due to the high frequency components present. As a result, coaxial cable was only useful for carrying high data rate traffic over a few kilometres. Initial research into using optical fibres proved frustrating as impurities in the glass also attenuated the light signal travelling down them. However, after intensive research, AT&T's Bell Laboratories and Korning produced a solution and patented a process for manufacturing pure glass for use in trunk telecommunications. From that point, the use of fibre as a viable communications medium was assured.

Overview of the Fibre Optic Medium

An optical fibre consists of two coaxial glass materials with differing refractive indices (the outer coating, or cladding, having the lower index). Light introduced into one end of such a combination will be internally refracted off the outer cladding and will propagate down the fibre. The light source which is used is usually a solid-state device such as a light emitting diode (LED) or a laser. The advantages of using such a transmission medium are several, and some of these are listed:

411

- increased bandwidth and hence, higher possible transmission capacity;
- lower signal attenuation over larger distances;
- immunity to electrical interference;
- it is almost impossible to 'eavesdrop' and hence, is secure;
- small physical size and is lightweight.

As with other cabling systems, the choice of optical fibre is not as straightforward as it may initially seem. The user is presented with a choice of fibre diameter (both for the core and the cladding), light-source wavelength, transmission mode and connector style. Commonly used core diameters are 50, 62.5 and 85 microns (although IBM also use 100 microns). These are generally surrounded by a cladding of 125 microns in diameter. A fibre is referred to using these two dimensions, for example, a 62.5/125 fibre. Each one of these types has different characteristics which are suited to certain applications. Table 14.1 compares the four types detailed above.

Table 14.1 Comparison of optical fibres

Core size (μm)	Attenuation (dB/km)	Bandwidth (MHz.km)
50	3	600
62.5	4	160
85	5	200
100	6	100

The attenuation factor is important, since this determines when a signal repeater must be introduced into the cable to ensure the integrity of the data. On a long link (such as under the Atlantic Ocean) this becomes a significant cost consideration. Using Table 14.1, it may be seen that the 50 micron fibre offers the lowest level of attenuation and also the highest available data rates. However, it must be remembered that light must be first introduced into the core of the fibre and that the smaller the core diameter the more difficult this is.

The light sources generally emit power which is outside the visible spectrum of the human eye; above 800 nm. Several wavelengths exist for emitters, the most common being 850 nm and 1300 nm (1500 nm is still under development). The figures shown in Table 14.1 are for a system using emissions at a wavelength of 850 nm. For a 62.5 micron cable, increasing the wavelength to 1300 nm has two advantages. First, the attenuation drops to only 2 dB/km. Secondly, the bandwidth increases to 500 MHz/km. However, 850 nm technology is well developed and, as such, is much cheaper to implement than 1300 nm. Therefore, the final choice depends on the individual user's needs and budget.

Two phases associated with fibre optic systems are *single mode* and *multimode*. Basically, with single mode (also referred to as monomode) systems, only one coherent light colour is employed at the source, whereas in multimode systems, several colours may exist. Without presenting a full mathematical breakdown of the theory, multimode systems are inferior in performance to single mode systems due to the mixture of colours. The different colours propagate at different speeds through the cable, with the result being that the received signal may arrive blurred once a certain data rate is reached. As a result, the data rate is limited.

Uses of Fibre Optics in Communications

The first area in which fibre optics technology was employed on a large scale was in the telephone trunk networks. Single mode cables are generally allocated two standard bandwidths, supporting 140 Mbits and 565 Mbits. This upper limit, however, is not the maximum rate at which such systems can operate, and the future will almost certainly see higher rates than this. A system running at 565 Mbits can carry the equivalent of 4000 simultaneous telephone conversations. The benefits of replacing standard coaxial and twisted pair systems with such a lightweight medium can be seen immediately.

The use of fibre to replace present telephone links has developed substantially with the advent of cable TV systems. The ability to transmit both TV and telephone signals down the same cable makes good economic sense for both the current carriers and the emergent cable TV companies. Several companies are working on passive optical networks (PONs) which enable fibre to be taken to individual line subscribers (or to units mounted in the street) at minimum cost. The possibilities of such a service are wide-ranging with systems including two-way interactive high density TV (HDTV) and videoconferencing. This leads to the concept of 'virtual mobility' whereby people are able to involve themselves in a wide range of external activities without leaving their homes.

Fibre Distributed Digital Interface (FDDI)

As increasing amounts of data was transferred across computer networks and the demand for faster access to such data became greater, fibre optics soon found a useful role in this area also. The most widely known of the optical networks is FDDI. The basic structure of the network comprises two fibre rings, with data travelling in opposite directions on each, to which the network nodes may be connected. The maximum data rate which is supported by each of the rings is 100 Mbits, therefore giving a maximum system rate of 200 Mbits.

Two classes of nodes are defined in the standard, Class A and Class B. Class A stations are connected to both primary and secondary rings, whilst the cheaper Class B stations only connect onto one of the rings. In the event of one of the rings failing, Class A stations can act in a supportive fashion by detecting the break and routing data between the two rings; in this instance, the system data rate falls to a maximum of 100 Mbits.

In order to achieve such throughput, several constraints must be met:

—   a maximum of 2 km between network nodes;

—   a total cable length of 100 km;

—   less than 500 stations attached to the ring;

—   a light source which operates at 1300 nm.

The overall network protocol is based on a token ring; only a station in possession of a free token may begin transmitting data. This system prevents erroneous frames being received due to collisions with frames from other stations. As a result, however, if one of the network nodes is switched off, the ring collapses as the free token will not be able to pass through that node.

Whilst FDDI is not yet implemented on a wide scale (since its standards are still in the process of being finalised), users may wish to implement a partial optical Ethernet system with a view to upgrading at a later date. It is usually more economical to install more cable than is at first necessary than to reinstall at a later date. The current versions of FDDI are designed to handle high-speed data traffic. A future version (FDDI-II) will allow mixed traffic to be transferred. Such traffic will embrace digital speech, data and vision. This mixture of traffic relies on a form of time-division multiplexing for its successful operation.

Presently, an FDDI network as described above would still transfer data from the network node to the user terminal via some form of copper; either coaxial cable or twisted pair. There is a growing need for the optical route to be taken all the way to the user and bypassing the copper route altogether, a process known as 'fibre to the desk', but due to the cost of such an approach and the general lack of such a high bandwidth requirement, it is felt that fibre to the desk will not be implemented on a large scale for at least four years. However, many organisations are taking the intermediate step of bringing fibre to the desk as part of their structured wiring policy, but leaving it unterminated — the so called 'dark fibre'. This approach makes economic sense since the other alternative is to install all the necessary ducting and have the cable 'blown' into place as required. The patent for the process of cable blowing is currently held by British Telecom and there are some worries about the cost of such a service in a potential monopoly situation, but since British Telecom have also licensed the technology to other companies, such as Sumitomo Electronics, this worry should diminish.

FDDI is not the only fibre-based network in existence. Some manufacturers are looking past the 100 Mbits barrier offered by FDDI and are going into next-generation products. *Fibrenet* is one such offering that supports a ring-based network capable of a data rate of 565 Mbits and is aimed at corporate customers and carriers in the UK and Ireland for splitting into smaller bandwidth allocations.

Summary

Initially, the benefits of utilising fibre optics were realised by the telephone carriers and this technology was soon implemented in the overhaul of the major trunk networks. This improvement of the telephone system is still in progress and it has been noted that future advances include the bringing of fibre to all line subscribers. There will be many benefits of such an approach, with the inclusion of cable TV, HDTV and data services being available to the subscriber.

Fibre optics is also playing a major role in the development of high speed computer networks. FDDI is an example of recent advances in such networks and its speed of 100 Mbits is already under threat from other systems in existence and under development. Without doubt, fibre has only just started its useful life cycle and advances in its use will be seen for several decades to come.

Fast Packet-Switching

Introduction

In general, today's telecommunication requirements are largely fulfilled by technologies which have been established, in one form or another, for a considerable time. For example, although digital voice transmission is often considered to be state of the art, this technology was well understood by the late 1930s. It is only the fact that the high speed devices required to switch digital circuits are now readily available which has allowed widespread implementation of the technology, improving both the quality and reliability of modern networks.

Packet-switching was first proposed in the early 1960s as a means of reliably delivering data across the then generally unreliable analogue networks. The concept of packet-switching is that a packet of data is delivered to the network with the destination address, source address and control data embedded in the packet. Between each switching node within the network, the address is re-examined to determine where to send the packet and techniques are employed to make sure that the packet is still intact. The route that each packet takes is then determined at each switching node, to take account of link failures and traffic congestion. The user of such a network does not need to know anything about this routing and checking, since the network guarantees delivery of the packet intact. This network is normally represented as a cloud. With this in mind, it is evident that the old approach of packet-switching is not necessary on a digital network; the links have become more reliable and speeds have risen to the extent that services such as ISDN can now deliver multiple 64 kbits links to the customers' premises.

Fast Packet-Switching — A New Concept

The concept of 'fast packet-switching' arose from the inherent reliability of a digital carrier network, and possesses a similar cloud status to the old packet-switching networks, ie the user does not need to know how packets are carried within the network, only how to connect to the cloud. Fast packet-switching works on the basis that within the cloud, the carriage of packets to the destination is almost certain to be error free. Hence, within the cloud, no re-routing or retransmission of packets is expected to take place. For this reason, the packet does not need to carry with it the overhead of addressing information. Similarly, the switching nodes do not need to store packets in transit as no re-transmission is expected. The route a packet takes within the fast packet cloud should always be the same for any one transaction, as the links within the network are not expected to fail.

Therefore, fast packet-switching uses the absolute minimum of control data in each packet, this being confined to a small header, usually limited to three bytes per packet. As each fast packet connection is established, the route is locked into the network to prevent searching for a connection and to ensure that packets are delivered in the order that they were sent. This eliminates the network overhead of re-routing and buffering of packets at each switching node, whilst removing the need to have packet sequence numbers assigned by the network. As no destination address is carried in the packet, the route through the network is set up at call initiation time and maintained for the duration of the session. The packet carries a virtual call indicator (VCI), which is used to identify to the network the source and destination of the call, but the interface must maintain a translation table to interpret these VCIs. Once a call is established, the network only has to check the network header for the VCI at the entry point to know where to send the packet.

As with any system, there are some uncertainties involved, and although the network may be reliable, there may still be instances when a packet is corrupted or does not arrive at its destination. However, this should be very rare and so the responsibility is placed with the end users for detection and correction of such a situation.

Re-routing between switching nodes in the event of a failure is simply accomplished by network control packets which are allocated all of the available bandwidth to establish a new route through the network. Because fast packet networks can effectively allocate all the network to control features in such an emergency, the time needed to establish an alternative route is 10 to 20 times less than re-routing over conventional time division multiplexer (TDM) systems. Typical times of less than two seconds result in no loss of voice or data connections and no time-out of sessions.

Several technologies have been proposed to implement fast packet-switching, *frame relay* and *cell relay* being two of the most common. Both of these technologies tackle the concept of fast packet-switching in slightly different ways. Frame relay is more akin to X.25, with software upgrades of existing switches often providing frame relay features. Frame relay uses variable length frames and is generally limited to speeds below 2 Mbits. Cell relay on the other hand, uses fixed length frames or cells and is aimed at much higher speeds than can be met with traditional software switching techniques. It is invariably implemented in hardware, with simple switch blocks built up into matrices to handle switching at 100 Mbits and over.

Figure 14.1 shows the relationship between the concept of fast packet-switching and the common technologies which implement it in frame relay and cell relay, together with the emerging standards that define the interfaces. CCITT I.122 is the Recommendation for link access protocol D (LAP-D) for multiplexing frames from different sources. CCITT Q.931 is the signalling specification for the frame relay interface. Asynchronous transfer mode (ATM) is the UK preferred implementation of cell relay technology, and IEEE 802.6 and distributed queue dual bus (DQDB) are the proposed standards for implementing metropolitan area networks (MANs).

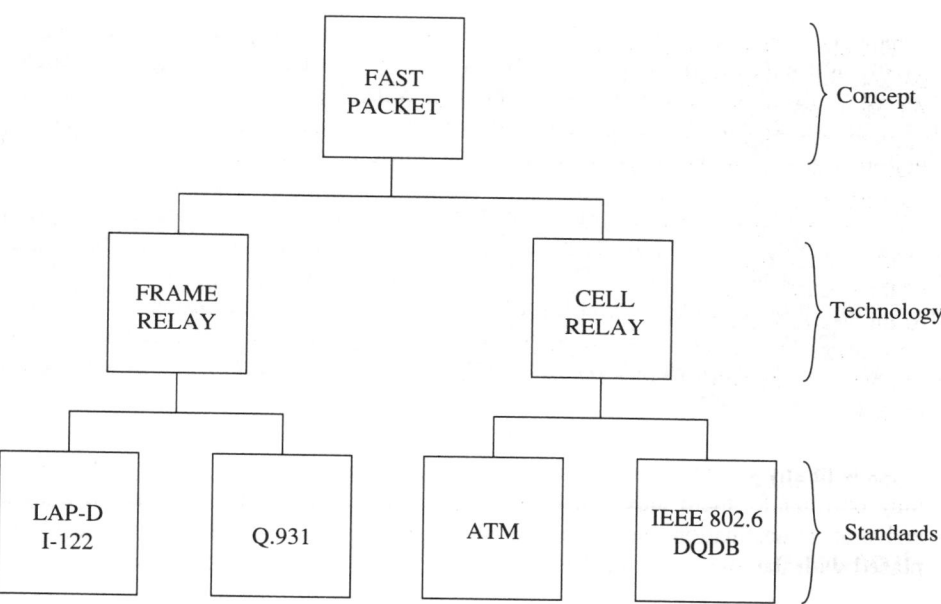

Figure 14.1   The fast packet-switching family

So far, the concept of frame relay has been pushed along quite rapidly by a market which demands ever-increasing bandwidths from its networks. Increasingly, communications requirements are calling for integrated voice and data networks, image transfer, and local area network (LAN) interconnection. This change in requirements over a relatively short time, means that a communications network must meet today's heavy demands yet be able to adapt to increasing data demands. For example, with the increase in LAN applications and ever more powerful client/server workstations, the high speed transfer of large amounts of data will become a critical requirement in the majority of corporate networks.

Fast packet-switching delivers the best of both worlds by bringing the economy and reliability of packet-switching to traditional applications such as voice, and the high throughput and low delays of circuit switching to data applications. One of the key aspects for fast packet-switching is the elimination of layer 3 protocols to provide protocol independent interfaces for switching equipment. Conventional packet-switching requires special protocol interfaces (such as X.25) for every device connected. However, with fast packet-switching, these special interface protocols are not required. As a result, fast packet-switching equipment should be totally compatible with existing communications networks, and could be used for any application where conventional circuit switching was used.

Summary

Fast packet-switching is one of the most efficient methods capable of integrating all current and future communications traffic over a digital network. It brings unparalleled benefits to the building and management of private digital networks. Voice activity detection, voice compression, redundant data discard and data frame multiplexing techniques should all allow for twice as much traffic on a network.

As up to 80 percent of the cost of a wide area network can be the cost of the carrier or bearer facility, reducing the number of private circuits in a network can substantially reduce the cost of the network. The key to efficiency in the topology of a network lies not simply in eliminating circuits from the network, but in reconfiguring these circuits to handle the traffic more efficiently. Fast packet-switching not only allows this, but also provides the means for standardising the introduction of both public and private digital networks of the future.

Frame Relay

Introduction

Contemporary network protocols used in a packet-switched environment, such as X.25, are designed to overcome deficiencies in the quality of the transmission medium by incorporating error control mechanisms. However, recent advances in technology means that it is possible to construct a network using high quality cabling systems, such as optical fibre which allow data to be transferred with very low error rates between network nodes. When existing protocols are used over such media, it is found that unnecessary delays are incurred in the process of error checking and correction procedures. Frame relay is a recent technique which enables fast networks to be deployed effectively over high quality media by removing these delays.

Existing Network Protocols

In order for data to be transferred efficiently across a network, it was found to be necessary to incorporate data validation and correction techniques at each node to counter the errors incurred during the transmission over the chosen medium. A typical network connection which could be established in an X.25 network is shown in Figure 14.2.

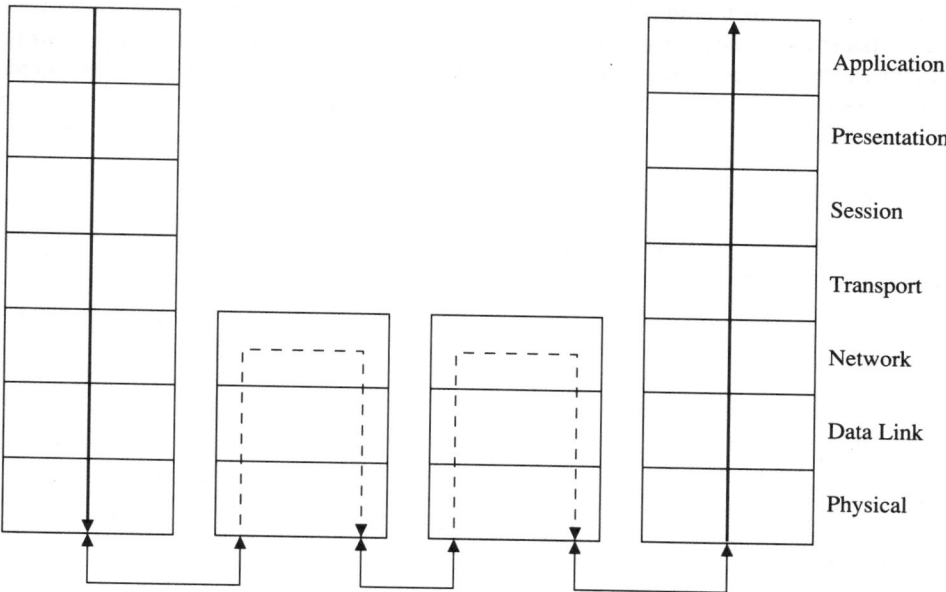

Application

Presentation

Session

Transport

Network

Data Link

Physical

Figure 14.2   X.25 Network connection

It can be seen that packets travelling across the network are processed by the physical data link and network layers in each of the nodes. The physical and data link layers are responsible for the transmission of valid signal levels across the connected medium and for the detection and correction of errors incurred. The network layer has several functions:

    —   to carry out end-to-end routing;

    —   to segment and re-assemble packets for transmission;

    —   to control congestion within the network.

These functions include overall frame validation and request for re-transmission in the event of any frame being corrupted.

Networks using the X.25 protocol may operate at speeds of up to 2 Mbits using suitable equipment and optimum configuration of packet size. However, a significant delay in passing through the node (the nodal transit delay) is introduced by the inclusion of these frame validation and re-transmission request techniques. Another consideration, although becoming increasingly less important is that this extra checking requires large amounts of buffer memory by the node switch in order to provide a satisfactory level of service.

As packet sizes become smaller, as would be the case in a network which experienced a lot of 'bursty' traffic such as local area network (LAN) interconnections, the effectiveness of such protocols becomes greatly diminished, as a greater percent of time is wasted in overheads. Also, with recent advances in high quality media, the process of frame validation and re-transmission techniques by each node should not be required, since node-to-node error rates will be very low. These two important factors have led to the development of the new protocol called *frame relay*.

Benefits of Frame Relay

Frame relay is basically a reduced version of the X.25 protocol standard. A typical network connection using this protocol is shown in Figure 14.3.

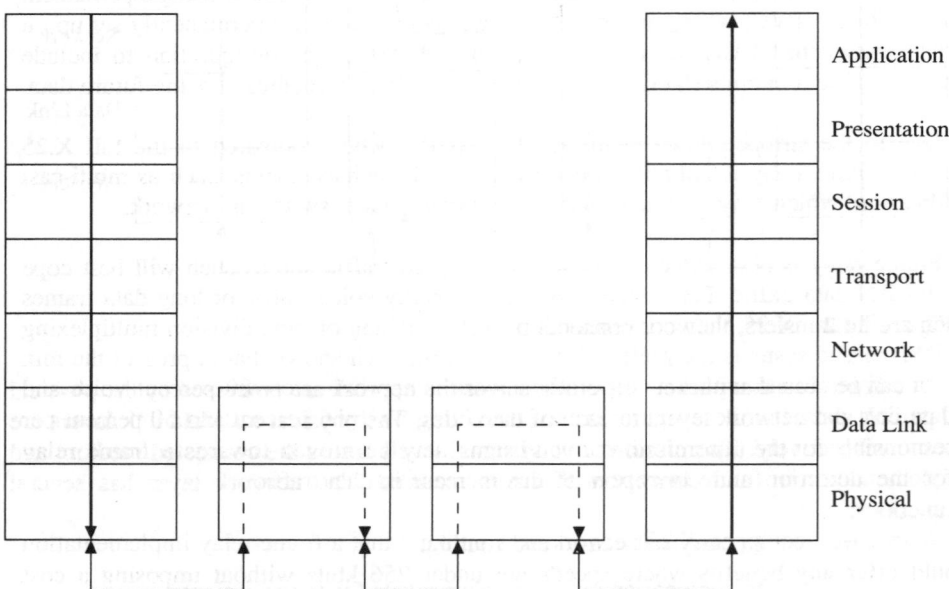

Figure 14.3 Frame relay network connection

It can be seen that this protocol only uses the bottom two layers of the network to transfer data through the node. The protocol does not incorporate any link set-up, error correction or flow control procedures. If any errors occur due to transmission or network congestion, then the system simply discards the corrupt frames. Detection of lost frames and subsequent recovery action is left to the end systems connected to the network. This is now felt to be a viable situation given the power available in modern personal computers and client/server workstations.

Two main benefits of such a system are:

— the nodal transit delay is reduced since no flow control or error recovery procedures are implemented;

— the overall system performance is increased since conventional switches have to be configured for the expected traffic characteristics, such as packet size. In practice, traffic does not always conform to this set-up and thereby reduces performance. Since frame relay does not get involved in this layer of the data transfer, performance will be at its optimum.

However, the frame relay protocol is still a fairly recent introduction into networks and has both advantages and disadvantages when compared with current techniques.

At present, frame relay services are defined for speeds of up to 2 Mbits, under the CCITT Recommendation I.122. Initially, such services will be offered using a permanent virtual circuit (PVC) where network end-to-end connections are permanently set up on a packet-switched basis. It is intended to extend this basic configuration to include switched virtual circuits (SVCs) using ISDN signalling procedures at some future data.

Due to the stripped-down nature of the protocol when compared to the full X.25 version, frame relay is not able to offer many of the enhancements, such as multi-cast addressing, which may be found in a conventional packet-switching network.

Frame relay is best suited to 'bursty', high speed traffic and as such will best cope with LAN data traffic. For systems which use mainly voice traffic or long data frames such as file transfers, then conventional packet-switching or time division multiplexing (TDM), based systems are preferable. However, research shows that at present the mix of data types found within a corporate network is approximately 80 percent voice and 20 percent data, but these figures are expected to grow by 6 percent and 50 percent per annum respectively. Therefore, many systems may be moved towards a frame relay implementation in future to cope with this increase in data traffic.

With a well equipped X.25 network, it is unlikely that a frame relay implementation could offer any benefits where speeds are under 256 kbits without imposing a cost penalty. Indeed, if the main concern is in the length of time taken to set up a call, then upgrading an X.25 network to include PVC support would be more straightforward than adding frame relay support. However, when network speed exceeds 256 kbits, frame relay provides greater system performance, due to lower nodal transit delays and reduced overheads.

Since frame relay switches do not perform any of the flow control and windowing operations associated with X.25 nodes, they may also be constructed with very small amounts of memory. However, the end user has to provide these flow control mechanisms and at present they are not yet in a standardised format and thus may cause problems during the initial implementation phase of such a system.

Frame relay systems are currently supported by several equipment vendors, usually as an upgrade to an existing packet-switched network. Sprint International and BT/Tymnet are among carriers who are in the process of launching international frame relay services. It is also thought that the next few years will see the same vendors offering upgrades from their frame relay products to support other fast packet protocols.

Summary

Current methods of node switching are designed to cope with dated transmission media which were found to be error-prone. Due to this fact, the protocols used had several layers devoted to error detection and recovery procedures. With recent advances in technology, it is now possible to construct a network utilising robust transmission media which yield a very low error rate. Use of contemporary protocols results in an unnecessary nodal delay, as error detection and recovery techniques are no longer required for a high percentage of the time.

Frame relay overcomes this wastage by stripping out such procedures and leaving it up to the end user to recover form any lost frames. However, due to the nature of such a technique it is only beneficial in networks which carry high speed, bursty traffic such as interconnection between LANs. It may therefore find a large market within public service networks which operate well above 64 kbits.

Asynchronous Transfer Mode

Introduction

Although the integrated services digital network (ISDN) concept is only just going into commercial service, recent advances, especially in fibre optics and microelectronics, have made it possible to consider an extension from the current narrowband integrated services digital network (N-ISDN) to a broadband ISDN (B-ISDN). The desire to do this is due to the limitations of N-ISDN. N-ISDN is essentially a public switched telephone network (PSTN) with digital lines extended into the customer premises rather than the historically 'conditioned' analogue telephony links. Consequently, it is restricted to some of the same parameters as the traditional voice network. Although the ISDN infrastructure offers unique functions over and above those in the traditional voice network, there are still a number of limitations with regard to the various services originally proposed for an ISDN.

Already, many subscribers have identified the requirement for a B-ISDN to meet their communication needs. The ever increasing use of LANs operating at speeds between 1 and 16 Mbits is generating a demand for high speed data transfer over the public networks. This demand will continue to escalate with the emergence of more powerful computers and workstations and high performance applications, including file servers, print servers, distributed processing and distributed databases.

It appears that there is a requirement for a new type of network, capable of transmitting all types of data whilst fulfilling some basic transport requirements, common to all services. Such a network would also be required to bring together the advantage of unified network access and transport mode with the possibility of using the same hardware for several logical networks. The introduction of a new transfer mode, asynchronous transfer mode (ATM), is an attempt to handle these different traffic types and requirements in a uniform way.

Overview of ATM Operation

In ATM networks, information is carried through the network in identical short, fixed length packets or cells. These cells also carry the required routing information. Voice, image, data and text are transparent to the network, which is a great step forward in the quest to reduce the number of standards used in the design of different types of communication equipment.

ATM has international recognition as being the transfer mode for the emerging B-ISDN, a fact that is validated through the numerous draft recommendations which are currently under development. In response to this backing by organisations such as the Consultative Committee for International Telephony and Telegraphy (CCITT), ATM products are expected to be available for a B-ISDN by 1995.

ATM appears, on first inspection, to be wasteful of bandwidth due to the header forming a significant part of the ATM cell. However, as the cells are only transmitted on user demand, then silent pauses in speech and LAN inactivity are ignored and hence, with careful engineering, a more efficient use of transmission and switching resources is possible.

ATM may be characterised by the following:

— information is sent in blocks of small predetermined fixed length;

— the principle is a basic function and involves no complex window type flow control on a link-to-link basis. Necessary error detection and retransmission is handled at higher layers and generally, on an end-to-end basis.

— routing is based on the information contained in the header that is added to each block of information. The header information routes the block to the appropriate output port in the ATM switch node.

The most important consideration in the design of the ATM protocol is the structure and functions of the cell header. A virtual connection is identified and characterised jointly by information carried within the header and by attributes carried to the switching device by signalling protocol.

At each network node, the information contained in the ATM header is used to invoke functions that are internal and specific to the individual node. In some cases, the ATM header is added to, or replaced by, further physical addressing information during the cell's passage through the node. It is suggested that a hierarchy of header types will emerge as ATM evolves from concept to product.

One of the main functions of the ATM header (see Figure 14.4) is to identify characteristics of the virtual channel on a communications link. The ATM header should contain only those functions associated with preservation, testing and integrity of the logical channel. The header is independent of services and transmission systems and will probably be subject to international standards for all switch interfaces. Upon entering an ATM switch the ATM header may be complemented or replaced by a switch header that provides physical routing through the switch. For all services transported on ATM, it is essential to identify the payload contents to enable service reconstruction at ATM endpoints.

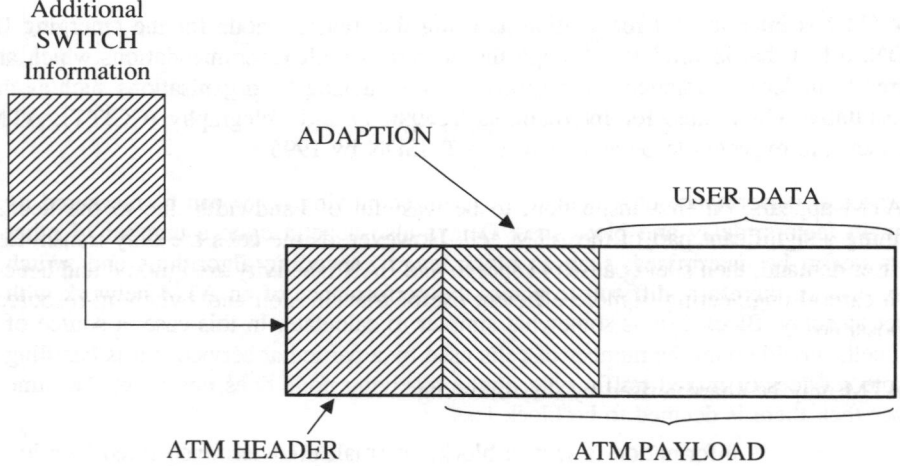

Figure 14.4 ATM cell structure

The multi-service nature of the B-ISDNs will require the development of distinct protocol sets serving control, user service, and transport functions and the overall partitioning of these functions has been defined by the CCITT (see Figure 14.5). The protocol structure is divided into three layers:

     — the service layer, consisting of the user plane (U-plane) and, the control plane (C-plane);

     — the adaption layer; and

     — transport layer (T-plane).

These all span the ATM transmission functions.

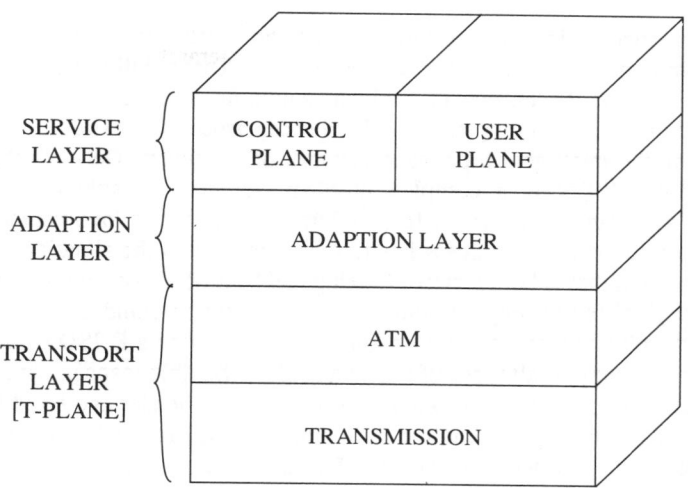

Figure 14.5 Broadband ISDN protocol hierarchy

The actual performance of a switching system employing ATM can be measured by two main factors: delay and block loss. Delays also depend upon a number of other factors yet to be determined, such as block length, encoding algorithms and switch structure. It is, therefore, difficult to assess the performance of an ATM network with respect to delay. Block loss is somewhat simpler to calculate. In this case, a source of ATM cells would count the number of cells used by a particular service if it is handling multiple services of mixed traffic type. If the receiving node does not count the same number then there is deemed to be block loss.

As mentioned earlier, one main advantage for the design of an ATM network is the ability to mix different traffic types. It is the frequency of transmission and the quantity of traffic which is of fundamental importance to the efficiency of an ATM network. Source 'burstiness', as it is known, is exploited by introducing statistical multiplexing in the basic bearer service. With statistical multiplexing, network performance will be affected by the load on the network and the mix and characteristics of the traffic. This load must be controlled in order to achieve an efficient ATM network. As such, the type of design and functions involved for such an ATM network are currently being researched by the RACE programme, a project funded by the European Commission and participating telecommunication companies.

The majority of the technical work being done surrounds the amount of distributed intelligence which will be required for the provision of varying levels and different types of service over the same logical network. This will mean that ATM network resources can be managed efficiently to obtain an optimum level of functionality.

Summary

The standards organisation are working against the clock. Pressure to provide high bandwidth communication over wider areas is growing. The inability to agree in a timely fashion will increase the chance of regional solutions. Reaching agreement on the fundamental transmission, multiplexing and switching aspects of B-ISDN is an important step in building the intelligent broadband network of the future. Defining the transport capabilities for B-ISDN is a complex problem, requiring a solution which takes transmission, switching and service factors into consideration. Involvement of many different organisations in this process has further complicated the situation. The task of achieving a solution that is both forward looking and backward compatible is not an easy one and the pay-off for coming up with a workable solution could be immense.

ATM could well be most plausible option for implementing a B-ISDN due largely to the fact that it will support all types of traffic and allow flexible management of network resources to allocate bandwidth on demand. The use of the header and fixed length cells lends itself very neatly to providing the means of standardising switching interfaces on a global scale. Only time will tell, but ATM certainly appears to be one of the major contenders that many of the public telephone operators throughout Europe are backing.

Synchronous Digital Hierarchy

Introduction

Synchronous digital hierarchy (SDH) is an alternative option upon which future broadband networks may be based. Whereas many users do not currently have access to bandwidths of 2 Mbits, SDH provides customer data rates in excess of 2400 Mbits. It also allows the customer to manage the network more efficiently. This section briefly examines the history of SDH and the way in which it operates.

History of SDH

Many years before digital techniques were incorporated into the switching centres of telecommunications networks, digital transmission techniques were being used to send data across the globe. The digital transmission lines were expensive commodities and ways were sought to increase their capacity to transmit information. In the 1960s, pulse code modulation (PCM) became the agreed method by which 4 KHz analogue signals were converted into 64 kbits digital signals. Using time division multiplexing (TDM), multiple 64 kbits digital signals could all be transmitted down a single twisted pair copper cable.

In Europe it was decided that 30 information channels, a single signalling channel, and a synchronisation channel (each carrying data at 64 kbits) should be multiplexed onto a single 2048 Mbits (often referred to as 2 Mbits) bit stream. However, it did not take long for the network operators to realise that although a 2 Mbits bit stream could cope with the link between two local exchanges, it was totally unsuitable for transmitting the large amounts of data which were being passed between trunk exchanges. They had two options: either to run multiple 2 Mbits bit streams or to multiplex each 2 Mbits bit stream onto a higher rate. The cost of laying multiple twisted pair cables was prohibitive and so the 2 Mbits bit streams were multiplexed onto 8.5 Mbits bit streams (normally known as 8 Mbits). As it became necessary to transmit more and more data, the hierarchy was extended to include 34 Mbits, 140 Mbits and 565 Mbits bit streams. The rates adopted in the USA were slightly different (1.5 Mbits, 6 Mbits and 45 Mbits) and the fact that there are two different hierarchies has caused some problems on the occasions when European and American systems have been interconnected.

Unfortunately, there is also a number of problems which occur even when the same bit streams are multiplexed to higher rates. The most troublesome is that each bit stream could have been generated by a different piece of equipment, at a different location and with a slightly different bit rate (the master clocks must not deviate from their nominal value by more than 50 parts per million). The constraint on the bit rate deviation is known as plesiochronous operation and the corresponding hierarchy (2 Mbits to 565 Mbits) is known as the plesiochronous digital hierarchy (PDH).

Another problem associated with multiplexing to higher rates arises when a single 2 Mbits bit stream has to be added (known as drop and insert) into an existing hierarchy. In this situation it is likely that a whole new multiplexing hierarchy has to be created at great cost.

To move away from these problems which are inherent in PDH, the American National Standards Institute (ANSI) developed a synchronous hierarchy of bit rates called the synchronous optical network (SONET). This was in 1984, and the concept was later expanded by the CCITT in Recommendations G.707, G.708 and G.709 and was simply named the synchronous digital hierarchy.

A Technical Overview of SDH

SDH has four important features:

- it is synchronous and therefore allows efficient drop and insert of channels, and cross-connect applications;

- it is an optical standard which allows fibres to be connected between equipment from different manufacturers;

- it makes full provision for network management channels which enables networks to be designed and controlled effectively;

- it can be introduced into existing networks in North America and Europe by interfacing directly with all currently used transmission rates.

The underlying philosophy of SDH is that any currently used transmission rate can be packaged into a standard sized container and located in a readily identifiable section within the multiplexed structure. All existing rates (such as the European 2 Mbits or the North American 1.5 Mbits) can be mapped into virtual containers (VCs). Once the existing rate has been mapped into a VC, multiple VCs can be placed together into standard formats. In this way, SDH can carry European 2, 8, 34 and 140 Mbits traffic, and the American equivalents. VCs originating from different existing rates can also be combined so that, for example, 2 and 1.5 Mbits traffic can be simultaneously carried within the same structure.

To allow the mapping process to be performed efficiently within a 140 Mbits structure with sufficient management overhead, the basic data rate for SDH has been set at 155.520 Mbits. This is labelled as the synchronous transmission module 1 (STM-1) and it forms the fundamental building block for higher rate traffic. Higher SDH rates are known as STM-4 (622.080 Mbits) and STM-16 (2488.320 Mbits) and these are formed by simple multiplexing of the STM-1 structure.

Another layer of control allows for network administration and is represented by the administrative unit (AU). The AU is the basic unit through which operators manage the network. There are several AU data rates which public telephone operators (PTOs) must choose from, and by sorting individual channels according to their outgoing destinations, all VCs are slotted into the chosen AU. In Europe the preferred AU rate is known as AU-31 which carries a 34 Mbits signal (or an equivalent capacity made up of lower rate channels) whereas North American networks tend to use the AU-32 rate which carries a 45 Mbits signal.

In order to provide a fast response to circuit failures, route provision and maintenance points, devices known as digital cross-connects are used. These devices form semi-permanent links between individual channels, and whereas conventional cross-connects require physical patching, a digital cross-connect is computer controlled.

Conventional networks have used cross-connects which have interfaced either 1 Mbit or 1.5 Mbits circuits by cross-connecting the constituent 64 kbits channels. SDH simplifies this switching situation by allowing easy identification of 2 Mbits channel within the STM-1 data stream (see Figure 14.6). The cost benefits of this attribute of SDH allows a network of high capacity cross-connects to be deployed in nodes throughout the network.

Several types of cross-connects are available within SDH, resulting in a hierarchy of cross-connect levels which replace the conventional hierarchy of multiplexers. The cross-connect levels include:

- 64 kbits cross-connects to interface at STM-n levels;

- VC cross-connects to allow switching of 1.5, 2, 6, 8, 34 and 45 Mbits traffic;

A) CONVENTIONAL CROSS CONNECT

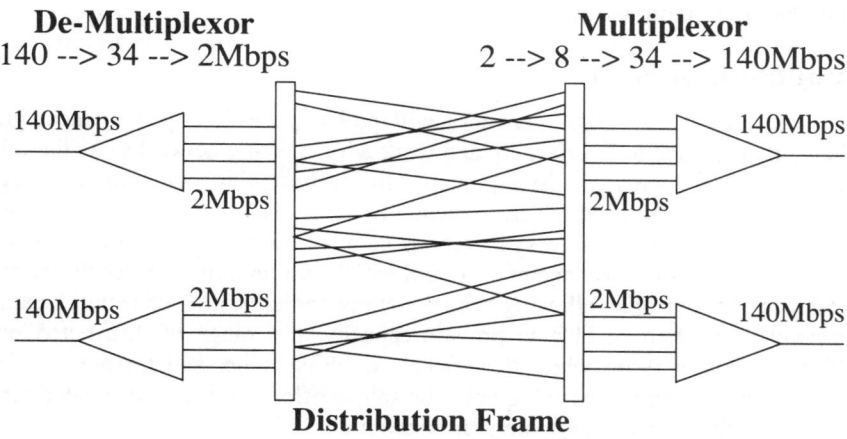

B) SDH CROSS CONNECT

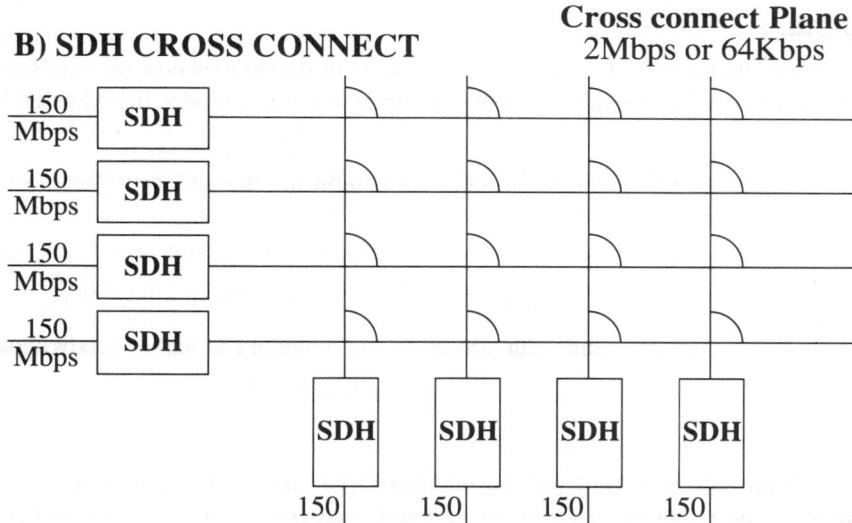

Figure 14.6 Conventional and SDH cross-connects

   —   AU cross-connects to work on major trunk nodes and to provide network protection;

   —   synchronous transmission cross-connects to switch purely STM-n data streams.

These cross-connects are used primarily because they are more cost-effective than current multiplexer hierarchies although they also allow for central management and transmission routing. Because of this, it is likely that a new approach to network topology design will be necessary, and as a result, conventional high-order multiplexing in core networks could disappear.

Introduction of SDH

During the introduction, SDH networks will probably be deployed to meet growth requirements and have to interface to, and work with, existing networks. As the roll-out of SDH networks increases, the arguments for restructuring contemporary network topologies will become stronger. Potentially, this restructuring will allow the development of a total managed synchronous network. Once this type of network is in place, there will be better management control which will, in turn, increase the network's versatility by, for example, allowing central routing and allocation of network capacity. The network will then be able to provide capacity to a range of distributed service providers, including those who offer packet-switching, telex and private circuits. In addition, operating companies will quickly be able to offer new centrally located services to geographically scattered customers throughout an SDH network.

Summary

SDH can simply be thought of as a hierarchical set of digital transport structures which will transport digital information over physical transmission networks. In particular, SDH offers:

- simplified multiplexing and de-multiplexing techniques compared with those used in conventional networks;

- easy insertion of lower speed channels into a high speed channel;

- payload transparency and transport of existing digital signals;

- flexible bandwidth allocation and evolution to higher bandwidths;

- greatly enhanced operation, administration and maintenance capabilities.

Once again only time will tell, but although SDH may exist alongside ATM, the pressure to standardise transmission systems and networks on a global basis could prove too much for one of the two options. SDH is firmly established in North America in the form of SONET but it has a long way to go before it can obtain a firm grip with in Europe.

Pan-European Cellular Radio

Introduction

Europe has a well-developed cellular radio infrastructure for its geographical size. Unfortunately, each country has adopted its own standard to suit its own requirements and there is no cooperation between individual countries when it comes to providing cellular radio services. This uncoordinated approach means that someone wishing to travel from, say, the UK to Italy, and who wishes to stay in touch with base whilst doing so, would require at least three cellular handsets, each registered for the relevant cellular network. Even though some countries use the same cellular standard (such as the UK and Italy), subscribers on one network would not be able to operate on the other as their handsets would not be validated.

In 1982 the Conference of European Post and Telecommunications Administrations (CEPT) founded a coordinating group to oversee the development of a pan-European cellular radio service. The broad aim of this work was to enable European travellers to be in telecommunications contact wherever they were in the EEC, using a single radio telephone instrument. The name given to the coordinating group was Group Speciale Mobile (GSM). In early 1990, the specification for the GSM system was frozen; it was approximately 6,000 pages long. The first GSM services were scheduled to be rolled out on 1 July, 1991, by all the cooperating countries (they had signed a Memorandum of Understanding to this effect in 1989). Other, non-European countries have since become interested in adopting this system and so its name has changed. GSM, instead of being the abbreviation for an obscure technical committee, is now deemed to stand for Global System for Mobile communications.

The GSM Standard

The basic architecture of the GSM system is similar to most cellular radio systems in existence (see Figure 14.7). The geographical area of interest is covered by cells of radio coverage. Each cell has a base transceiver station (BTS) responsible for providing the radio link between subscribers and the network. Adjacent BTSs use different sets of radio frequencies so they do not interfere with each other. The frequency re-use patterns on GSM are similar to those employed by the UK total access communications system (TACS).

BTSs are grouped together under a base station controller (BSC) which is responsible for manipulating radio frequency (RF) power levels and for controlling handover between BTSs. Groups of BSCs are controlled by mobile switching centres (MSCs). These MSCs are the heart of the GSM system. They manage call set-up and termination, routing, billing and accounting and provide the interface to public voice and data networks. MSCs have access to two databases; the home location register (HLR) and the visiting location register (VLR). The HLR contains the details of all subscribers normally resident in each particular cell, including the required challenge-response pair of codes to enable a subscriber to gain access to the network. The VLR contains the details of which cells subscribers were last known to be active in. Both of these are used in the authentication of subscribers and their equipment when they gain access to the network.

The subscriber equipment contains a subscriber interface module (SIM) which is either resident in the mobile equipment or takes the form of a *smart-card* which can be inserted in any GSM compatible phone, allowing registered users to use any GSM phone in any participating country. The SIM contains authentication information such as the challenge-

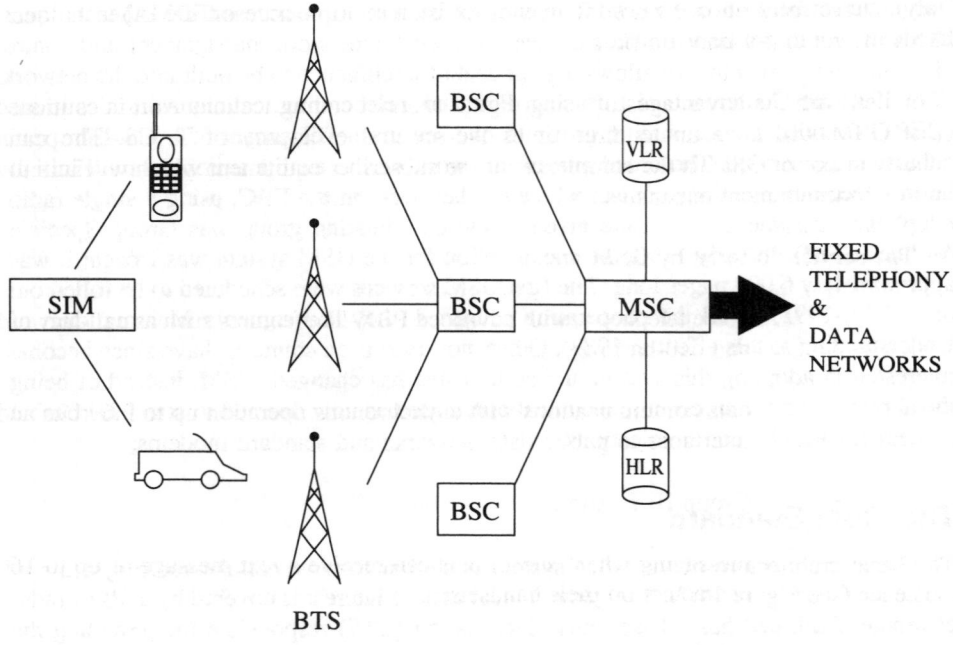

Figure 14.7 GSM architecture

response information checked by the HLR and details of the services that the user is registered for. It also contains the international mobile subscriber identity (IMSI) which is unique across the whole of Europe for a given subscriber.

GSM will be the world's first digital cellular radio system. The current UK system, TACS is an analogue system and GSM has several advantages over such analogue systems. The main advantage of GSM is that it provides greater spectral efficiency, which means that it can support more subscribers in a given bandwidth over a given geographical area. This is achieved through the following:

    —   the use of low bit-rate speech coding techniques, to reduce the spectral requirements for a telephone call;

    —   the use of advanced digital signal processing algorithms to make the system more resilient to signal dispersion and interference problems, thus enabling the distance between cells using the same frequency sets to be reduced.

Furthermore, GSM uses time division multiple access (TDMA) unlike analogue system which have conventionally used frequency division multiple access (FDMA) techniques. This means that vacant traffic slots can be used for network management and control information, which in turn allows a great deal of intelligence to be built into the network. Coupled with the advantages of using digital channel coding techniques it is estimated that GSM will have up to three times the spectral efficiency of TACS. The main disadvantage of GSM is the complexity of the subscriber equipment which will initially make handsets more expensive.

The services offered by GSM are:

—   voice telephony, with advanced PBX-like features such as call-forward and call-barring;

—   data communications with asynchronous operation up to 9.6 kbits and interfaces to public data networks and standard modems;

—   Group III facsimile;

—   messaging where subscribers can receive a text message of up to 160 characters on their handsets;

—   cell broadcast, where short messages can be sent to all users in a particular cell for public service announcements such as: traffic conditions, or financial news.

Speech coding on the system will initially be at 13 kbits. A half-rate coder is under development and will further increase the spectral efficiency of the system, allowing it to support a greater number of users. the frequency bands to be used by GSM are 890-915 MHz (mobile transmit) and 935-960 MHz (base station transmit).

As with TACS there are different classes of handset as shown in Table 14.2 below.

Table 14.2 GSM handsets

Class	Description	Power Output (Watts)
1	Vehicle/Portable 20	
2	Vehicle/Portable 8	
3	Hand-held	5
4	Hand-held	2
5	Hand-held	0.8

GSM Roll-Out

As previously mentioned, the nominal start date for the GSM service was 1 July 1991. The participating countries (those who are members of the EEC and EFTA) signed a Memorandum of Understanding to the effect that the network operators in each country would provide a basic set of GSM services, without internetwork roaming by that date. However, the deadline date passed with only one operational network being in place. This was the 'radio linja' system in Finland. Other countries are still putting network infrastructure together and trial testing various parts of it.

The UK has the most well-developed analogue system in Europe, with a level of market penetration second only to Norway and Sweden. The network operators in the UK are understandably reluctant to upset their userbase by shutting down the analogue networks and forcing everyone to migrate to the initially more expensive GSM network. It is estimated that as few as 10 percent of the UK cellular userbase want the extra facilities provided by GSM (Europe-wide roaming and PBX-like services) and they would be reluctant to pay extra for them. It is not surprising then, that progress with GSM has been slow in the UK.

Cellnet are aiming for a full commercial launch of their GSM network in mid-1992, covering the main population centres in the UK. It will be extended to other areas subject to demand. By 1994 Cellnet expect to be able to offer a GSM service to 98 percent of the UK population. Vodafone are aiming to cover 90 percent of the UK population by the end of 1992. They will extend this coverage as demand dictates.

Options for European Mobile Communications

GSM is not the only European mobile communications system available, but it will, when fully implemented, be the most flexible. Satellite communications, such as those offered by Inmarsat and Eutelsat provide not only European but worldwide coverage. Inmarsat, administered in the UK by BT provide a voice and low-speed telegraphy service. The subscriber equipment fits into a suitcase-sized box and it provides global communications via a network of geostationary satellites. Access to fixed voice and data networks in the home country is also provided. These services are used most by agencies, international road haulage companies and disaster relief organisations.

Radio paging is available on a European scale with the *Euromessage* service. This is a cooperative venture between several European national paging operators and it allows messages to be sent to subscribers anywhere within the coverage of a cooperating paging operator. Furthermore, the European radio message service (ERMES) is due to be launched in 1993. This is the first pan-European paging system and will provide a more advanced service than current national paging systems. The coverage of this system will be limited to EEC countries.

Summary

Many European countries have developed their own cellular networks independently of each other which makes it difficult for the European traveller to stay in touch with base. The introduction of GSM is intended to overcome these limitations and provide additional services. However, as previously noted only one of the participating countries met the agreed commercial launch date of 1 July 1991. The reasons for this are due to many factors, not least of which is the lack of standards and support in the area of equipment manufacture. Coupled with the expense of implementing a new infrastructure and the risk of upsetting existing customers, it may well take a long time for GSM to become firmly established. In fact, a viable service is not expected until well into the 1990s and many European countries will probably offer GSM alongside their existing analogue systems in the future.

Finally, some of the alternatives, such as satellite communications and pan-European paging may also affect the roll-out of GSM. Even though none of them will be as flexible as GSM, they may very well satisfy the needs of those who require a less sophisticated service.

Cordless Telephony

Introduction

Cordless telephones allow people to move around inside a predefined zone, whilst making or receiving a telephone call. The telephone instrument is not connected to the exchange solely via wires; instead a combination of radio waves and cables is used to provide the communications link.

Figure 14.8 shows how cordless telephones might be implemented inside an office building. Radio waves provide the connection from the cordless handset to the radio base station (RBS). Standard cabling may in turn provide the communications link from the RBS to the PBX. The PBX then allows switching between exchange lines in the normal way. Additionally, the PBX needs to keep a record of the location of cordless handsets for incoming calls.

It is thought that the higher initial costs of cordless telephony may be recouped through lower running and cabling maintenance costs and reduced telephone charges for returned calls. There are no case studies to prove this and no accurate estimates of payback periods are available, as there are currently no installed cordless exchanges, but it is easy to see how this could be the case. However, the decision to use cordless telephony should not be based solely on cost. The nature of an organisation's business should decide whether the convenience of being able to contact someone, wherever they

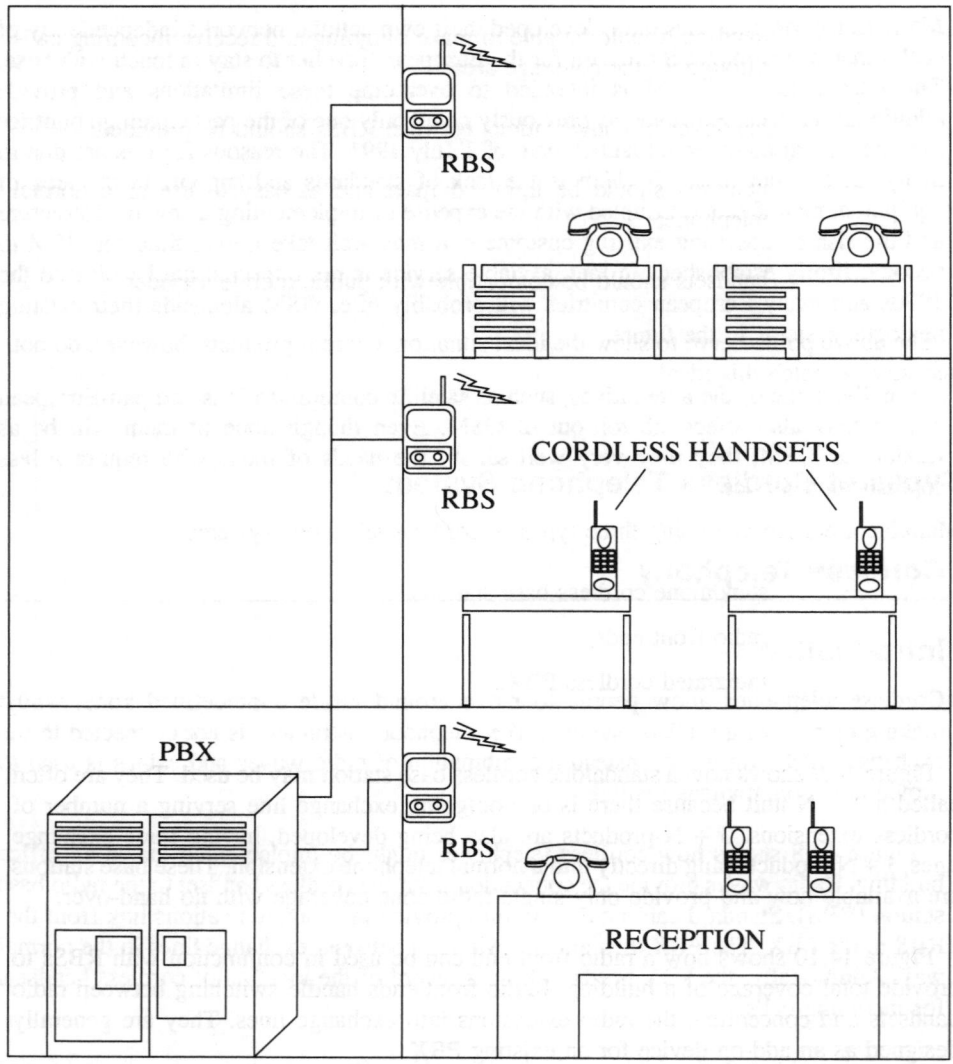

Figure 14.8 Cordless telephony at work

are in the building, is important. This may be crucial for businesses who have, for example, maintenance staff and sales personnel who need to be contactable throughout the working day.

What Services Are Desirable?

In order for a cordless telephone system to be beneficial, it should take into account the following aspects:

- the whole of the office premises should have radio coverage;

- handsets should be able to make outgoing and receive incoming calls throughout the coverage area;

- handover of conversations between RBSs should be provided;

- handsets should be light, compact and as easy to use as a normal telephone;

- handsets should be compatible with public mobile services.

The above points serve to show the ideal situation. Current products, however, do not necessarily match this ideal.

Types of Cordless Telephone System

Manufacturers are proposing three types of cordless telephone systems:

- standalone cordless base stations;

- radio front-ends;

- integrated cordless PBXs.

Figure 14.9 shows how a standalone cordless base station may be used. They are often called a 1 + N unit because there is one outgoing exchange line serving a number of cordless extensions. 3 + N products are also being developed, having three exchange lines. 1 + N products plug directly into a normal telephone extension. These base stations are available now and provide only single radio zone coverage with no hand-over.

Figure 14.10 shows how a radio front-end can be used in conjunction with RBSs to provide total coverage of a building. Radio front-ends handle switching between radio handsets and concentrate the radio extensions into exchange lines. They are generally designed as an add-on device for an existing PBX.

Figure 14.11 shows how an integrated cordless PBX may be used. They provide the same facilities as radio front-ends but incorporate the cordless functions into the exchange itself. However, it is not expected that such products will be commercially available for use in Europe until 1993.

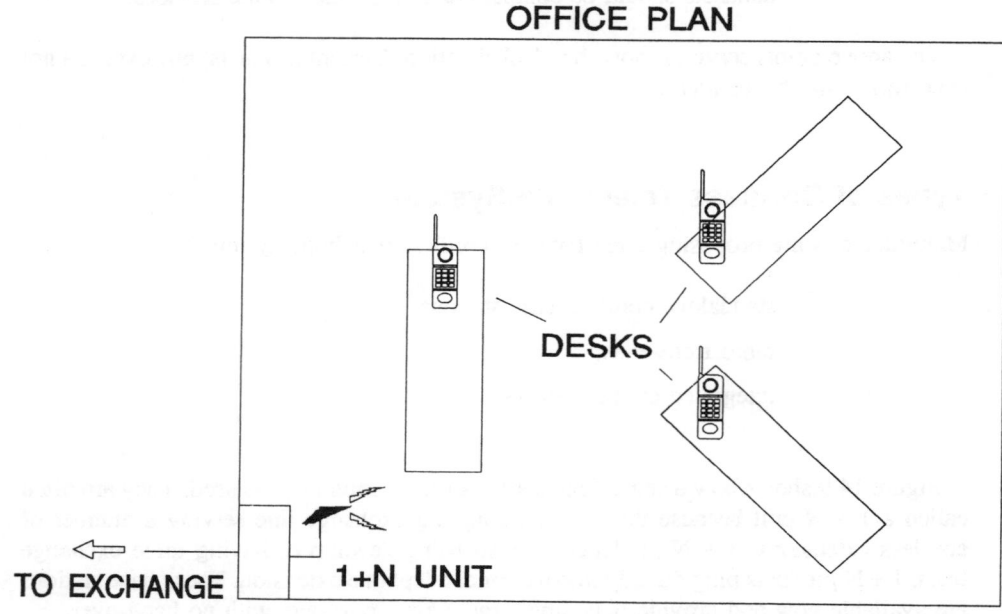

Figure 14.9 Standalone cordless base stations

Cordless Telephony Standards

It is impossible to read any of the manufacturer's literature without coming across several acronyms for cordless telephony standards. Each standard has different capabilities and the major ones are often named CTn, where CT stands for cordless telephone and n is a number denoting the type.

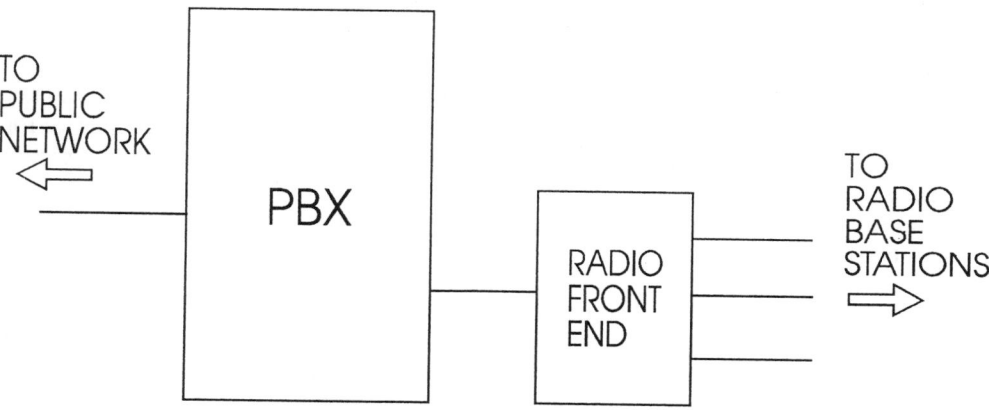

Figure 14.10   Radio front-ends

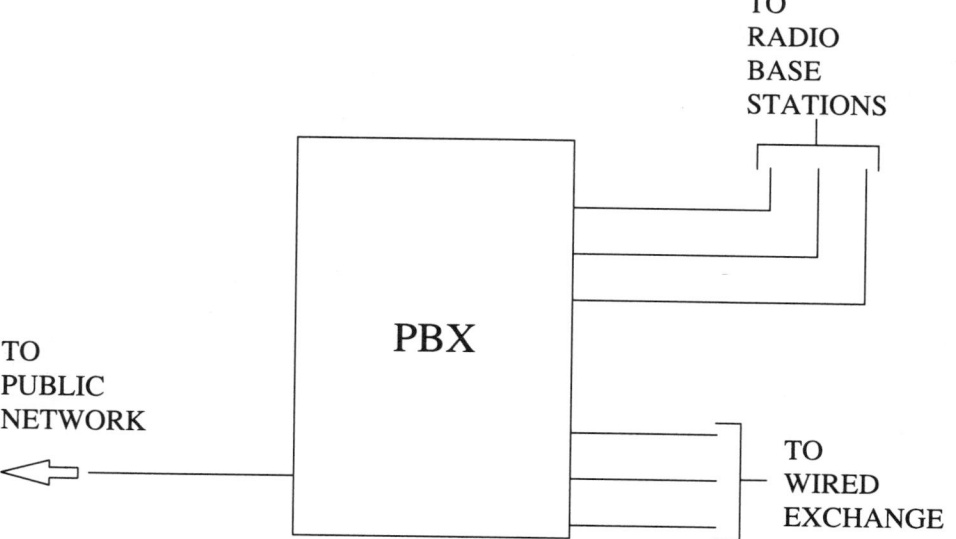

Figure 14.11   The integrated cordless PABX

CT0 is the standard upon which the domestic cordless telephones in the UK were originally based. These systems have few radio channels available to them so interference from other sets in the neighbourhood is common. It is easy to eavesdrop on conversations; a radio receiver covering the right frequency band is all that is needed. CT0 was not designed for the high densities of users experienced in an office environment.

CT1 has similar characteristics to CT0. Again, it was designed for home handsets, but was mainly used in the rest of Europe.

CT2 is also known as common air interface (CAI). This has been adopted by the European Telecommunications Standards Institute (ETSI) as an interim European standard and the public 'telepoint' services in the UK also operate to this standard. In the office, CT2/CAI can provide facilities for making and receiving calls, but it will not support handover of conversations between radio zones. It is therefore more suitable for smaller offices, where single-zone coverage is sufficient. CT2/CAI was the first digital standard and as such, it is more secure than CT0 and CT1. Each base station can generally support 40 two-way conversations, with a range of 50 metres in an office with partition walls.

The digital European cordless telecommunications (DECT) standard was set by ETSI and is the recognised European standard for cordless telephony. DECT systems support 132 simultaneous, two-way conversations per base station and provide handover facilities. It is more difficult to eavesdrop on conversations than with CT2 because speech transmissions are encrypted. Their range of operation is similar to CT2 but ETSI chose DECT to be the European standard as it provides all the main facilities that should ideally be offered by a cordless PBX.

CT3 was developed by Ericsson (a Swedish telecommunications manufacturer) to enable them to have experience with the type of technology required for DECT, and thus get ahead of the competition with their products. As a result, only Ericsson support CT3 products. CT3 was designed for a high density of users, with each base station being able to support up to 64 simultaneous, two-way conversations. It provides handover facilities and has a range similar to that of CT2 but has similar security features to DECT.

Other communications systems which could be used in a cordless office environment are personal communications networks (PCNs) and the pan-European digital cellular network, known as GSM. The PCN standard is termed DCS1800 and the one for GSM is GSM900. Both of these services are designed for use in a wider geographical area than inside office buildings. However, it is feasible that they could be used in, say, a business park to provide centrex-type services.

Products and Services

Currently, only standalone base station devices are available, these being provided by the public telepoint operators such as *Phonepoint*. Their base station, for example, can support up to eight handsets but they will have to share one exchange line. It can be configured so that, either all eight handsets ring when there is an incoming call, or one rings and is then able to pass the call on to one of the other seven after answering it. This base station could even be used in the home providing each member of the family with a handset. The handsets used with this base station are fully CT2/CAI compatible and can be used with the public telepoint services, provided that the user is registered with a telepoint operator.

These products are most suitable for small offices such as estate agents, solicitors and small 'satellite' type offices. Phonepoint also plan to introduce a new product which combines a radio pager and a CT2/CAI handset into one hand-held terminal. This will allow people to be contacted wherever they are and give them the freedom to respond when it suits them. Other telecommunications manufacturers and telepoint operators will also be providing standalone base stations products in the future.

Radio front-ends form the basis of the products to be released in the next two years and suppliers of PBX equipment will be providing them as add-ons for their existing customer base. Integrated radio PBXs are expected to arrive in 1993. Suppliers are cautious about investing vast amounts of resources into developing these products when the market is not yet certain. Table 14.3 summarises the intentions of the major telecommunications manufacturers with regard to cordless telephony products.

As can be seen from the table, information is scarce and few products are ready for commercial launch. The Japanese manufacturers are adopting a 'wait-and-see' attitude, and they will enter the market at a convenient point in the future.

Summary

Current systems cannot meet all of the requirements identified. However, steps are being taken to reach the ideal. Public telepoint services are due to be relaunched in London. Handsets on these systems should be compatible with early single-zone base stations, but these do not provide handover for large multi-base station applications.

Cordless telephony systems with handover are some way off. DECT-based solutions will be the first ones to meet most of the requirements, but the public telepoint services will need to change to be compatible with these if the goal of 'one handset to do everything' is to be reached.

The future is somewhat unclear. However, it is hoped that, for the sake of the industry and consumers alike, one standard for domestic, public and commercial systems is adopted. This will produce a large unified market bringing benefits to the consumer through lower equipment prices and greater choice. It should also benefit the manufacturers and the public network operators through, respectively, greater sales revenues and increased air-time revenues.

Cordless Local Area Networks

Introduction

Cordless LANs (CLANs) allows computers and computer peripheral equipment to be interconnected without the need for wires. The communications link is provided either by radio waves or by line-of-sight optical transmission. The major advantage of CLANs is that they allow computers and terminals to be moved with the minimum of fuss and expense.

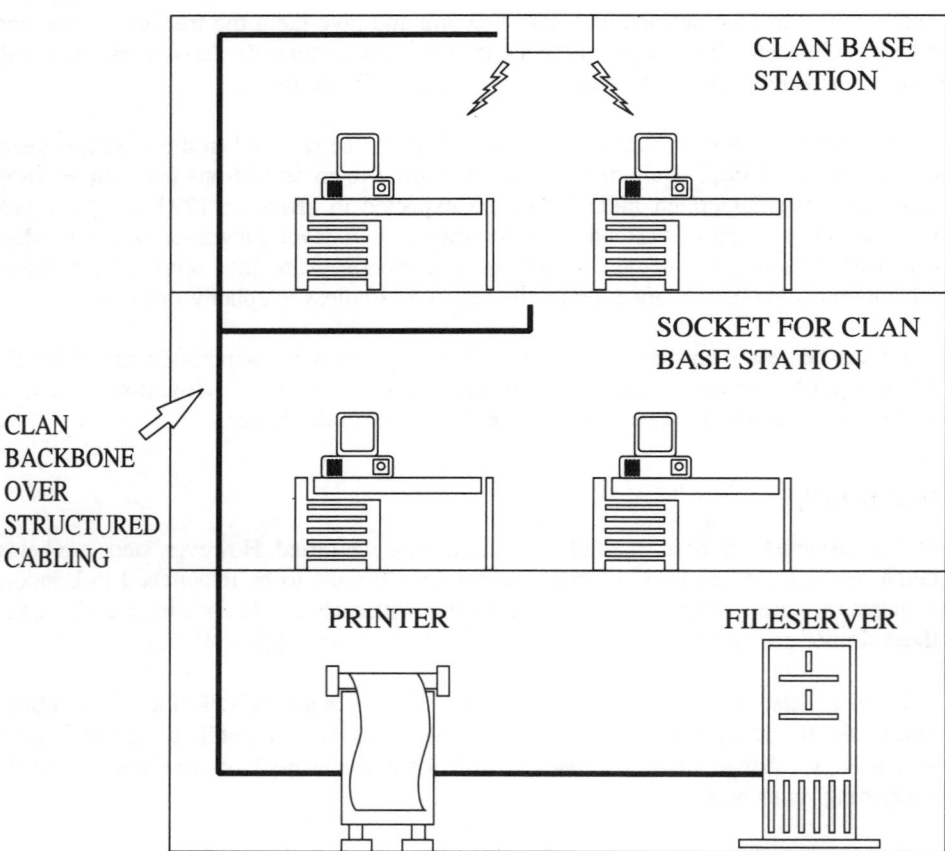

Figure 14.12 CLANS in action

Figure 14.12 shows how CLANs might be used inside a multi-storey office building. Ideally, access should be provided throughout the whole building so people and equipment can be moved to any location. Obviously, more than one radio coverage zone is needed to provide coverage of the entire building. This has a spin-off benefit in the fact that less radio spectrum is needed to cover the building; frequencies can be re-used in different zones.

Table 14.3 Planned cordless telephony products

Company	Product
Alcatel	Cordless PBX based on DECT. Launch date unknown
Bosch	Cordless PBX based on DECT. Launch date unknown.
Ericsson	They are producing a radio front-end for their MD110 PBX, based on CT3. This is undergoing trials. Commercial launch of a radio front-end, based on either CT3 or DECT will be early 1992.
GPT	They are producing a radio front-end for the ISDX and MDSS PBXs. This will be CT2/CAI-based and will be commercially available in mid-1992. They are also producing standalone base stations, based on CT2/CAI, in both 1 + N and 3 + N format.
Matra	They are working on CT2 and DECT PBX products but their launch dates are unknown.
Northern Telecom	They are developing the CT2 plus standard, which is effectively CT2 with handover. This unlikely to appear in Europe. They are also working on CT2 and DECT products, but there are no firm dates for launch yet.
Philips	They are working on CT2 products, which will be available in 1992 and on DECT products, to be ready in 1995.
Siemens	They are working on DECT-based products, but there are no details available of launch dates.

General Types of CLAN

CLANs use either radio waves or infra-red light to provide the communication link. The radio-based CLANs in the USA work at around 915 MHz and also the microwave bands using frequencies around 18 GHz. Many of these products would be unsuitable for use in the UK as they would directly clash with the spectrum used by the UK cellular radio service. The Radiocommunication Agency (RA) of the DTI has submitted a draft proposal to the European Radio Committee (ERC) which suggest the use of 2445-2475 MHz, 5.8, 17, 24 and 60 GHz for CLAN use in the UK. In view of this, it will be well into 1992 before a full range of radio CLAN products is available in the UK.

It is expected that CLANs working in these bands will be exempt from licensing on a site-by-site basis in the UK (the CLAN manufacturer will be granted a 'cover-all' licence, provided the product meets the required specification), although individual site licences will probably be needed for CLANs in Northern Ireland. Some radio CLANs employ spread-spectrum techniques.

Spread-spectrum techniques were developed in the Second World War to make allied radio transmissions immune to interception. Instead of the energy in a radio transmission being concentrated into a narrow frequency band, the signal is 'squashed' so that it occupies a larger portion of the radio channel, but at a much lower amplitude level (see Figure 14.13). Interference to other radio users is minimised, as spread-spectrum signals are similar in nature to natural radio noise. Interference between different spread-spectrum systems occupying the same portion of radio spectrum is handled by the channel coding techniques used. They are also inherently difficult to eavesdrop on, making them ideal for use in financial institutions and government departments.

Optical CLANs do not require a licence for operation. They are suitable only for applications where there is a line-of-sight between the equipment requiring interconnection. Obviously, optical CLANs are unable to penetrate partition walls, so they can only be used where this would not be the case. Radio CLANs, in the bands suggested for use in the UK, have a range of approximately 100m in an office with partition walls and approximately 200m in an open environment.

The CLANs based on higher microwave frequencies such as 60 GHz, are able to penetrate concrete but they tend to have less overall range. Radio CLANs could be of most use in a traditional office environment where there are many walls dividing offices and work areas.

Radio CLAN coverage is also influenced by office furniture, such as steel filing cabinets. As a result, care needs to be taken when deploying a radio CLAN: radio dead-spots, where there is little or no radio signal, can be overcome by moving radio terminals around. Dead-spots can also be overcome by use of space diversity techniques. In other words, two antennas feeding the ratio terminal, with the distance between them greater than a wave-length of the radio wave in use. With space diversity, it is unlikely that both antennas will be in a dead-spot, so by selecting the one with greatest signal strength, the radio terminal can guarantee radio access. The disadvantage with using any diversity technique is the cost of the extra hardware needed to provide it.

In general, the advantages of a CLAN can be summarised as follows:

— costs associated with moving are reduced;

— the psychological barriers to moving are reduced;

— the expansion of the LAN is made easier.

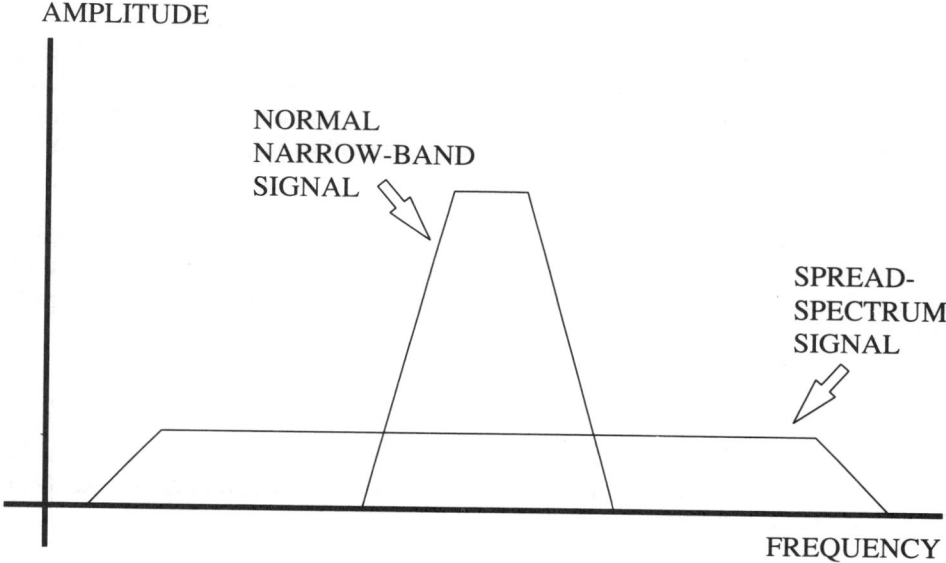

Figure 14.13 Spread spectrum signals

It can cost somewhere in the region of £100-£200 to move a wire or fibre optic LAN connection. Cable-free LANs will reduce this cost, provided that adequate CLAN coverage is available and mains power access is adequate in the new location. It is therefore easier to move terminals and computers so that people will be less reluctant to do so, and as a result they will be more willing to optimise office space usage. Adding new nodes is easier with CLANs than with ordinary LANs as there are fewer cabling considerations.

However, the major disadvantage of a CLAN is the high initial cost. It is estimated that a CLAN connection costs 2-3 times as much as an ordinary LAN connection to install. On the other hand, maintenance and running costs for CLANs will be much lower. If people move office frequently then it may well be worth paying more initially to make bigger savings in the future.

Additional Requirements of a CLAN

It is worth considering what the major requirements of a CLAN should be to make it truly beneficial. Armed with this information, the current product offerings can then be viewed with some objectivity. Obviously, the definition of 'useful' will vary between organisations. However, as with cordless telephones, there are some features which are, in the main, essential for a CLAN to be of practical use.

First of all, CLAN products should be compatible with existing LANs. This would allow existing LANs to be extended, or portions of them to be replaced, with minimum disruption of users and applications.

Secondly, CLAN products should be compatible with mobile data networks. This is desirable, as there are people armed with lap-top computers who want access to a LAN whilst they are both in and out of the office. However, it would be difficult for both this and the previous requirement to be satisfied, as the current mobile data networks do not use any of the current LAN standards.

CLAN interface devices should be physically small. A CLAN access device which occupies half of a desktop is obviously not going to be very practical. They should be small and unobtrusive, perhaps fitting into the expansion slots of a personal computer or be capable of being mounted on the edge of a desk.

CLAN Products

The product details listed in Table 14.4 are for CLANs which are currently available in other countries. It remains to be seen which, if any, will be re-engineered to make them suitable for the UK market.

As can be seen by comparing Table 14.4 with the frequency bands for CLANs in the UK, noted earlier, most of the systems will require re-engineering if they are to be used in the UK. However, there are some such as the radiolink product from IT Security, that will need no further work and these are currently available.

The speeds quoted are somewhat lower than most cabled LANs, which generally operate at around 10-16 Mbits. Optical LANs are nowhere near this speed yet, although BICC estimate that they will be producing an optical CLAN, running at over 10 MHz within the next five years. Motorola's radio CLAN is Ethernet compatible and runs at 'Ethernet-like speeds'. This should make it very appealing to organisations who want to take advantage of CLANs and still need to use cabled LANs as well. The level and type of traffic on a network will dictate whether low transmission speeds will cause unacceptable network response times, but some organisations with low traffic requirements may still find low-speed CLAN products attractive.

Summary

Current CLAN systems are much nearer to the ideal situation than cordless telephones systems for the office. However, no CLAN products have yet been produced which can interface properly with current mobile public data networks. This requirement is somewhat less important than a CLAN being compatible with a standard commercially available wired LAN, and therefore not be seen as too much of a disadvantage.

Most CLAN products are fairly small, although they look large when compared with a laptop or notebook PC. Smaller products, which could be mounted inside personal computers would be greatly welcomed but may not appear for some time.

Table 14.4 CLAN products

Supplier	*Product*	*Technology*	*Speed*
Agilis	21CSP	Spread spectrum	250 Kbits
BICC	—	Infrared	4 Mbits
IT Security	Radiolink	Spread spectrum	250 Kbits
Motorola	Altair	Microwave (18Ghz)	15 Mbits
NCR	Wavelan	Spread spectrum	2 Mbits
Photonics	LocalTalk	Infrared	1 Mbit
Telesystems	Arlan100	Spread spectrum	200 Kbits
	Arlan600	Spread spectrum	1 Mbit

It remains to be seen how many of the radio products listed in Table 14.4 will be re-engineered for the UK market. If they are reworked and the standards are finalised, then these products should start to appear by the end of 1992.

Electronic Data Interchange

Introduction

Electronic data interchange (EDI) should not be viewed as a technology, but rather as a solution to a business need. This implies that the business case for EDI will vary between companies depending on the different needs that EDI is to address. EDI is not an end in itself, nor is it a gimmick. It is a tool which many companies have used dramatically to restructure their internal operations and the way in which they relate to customers, suppliers and authorities. Although the business case for EDI will differ between companies, several key reasons (improved customer service, reduced operating costs and responsiveness to market demands) continually recur.

Is EDI For You?

The first participants in the EDI revolution were mainly large companies, but this is no longer the case. Certainly, the active involvement of a number of large companies was required to stimulate interest in, and encourage the uptake of EDI. However, smaller companies are also now seeing potential benefits. This point was made strongly at a European chemical industries conference when the chairman of the EDI project made the following comment:

> "EDI is not for the big companies alone, it is not a goal in itself, it is
> a tool, a new way of doing business, for the big and the small."

Anyone browsing through the various EDI publications and press articles looking for a precise and simple statement of what constitutes electronic data interchange, will almost certainly encounter a range of definitions. Each definition will have been put forward in an attempt to encapsulate the essential features of EDI, whilst trying to exclude (or in some cases even include) other related technologies. One statement is that EDI is:

> 'The transfer of structured data, by agreed message standards, from
> one computer system to another, by electronic means.'

In this context, 'structured data' refers to a precise, recognised and accepted method of assembling data. Such data items as product number, customer name, and unit price may be structured into a purchase order or invoice for example. The definition also uses the phrase 'from one computer system to another'. This may well include intra-company and inter-company communications provided that the transactions are between trading partners (usually between supplier and customer).

The phrase 'by electronic means', implies a direct computer-to-computer link for electronic data transfer. EDI is certainly striving to achieve this and without doubt this linkage is essential to realising fully the benefits offered by EDI. However, some 'paperless trading' is currently practised using magnetic tape as the transfer mechanism, and this form of EDI is certainly included within the definition.

The definition makes no reference to the timing factors involved in EDI, although it should be apparent that the trading cycle will be dramatically altered. Most current EDI applications use some form of batch transfer mechanism, whereby the data is stored by a third party prior to the recipient retrieving the information. However, some business transactions require a more immediate and conversational method of exchanging information, where the two parties actively cooperate at the time of the exchange of data. An example of this is in the travel and leisure industry, where a holiday booking may necessitate a 'conversion' between tour operator, travel agent and ferry company in order to ensure confirmation of the holiday. This interactive facet of paperless trading or EDI is not yet as fully developed as the batch, or store-and-forward, method.

EDI covers a wide variety of business applications and as a further aid to clarity, it may be useful to highlight some of the essential differences between EDI and electronic mail. These differences include:

- *Interchange agreement.* EDI trading commonly involves an agreement between trading partners concerning the types of information to be transmitted and providing legal status to the electronic documents. Even where no formal agreement exists, the EDI partners have a much higher level of expectation over what will appear and when it will appear, than do standard electronic mail users.

- *Structured data.* As stated, EDI is concerned with specific presentation of trade data, to defined standards. Although data may be structured for electronic mail, for example, in telex systems, this is not a cornerstone of the technology.

- *Personal/company mailbox.* EDI systems are likely to have company or functional electronic mailboxes for aspects such as invoicing or order processing as opposed to the personal mailboxes prevalent in electronic mail systems. This distinction is associated with the need for user intervention when processing electronic mail, whereas the basis of EDI is that the information may be processed automatically by computer systems. In other words, the rules for processing invoices, purchase orders and so on, may be fixed and therefore programmable.

Banks have, for many years, been at the forefront of using technology as a means of automating their business. Their interest and participation in the development of EDI is no exception. Perhaps their most well-known use of a specialist area of EDI is in the banks' use of electronic funds transfer (EFT). A number of systems are currently being used by banks to carry out EFT both nationally and internationally.

The EDI initiatives taken by individual banks have now been implemented and both the National Westminster and Barclays have announced commercial EDI payment services, these being 'BankLine Interchange' and 'Trading Master', respectively. The services are seen to offer new opportunities and benefits to the banks' corporate customers. Completing the EDI trading cycle with a payment service has the unique effect of integrating trading activities with settlement procedures and financial control. It is claimed that through improved financial control, the services provide competitive advantages and new opportunities when planning a business strategy. Figure 14.14 illustrates the activities and information flows involved in making an EDI payment as outlined in a joint announcement (January 1991) by five of the UK clearing banks committed to this new payment mechanism.

Potential Savings and Benefits

Traditionally, many companies have viewed EDI as providing the means to eliminate non-value-added activities, improve business efficiency and reduce costs. Initially, the majority of benefits quoted were in the area of cost reduction through postal savings and the time taken to process orders and invoices. Although these cost savings are still important it is now apparent that the driving force behind EDI is its ability to provide significantly better levels of customer service, and consequently, a competitive edge over rival companies. In some cases, there may not be any savings as such, but it is a cost to be carried in order to remain a 'preferred supplier'. Furthermore, it is frequently predicted that in the not-too-distant future, companies will look to EDI to provide totally new business opportunities, and indeed, this is beginning to happen within some industries.

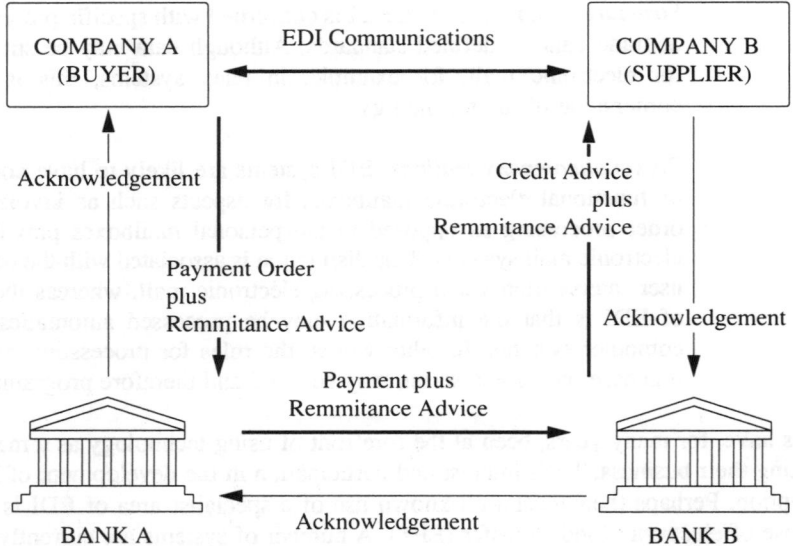

Figure 14.14 Use of EDI to close the trading loop

EDI benefits may be conveniently separated into three areas: strategic benefits, operational benefits, and opportunity benefits.

Strategic benefits include:

—   a faster trading cycle;

—   'just-in-time' manufacturing;

—   terms of trade dictated by bargaining power;

—   a need to respond to highly competitive market entrants.

Operational benefits include:

—   reduced costs;

—   improved cash flow;

—   security and error reduction;

—   acknowledged receipt.

Opportunity benefits include such factors as: enhanced image, competitive edge, which, although perceived as beneficial, are difficult to quantify. The appearance of screens on travel agents' desks, allowing a holiday to be booked electronically, is one such example. These factors give rise to new business opportunities, resulting from an improved service given to trading partners. As many companies begin to insist on EDI trading, so the company offering this service will enhance its chances of securing a wider choice of trading patterns.

Summary

EDI is about 'better business practice', both in the way a company manages its internal operations and, more directly, in the way it manages its relationships with customers and suppliers. EDI encourages, or more often forces, companies to re-examine business processes and trading relationships; the outcome is invariably a more productive and competitive organisation, able to rely on a responsive supply line, and delivering a much higher level of customer service. EDI depends upon high-quality information ('getting it right first time') leading to reduced errors, fewer misunderstandings, the elimination of double-handling, and a vastly improved image to customers and suppliers.

The major challenge for any company wishing to establish EDI as a means of trading is not a technical one. The development and installation of EDI hardware and software can be achieved in a relatively short time. The EDI challenge is one of effecting cultural change within organisations, building trading relationships, and an understanding between trading partners. EDI is providing many organisations with a corporate weapon of immense value as they strive to become more lean and effective in a volatile marketplace. The EDI business strategy requires not only a realistic assessment of benefits, but must seek to quantify potential risks, and appreciate the likely impact on the organisation. EDI is seen to be today's strategic investment which will ensure that a company remains competitive throughout the 1990s and into the next century.

Appendix 1
Glossary

Amplifier	A device which increases the electrical or optical power of a signal.
Analogue Signal	A signal whose characteristic quantities follows continuously the variations of another physical quantity.
Analogue Transmission	The transmission of a continuously variable signal (such as a speech signal) as opposed to a discretely variable signal.
Attenuation	A decrease in magnitude of current, voltage or power of a signal in transmission.
Bandwidth	The range of frequencies or data rates available for signalling in a communications channel. The difference between the maximum and minimum frequencies or data rates on that channel.
Baseband	A signal concentrated at low frequencies with bandwidth depending on the type of signal (for example, speech).
Basic Rate Access (BRA)	A method or structure for accessing ISDN using a maximum data rate of 144 kbits broken down into two 64 kbits user channels (B-channels) and a single 16 kbits signalling channel (D-channel).
Bearer Service	A type of telecommunication service that provides the capability for the transmission of signals between user-network interfaces.
Bit (Binary Digit)	One of the two figures (that is, 0 or 1) used in the representation of numbers in binary systems.

CCITT	An abbreviation of Comité Consultatif International Téléphonique et Télégraphique, the committee of the International Telecommunications Union (ITU). The ITU is an agency of the United Nations.
Common Channel Signalling	A method of signalling in which signalling information relating to a number of circuits is conveyed over a single channel by addressed messages.
Crosstalk	Interference from one cable manifesting itself in another cable.
Decibel	A logarithmic measure used for comparing two signal power levels.
Digital Access Signalling System 1 (DASS1)	British Telecom's original signalling system developed both for single line and multiline IDA, but only used for single line IDA in the BT ISDN pilot service.
Digital Access Signalling System 2 (DASS2)	British Telecom's message based signalling system which provides common channel signalling via digital circuits for both single line and multiline IDA between digital ISPBXs and local ISDN exchanges.
Digital Private Network Signalling System (DPNSS)	A signalling system which provides common channel, inter-ISPBX signalling across private digital networks. DPNSS was formulated by British Telecom and PABX manufacturers.
Digital Signal	A discretely timed signal in which information is represented by a number of well defined discrete values.
Digital Transmission	The transmission of a discretely variable signal in which the message is coded into separate pulses or signal levels.
Encoding	The generation of a code word to represent a quantised signal value.
Equaliser	A device which improves circuit quality by equalising the different levels of distortion or interference.
Filter	An electronic circuit which blocks the transmission of certain frequencies whilst letting others pass unhindered.

Frequency Response	A measure of the attenuation in a transmission line as a function of the signal frequency.
High-level Data Link Control (HDLC)	An international standard describing a link level protocol using a frame and bit structure as opposed to a character structure.
Integrated Digital Access (IDA)	The means of providing digital access for subscribers to the BT ISDN service. The two forms of IDA are single line (a pilot service soon to be superseded by ISDN 2) and multiline (a commercial service).
Integrated Digital Network (IDN)	A set of digital nodes and digital links that uses integrated transmission and switching to provide digital connections between two or more defined points.
Integrated Services Digital Network (ISDN)	A network that provides or supports a range of different telecommunications services and provides digital connections between use-network interfaces.
Integrated Services Private Branch Exchange	A digital PABX which is used to terminate an ISDN and thus provide access to private ISDNs and direct access to public ISDN.
Interference	The effect of unwanted signals (such as noise) on a transmitted signal.
International Standards Organisation (ISO)	A major standards making body which exists to promote the development of world standards. Membership consists of national organisations which are the most representative of standardisation within their country.
Laser	A device which amplifies light into an extremely narrow and intense monochromatic beam (light amplification by stimulated emission of radiation).
Light Emitting Diode (LED)	A semiconductor device which produces incoherent light when a voltage is applied to it.
Local Area Network (LAN)	A network which spans a limited geographical area (usually within one building or site) and interconnects a variety of computers and terminals, often at very high data rates (1 to 50 Mbits).
Local Loop	The transmission line which runs between a local telephone exchange and a subscriber's premises.

MegaStream	British Telecom's 2.048 Mbits digital point-to-point communications service.
Nyquist Rate	The minimum sampling rate for complete sample representation of an input signal.
Open Systems Interconnection (OSI) Model	A seven layer hierarchical structure which is used to design, specify and interrelate communications protocols. The structure was developed by ISO. OSI is not a standard or a protocol in itself.
Postal, Telegraph and Telephone Authority (PTT)	The name given to telecommunications authorities in Europe and elsewhere, which act as common carriers for telecommunications.
Predictor	A device that provides an estimated value of a sampled signal derived from previous samples.
Primary Rate Access (PRA)	A method or structure for accessing ISDN using maximum data rates of 2.048 Mbits (in Europe) or 1.544 Mbits (in America and Japan).
Private Automatic Branch Exchange (PABX)	An automatic, user-owned telephone exchange for voice and/or data. It may also perform analogue to digital signal conversion for voice transmission.
Public Telephone Operator (PTO)	The generic name given to a company or organisation which is licensed to operate a public telephone system. In Europe these companies are known as PTTs (Postal, Telegraph and Telephone Authority).
Pulse Code Modulation (PCM)	A method of transmitting analogue speech in a digital form over a transmission link. The speech bandwidth is converted into a 64 kbits bit stream.
Quantisation	A process in which a continuous range of values is divided into discrete and adjacent intervals.
Sample	A representation value of a signal at a chosen instant, derived from a portion of that signal.
Signal-to-Noise Ratio	A logarithmic measure of the ratio of the power of noise present in a transmitted signal.
Stored Program Control	A method of switching in a telephone exchange by means of a computer running a program.

Supplementary Service	A service offered by an administration to its customers in order to satisfy a specific telecommunication requirement.
Synchronisation	The process by which a receiver is brought *in step* with a transmitter, usually achieved by having a constant time interval between successive bits, by having a predefined number of start, information and stop bits, and by using a clock.
Teleservice	A type of telecommunication service that provides the complete capability, including terminal equipment functions, for communication between users according to established protocols.
Time Division Multiplexing (TDM)	A process in which several signals are interleaved in time for transmission over a common channel.
White Noise	A mixture of sound waves covering a wide frequency bandwidth and which manifests itself as 'hiss'.

Appendix 2
List of Abbreviations

ADM	Adaptive Delta Modulation
ADPCM	Adaptive Differential Pulse Code Modulation
APC	Adaptive Predictive Coding
APD	Avalanche Photodiode
BABT	British Approvals Board for Telecommunications
BRA	Basic Rate Access
BSGL	British Telecom
CUG	Closed User Group
CVSD	Continuously Variable Slope Delta
DASS2	Digital Access Signalling System No.2
DM	Delta Modulation
DPCM	Differential Pulse Code Modulation
DPNSS	Digital Private Network Signalling System
DSS1	Digital Subscriber System No.1
DTE	Data Terminal Equipment
FDDI	Fibre Distributed Data Interface
HDLC	High-Level Data Link Control
Hz	Hertz (cycle per second)
IDA	Integrated Digital Access
IDN	Injection Laser Diode
ISPBX	Integrated Services Private Branch Exchange
ISO	International Standards Organisation
LAN	Local Area Network
LAPD	Link Access Procedure for the D-Channel
LASER	Light Amplification by Stimulated Emission of Radiation
LED	Light Emitting Diode
LPC	Linear Predictive Coding
NT	Network Termination
NTE	Network Terminating Equipment

OFTEL	Office of Telecommunications
OSI	Open Systems Interconnections
PABX	Private Automatic Branch Exchange
PAM	Pulse Amplitude Modulation
PCM	Pulse Code Modulation
PIN	Positive-Intrinsic-Negative Photodiode
PPM	Pulse Position Modulation
PRA	Primary Rate Access
PSDN	Public Switched Data Network
PSPDN	Packet-Switched Public Data Network
PSTN	Public Switched Telephone Network
PTO	Public Telephone Operator
PWM	Pulse Width Modulation
RMS	Root Mean Square
SNR	Signal-to-Noise Ratio
SNR$_q$	Signal-to-Quantisation Noise Ratio
SPC	Stored Program Control
TA	Terminal Adapter
TDM	Time-Division Multiplexing
TE	Terminal Equipment

Appendix 3
List of Standards

CCITT E.110 — E.333, E.164

CCITT Recommendation G.711

CCITT Recommendation G.721

CCITT I-series

CCITT Q.931

CCITT V-series

CCITT X-series

ISO 3309, 4335

Index